Climate Smart Agriculture: Concepts, Challenges and Opportunities

Climate Smart Agriculture: Concepts, Challenges and Opportunities

Erin Stuart

New York

Published by Syrawood Publishing House,
750 Third Avenue, 9th Floor,
New York, NY 10017, USA
www.syrawoodpublishinghouse.com

Climate Smart Agriculture: Concepts, Challenges and Opportunities
Erin Stuart

International Standard Book Number: 978-1-64740-352-2 (Hardback)

Cataloging-in-Publication Data

Climate smart agriculture: concepts, challenges and opportunities / Erin Stuart.
p. cm.
Includes bibliographical references and index.
ISBN 978-1-64740-352-2
1. Crops and climate. 2. Agriculture--Environmental aspects.
3. Agricultural innovations. 4. Agriculture. I. Stuart, Erin.

S600.5 .C55 2023
630.251 5--dc23

TABLE OF CONTENTS

PREFACE

Every book is initially just a concept; it takes months of research and hard work to give it the final shape in which the readers receive it. In its early stages, this book also went through rigorous reviewing. The notable contributions made by experts from across the globe were first molded into patterned chapters and then arranged in a sensibly sequential manner to bring out the best results.

Climate-smart agriculture (CSA) is an integrated approach that helps steer actions to transform agri-food systems towards green and climate resilient practices. This approach seeks to manage landscapes such as cropland, livestock, forests and fisheries that address the interlinked challenges of food security and climate change. The negative impacts of climate change can be observed through weather variability, shifting agroecosystem boundaries, invasive crops and pests, and more frequent extreme weather events. CSA endeavors to simultaneously tackle three main objectives. The foremost aim of CSA is to increase the agricultural productivity and incomes in a sustainable manner. Secondly, it seeks to enhance resilience by reducing the vulnerability to drought, pests, diseases, and other climate-related risks. The final objective of CSA is to reduce emissions caused due to greenhouse gas emissions, avoiding deforestation, and identifying ways to absorb carbon out of the atmosphere. The topics included in this book on climate-smart agriculture are of utmost significance and bound to provide incredible insights to readers. It will serve as a valuable source of reference for students, policy makers, and researchers studying the impact of climate change on agriculture and agricultural sustainability.

It has been my immense pleasure to be a part of this project and to contribute my years of learning in such a meaningful form. I would like to take this opportunity to thank all the people who have been associated with the completion of this book at any step.

Erin Stuart

1

Climate-Smart Agriculture: An Overview

Todd S. Rosenstock, David Rohrbach, Andreea Nowak, and Evan Girvetz

1.1 Tracking Progress

In 2001, the Third Assessment Report of the Intergovernmental Panel on Climate Change highlighted the potential impacts of a changing climate on global agriculture. The Report stated that rising temperatures and drought could lead to significant declines in yields for many of the world's poorest nations, including Africa. This stimulated a new set of global commitments to research and promote agricultural practices that are more climate-smart. Since then, almost USD 1 billion has been committed to climate-smart programming in Africa, with more likely to follow (Fig. 1.1). Most African governments have formed climate-smart agriculture task forces. New transnational partnerships, such as the East African Regional Climate-Smart Agriculture Alliance, have linked government efforts to support regional change. In 2015, these commitments were reinforced by the adoption of a Statement

T. S. Rosenstock (✉)
World Agroforestry Centre (ICRAF), c/o INERA,
Kinshasa, Democratic Republic of the Congo

CGIAR Research Program on Climate Change, Agriculture and Food Security,
Kinshasa, Democratic Republic of the Congo
e-mail: t.rosenstock@cgiar.org

D. Rohrbach
Consultant, Washington, D.C., United States of America

A. Nowak
World Agroforestry Centre (ICRAF), c/o INERA,
Kinshasa, Democratic Republic of the Congo

E. Girvetz
CGIAR Research Program on Climate Change, Agriculture and Food Security,
Kinshasa, Democratic Republic of the Congo

International Center for Tropical Agriculture (IITA), Nairobi, Kenya

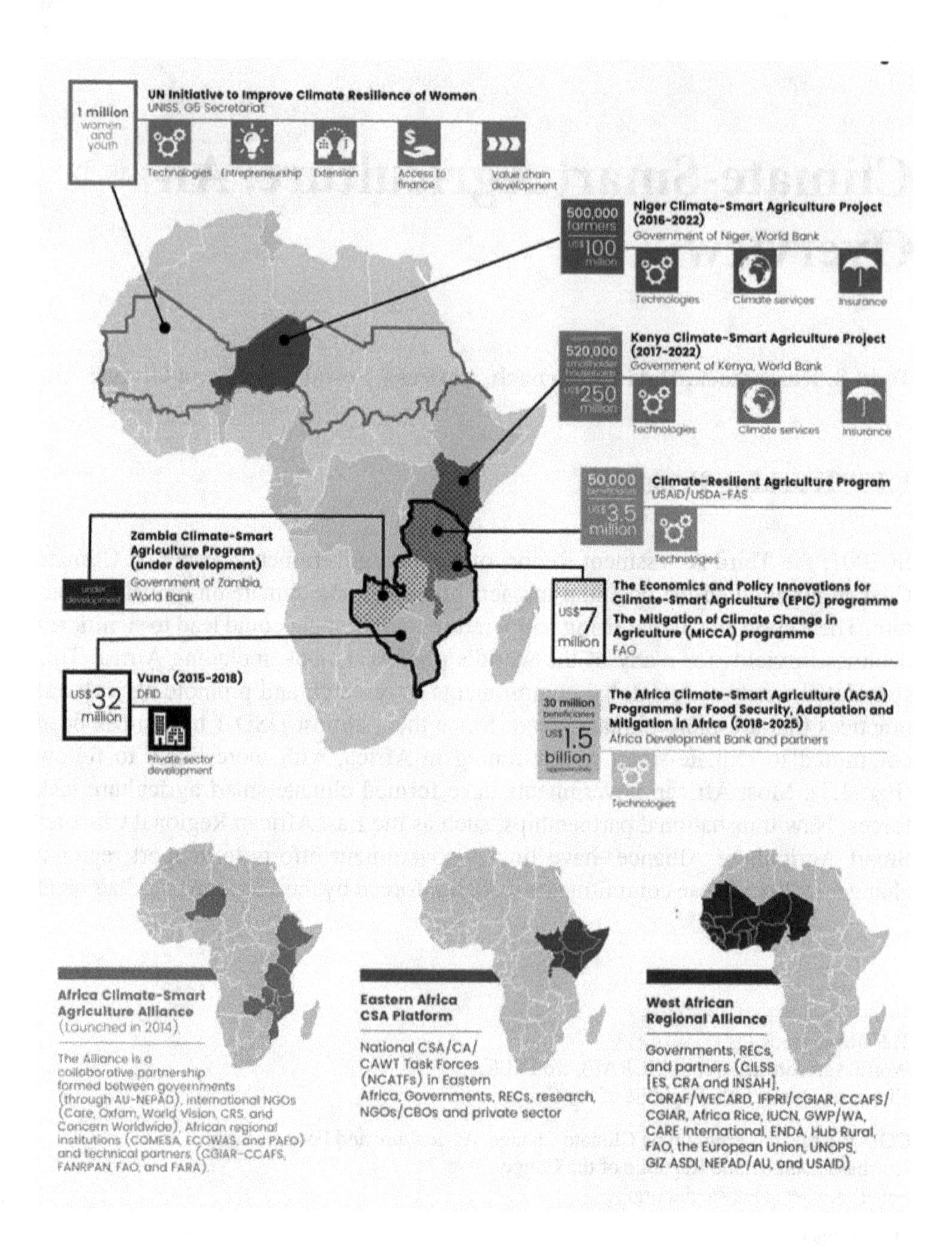

Fig. 1.1 Investments and alliances promoting CSA in Africa. (Source: Authors)

of Shared Ambition for climate-smart agriculture and a subsequent Action Plan by the corporate members of the World Business Council for Sustainable Development (WBCSD 2015). What's more, non-governmental and some civil society organisations have formed complementary advocacy groups, such as the Alliance for Climate-Smart Agriculture in Africa (ACSAA) that includes international non-governmental organizations, policy institutions, technical partners and farmers groups.

The responses to these large commitments of strategic and financial support have been substantial. Hundreds of technological solutions have already been identified as climate-smart because they mitigate the effects of rising temperatures and variable rainfall; or contribute to a reduction of greenhouse gas (GHG) emissions; or accumulate carbon in biomass or soils (Table 1.1). New varieties of crops with greater drought tolerance are already in production, and more are being released each year (Challinor et al. 2016). Many land management systems conserve soil moisture (Thierfelder et al. 2017). Agroforestry systems reduce the ambient temperature of nearby crops and livestock (Lin 2007; Barton et al. 2016). Feeding strategies that increase productivity and reduce GHG emissions from livestock are well known (Bryan et al. 2013; Thornton and Herrero 2010). And information delivery systems help farmers to plan the right period(s) to plant.

Indeed, 'climate-smart' has bordered on becoming a brand. Carried to the extreme, today there are now climate-smart extension systems, climate-smart finance, climate-smart landscapes, climate-smart livestock, climate-smart soils, and climate-smart varieties etc. (Gledhill et al. 2012; Graefe et al. 2016; Minang et al. 2014; Paustian et al. 2016; Sala et al. 2016).

Whereas many technologies are available to help farmers better cope with climate risks, improving farmers' access to these technologies, while strengthening incentives around their adoption, remains the more significant challenge. Despite millions of dollars of investment, adoption rates of new agricultural technologies in much of eastern and southern Africa remain low (Giller et al. 2009; Asfaw et al. 2016). The majority of farmers continue to struggle with the costs and risks of new technologies. Increasing climate risks simply make these efforts more difficult.

This volume highlights current efforts being made by scientists in eastern and southern Africa in developing and disseminating climate-smart agriculture (CSA) technologies. Emphasis was placed on getting previously unpublished data written up and presented. Unlike many edited volumes, the book started with an open call for chapters on five key topics. More than 70 applications were submitted and evaluated against the criteria, which included: relevance of the topic, whether new data were being presented, and the quality of the science. Twenty-three applications were selected to move on to full chapter development. Twelve specific contributions were then commissioned by the book's editors to fill gaps in the discussion. After at least two technical reviews and multiple rounds of revision, 25 of these papers were accepted for publication within this volume. Unpublished chapters, which still contain important content for development, can be found on the webpage that accompanies this book.

Table 1.1 Different climate hazards and associated field- and farm-level adaptation interventions, by broad agricultural land use

Land use	Intervention	Climate hazard			
		Temp	Water	Variability	Flooding
C M	Improved crop varieties: dual-purpose, higher-yielding, stress tolerance (heat, drought, salinity, pests)	√	√	√	√
C M	Change crops: new mixes of crops of different characteristics (heat-, drought-tolerance), crop rotations	√	√	√	√
C M	Crop residue management: no till/minimum tillage, cover cropping, mulching		√	√	
C M	Crop management: modified planting date/densities, multicropping with legumes, agroforestry species	√	√	√	√
C M	Nutrient management: composting, appropriate fertiliser and manure use, precision nutrient application		√	√	
C P M	Soil management: crop rotations, fallowing (green manures), conservation tillage, legume intercropping		√		
C P M	Improved water use efficiency and water management: supplemental or reduced irrigation, water harvesting, modifying the cropping calendar, flood water control		√	√	√
P M	Change livestock breed: switch to more productive/smaller/more heat- and drought-resilient breeds	√	√	√	
P M	Change livestock species: switch to more cash-fungible species, use more drought- and heat-tolerant species	√	√	√	
P M	Improved livestock feeding: diet supplementation, improved pasture species, low-cost fodder conservation technologies, precision feeding			√	
C M	On-farm pond aquaculture as a low-emissions adaptation and livelihood diversification strategy	√	√	√	
P M	Improved animal health: disease surveillance, vaccination, disease treatment			√	
P M	Grazing management: adjusting stocking densities to feed availability, rotational grazing, livestock movement		√	√	
P M	Pasture management: use of sown pastures, setting up of fodder banks and other strategic dry-season feed resources		√	√	
P M	Manure management: anaerobic digesters for biogas and fertiliser, composting, improved manure handling, storage and application techniques			√	

(continued)

Table 1.1 (continued)

Land use	Intervention	Climate hazard			
		Temp	Water	Variability	Flooding
M	Alter system integration: alter animal species and breeds, alter the ratio of crops to pasture, or crops to livestock and/or to fish		√	√	
C P M	Use of weather information: seasonal forecasts for agricultural planning, and short-term forecasts for early warning of extremes such as high temperatures and heavy rainfall		√	√	√
C P M	Index insurance for crops and livestock		√	√	√

C cropland, *P* pastureland, *M* cropland and pastureland
Source: CCAFS and GCF (2018)

1.2 Overview of the Chapters

The 25 chapters in this book have been divided among 5 themes. Four chapters explore issues around climate change, including impacts and risks. Six investigate mechanisms in seed and crop germplasm delivery systems. Six examine various perspectives and lessons learned on technologies and practices through a CSA lens. Five more examine the resilience to climate change of value chains; and four look at financing, extension and other mechanisms to reach scale. Each chapter reflects on a fundamental question: how to make complex crop and livestock systems more climate-smart? Each chapter ends with messages on the implications for development practitioners to inform future decision-making.

The chapters that explore climate change, along with its impacts and risks, include one on future projections, two on impacts and one on systems. Girvetz et al. investigate the certainty and uncertainty of future climate change in sub-Saharan Africa. This involves longer term and near-term predictions of traditional indices, such as temperature and precipitation, as well as new bioclimatic indicators that help make forecasts relevant to the risks faced by agricultural systems. The authors use a freely accessible online tool known as Climate Wizard (www.climatewizard.org), which is available to practitioners to help them incorporate climate information in programme and policy design. Bett et al. describe two cases of how predicted climate change will affect the occurrence of livestock pests and diseases that already cause significant damage to livelihoods and economies. The authors' concrete recommendations around mitigating future impacts support the notion of taking action today to prepare for the challenges of tomorrow. Hunter and Crespo analyse the climate risks and impacts for both staple (maize and cassava) and cash (coffee) crops at the subnational level in Angola. The authors' findings demonstrate a clear need for future investments—for example, in long-lived coffee—despite the inherent uncertainty in climate models. Lastly, Masikati et al. look at the likely responses of maize and groundnut under climate change using common crop models. Their findings suggest that improved soil management can help mitigate future negative risks to productivity. Taken together, these chapters illustrate why this type of research is critical in moving beyond projections of the future to concrete action that can be taken today. Nonetheless, commissioning this section of the book was not unproblematic, pointing to an urgent need for more information around climate change impacts and risks—detail that is instrumental for initiating meaningful conversations on CSA.

The next set of chapters describe the challenges and opportunities around improving the delivery of quality crop germplasm to farmers. Improved planting materials are typically among the first suggested responses to climate variability—whether today's or tomorrow's—and this section explores some of the limits of this conventional wisdom. Das et al. open with a private-sector perspective on seed systems. The authors describe bottlenecks in the delivery of cereal seeds along with the necessary changes—such as public–private partnerships—they feel are needed to make investment opportunities more conducive to the private sector. Ertiro et al. bring fresh evidence in support of the development of drought-tolerant maize in Ethiopia. Droughts are already an every-year occurrence in the country under climate change, and this case highlights a suite of actions needed to move from breeding to widespread use of new varieties. Cramer focuses on one specific link in the seed system chain—early generation seeds. By comparing a successful case with an unsuccessful one, the author identifies a few key stumbling blocks that extend the time taken in breeding, delivery and adoption. Parker et al. illustrate that many of the issues presented for cereal crops are also applicable to roots, tubers and banana—staples for 300 million people in the humid tropics of sub-Saharan Africa (SSA). However, the solutions recommended by the authors differ markedly to those of previous chapters due to the structure and development of the system that delivers vegetable planting-materials. Many of the challenges discussed—e.g. the lack of development and long generation time—are also presented in Dawson et al. In this chapter, authors discuss the contributions of trees and orphan crops to resilient food systems and make recommendations for investments that develop this system in future. Faddha and van Etten close this section by presenting a cost-effective participatory approach to evaluate varieties under farm conditions using novel material from national gene banks or plant breeding. Using a case study, the authors argue that this triadic comparisons of technologies (tricot) approach has the potential to contribute to making seed systems more dynamic when demand and supply are linked and more diversified, as more varieties per crop will be delivered in a location-specific way. Together, the chapters in this section of the book present a sobering picture of the current germplasm delivery systems; with a low penetration of improved varieties within agricultural systems (20%), even for most well-developed breeding programmes, and a long development time (13–30 years). This may signal a massive development opportunity for the seed sector within CSA.

Subsequent chapters present perspectives on the climate-smartness of various technologies and management practices. In particular, they unpack the evidence and lessons learned on what makes a technology climate-smart. Rosenstock et al. conduct a systematic map —a rigorous and structured analysis of the available data—to examine the impact of 73 technologies on indicators of productivity, resilience and mitigation. They identify a significant skew in the available peer-reviewed literature towards maize-based systems, productivity outcomes and on-farm trials. This suggest that anyone interested in creating evidence-based programmes and plans will find many gaps in the scientific knowledge. A complementary quantitative approach towards assessing the multidimensionality of agricultural technologies can be found in Kimaro et al. Here, the authors collect agronomic data on the performance of technologies across the three pillars of CSA (productivity, resilience and mitigation) in three agroforestry systems of Tanzania (shelterbelt, intercropping and border plantings of fuelwood and food crops). Their findings highlight the perspective and flexibility needed to understand whether a technology is climate-smart or not. Performance assessments, however, only provide part of the evidence. Manda et al. design and pilot a participatory framework to evaluate practices against farmer-selected criteria of productivity and resilience. This qualitative approach can help fill gaps in knowledge-which other chapters of the book have pointed towards-while being farmer-centric. Mwungu et al. present an analysis of barriers to the adoption of a technology. Specifically, the authors investigate drivers behind the adoption of improved varieties in rural, post-conflict Uganda. They find that household size and information networks influence adoption, with results pointing towards both general and context-specific rules on the adoption of CSA technologies. For example, while household size is typically positively correlated with adoption, trust in information networks may be increasingly important in some contexts, such as post-conflict zones. Davies et al. analyse how culture and spirituality can affect the adoption of technologies. The introduction of culture as a determinant of adoption is unique in most discussions of technologies in general and of CSA in particular. This concern may be acutely pertinent for technologies aimed at addressing climate risks, given that weather—good or bad—is often viewed as a manifestation of divine intervention. Together, the chapters presented in this section of the book provide insights into the social considerations and scientific approaches that inform the adoption of CSA.

Because technologies are only part of the food system, the fourth set of chapters explores how value chains contribute to the climate-resilience of smallholder farmers and how climate risks to these value chains can be reduced. Barzola et al. focus on farmers and test the hypothesis that farmer entrepreneurship—the innovative use of agricultural resources to create opportunities for value creation—as well as engagement in the value chain facilitates the adoption of CSA technologies. The study found that farm size influences entrepreneurial innovativeness in a surprising way—with smaller farms more likely than larger ones to engage in all forms of innovation. Actors seeking to promote innovation, including the adoption of technology, might therefore consider investing in programmes that help farmers to develop a more entrepreneurial outlook. Hammond et al. further explore farmer participation and climate resilience. The authors use an innovative survey tool, the Rural Household Multi-Indicator Survey, to investigate how participation in *Shea* value chain activities benefit poor farmers. Shea trees serve as a buffer against desertification, accumulate carbon in the landscape and protect soil and water resources, while processing activities (more specifically, shea butter production) can increase farmers' adaptive capacity by boosting incomes. In contrast, Sloan et al. examine how private-sector firms, in different parts of the supply chain, view, understand and engage with climate change and the promotion of CSA technologies. The key factors influencing the readiness of companies to incorporate CSA into their strategies were found to be specialised staff and a track record of actively promoting sustainability within the company. The scientific community therefore needs to provide actionable information to incentivise companies' investments in CSA; particularly emphasising returns on investment and the cost of inaction. Mwongera et al. discuss the need to link climate change analyses with value chain approaches in designing CSA interventions. Using a case study from Nyandarua County in Kenya, the authors illustrate how the climate risk profile (CRP) approach supports identification of major climate risks and their impacts on the value chain, identifies adaptation interventions, and promotes the mainstreaming of climate-change considerations into development planning at the subnational level. They conclude that the magnitude of a climate risk varies across value chains. Allen and de Brauw take an even broader perspective to explore mechanisms that promote nutrition-sensitive value chains, as part of efforts to manage climate risks and increase resilience through diversification. Access to improved, biofortified seeds, reducing post-harvest loss (for example, through adequate storage and the transportation of perishable crops), and diet diversification are key value-chain interventions for improved nutrition. Vermulen considers the very big picture, describing recent private-sector progress towards realising CSA targets. The author looks at the Climate-Smart Agriculture Initiative of the WBCSD and shows that the global agri-food sector is exceeding WBCSD targets for global food production, but falling short on emissions reductions, and failing to track outcomes for farmers' livelihoods. There are major gaps in information, monitoring, reporting and verification which need to be tackled if the ambitious CSA targets are to be met. Overall, this section of the book highlights the instrumental role of systemic, collective action for promoting climate-smart value chains.

In order to meet global food security ambitions, CSA technologies need to be accessible and accessed by farmers. The final section of the book discusses mechanisms for bringing CSA to scale. Franzel et al. explore farmer-to-farmer extension systems and find that these approaches can significantly increase the pool of farmers adopting CSA practices, but that this varies across practices and contexts. Their chapter suggests that this innovative advisory approach should not replace traditional, low-performing extension services, but rather complement existing approaches (such as extension campaigns, farmer field schools or information and communication technology). Acosta et al. study the role of multi-stakeholder platforms in promoting an enabling policy environment for climate action. These platforms can create ownership, knowledge and science-policy dialogue at various scales. In a similar vein, Kadzamira et al. discuss the role of different partnership arrangements in scaling CSA in Zambia, Zimbabwe and Malawi. Accordingly, successful partnerships for scaling build on existing structures and mechanisms, bring mutual benefits for all stakeholders, and ensure transparency in decision-making processes. Finally, Ruben et al. investigate the different rural financial instruments that are available for promoting CSA. While the adoption of practices and technologies may be stimulated through interventions that address very specific resource constraints (through credit, insurances, and input provisions, for example), scaling CSA requires more systematic investments (for example, blended mechanisms) that allow for increases in farm income while minimising risks.

1.3 Implications for Development

This book highlights a wide cross-section of effort to design and disseminate agricultural technologies and approaches that help farmers better cope with climate risks. During a review of the chapters, however, several common gaps were identified that may merit attention in future research.

The main climate risk considered in these pages is drought—an obvious choice, given the long history of efforts to identify technologies suitable for drought-affected regions of Africa. Drought is already endemic in large parts of eastern and southern Africa. A principal concern is that these areas will expand as the climate continues to change. However, there is relatively little discussion about the variation in drought across the region and how this is expected to alter over time. This is based on the assumption that current drought risks are indicative of weather patterns under a changing climate. Yet it is not obvious that current drought risks will simply expand spatially. Over the next generation or two, the types of drought may change (cf. Chavez et al. 2015). A larger proportion of farmers may find that the rains start late or end earlier, or that the seasons simply shorten. In some areas, mid-season dry spells affecting flowering may become more common. This points to a need to better characterise how drought risks are likely to change over time and, more explicitly, account for this in technology design.

While rising average temperatures are linked with the likely spread of drought, the chapters in this book suggest that comparatively little work has been completed on solutions to these temperature changes. This is surprising given the irrefutable evidence that temperatures are rising in line with the growth of GHGs, and may be rising faster in sub-Saharan Africa than in other parts of the world. Higher average temperatures are widely expected to shift the incidence of pests and diseases affecting crop and livestock production (Bett et al. 2017). However, models tracking the speed and incidence of this change remain rudimentary. Observers note that rising temperatures may also affect plant flowering and fruit production, as well as the timing and severity of drought. But the thresholds for these changes do not seem to be well defined in applied technology development programmes. If scientists remain uncertain about the levels, spatial distribution and timing of changing temperatures, designing technology suitable for the diverse farming systems of eastern and southern Africa will continue to be challenging.

Similarly, solutions to the endemic and possibly worsening climate risk of flooding are almost totally absent in this collection of studies. This includes the need to develop varieties that are more tolerant of water-logging and to strengthen water management and control systems.

Most of the chapters concentrate on the improvement of technologies and management strategies for coping with today's climate risks. This is understandable, given the pressing need to improve the productivity of farming systems. The emphasis is, therefore, on technologies that help farmers to better cope with today's risks, which are also likely to support larger numbers of farmers who may be affected by a changing climate in the future. As such, this body of work may be better characterised as 'climate-risk management' rather than 'climate change management'. But given that it can take several decades to develop a new crop or livestock variety, some investment needs to be allocated to coping with changes likely to occur over the next generation or two. And the possible differences between today's climate risks and the probable changes in these risks over time needs to be more consistently acknowledged.

Finally, while the focus of these papers has been on climate risks, greater attention needs to be directed to the trade-offs in household decision-making that may lead many farmers to identify climate risks as secondary. Indeed, market risks—such as price and quality—may be more important than climate risks in regions benefiting from the expansion of commercial opportunities. Even in drought-prone regions, such as those growing cotton or sunflower or livestock, farmers may be willing to adopt technologies offering moderate risks and larger potential returns. Similarly, efforts to reduce market risks may allow farmers to experiment with a wider range of productivity-enhancing technologies. Ideally, every new technology will offer higher yields as well as lower risks, including climate risks. In practice, the distribution of technology traits will continue to vary for different environments.

Ultimately, these papers highlight the increasing attention being given by agricultural research and extension officers operating in eastern and southern Africa to problems of climate risk and the threats of climate change. Most of the chapters in this book emphasise concerns around technology targeting, dissemination and scaling up needed to speed the adoption of improved practices. The challenge remains to achieve faster gains on the ground. More evidence-based examples of scale up are therefore needed, along with greater attention on documenting and sharing lessons from successful and unsuccessful practices.

Next year, CSA will turn ten. The development community must face the existential question of whether it will be time to celebrate? Only 2 years remain before countries need to report progress towards implementing their Nationally Determined Contributions that are at the heart of the Paris Agreement—virtually all of which identify improving agricultural practices in Africa as a priority under climate change. And only 12 years remain before the 2030 deadline set by the United Nations Framework Convention on Climate Change to have 500 million climate-smart farmers. How can we best combine our future efforts to achieve this target?

References

Action Aid (2014) Clever name, losing game? How climate smart agriculture is sowing confusion in the food movement. Available at: http://www.actionaid.org/publications/clever-name-losing-game-how-climate-smart-agriculture-sowingconfusion-food-movement

Asfaw S et al (2016) What determines farmers' adaptive capacity? Empirical evidence from Malawi. Food Secur. Available at: http://link.springer.com/10.1007/s12571-016-0571-0 8, 643–664

Barton DN et al (2016) Assessing ecosystem services from multifunctional trees in pastures using Bayesian belief networks. Ecosyst Serv 18:165–174. https://doi.org/10.1016/j.ecoser.2016.03.002

Bett B et al (2017) Effects of climate change on the occurrence and distribution of livestock diseases. Prev Vet Med 137(November 2015):119–129. https://doi.org/10.1016/j.prevetmed.2016.11.019

Bryan, E. et al. (2013) Can agriculture support climate change adaptation, greenhouse gas mitigation and rural livelihoods? Insights from Kenya. Clim Chang, 118, 151–165. Available at: http://www.springerlink.com/index/10.1007/s10584-012-0640-0 [Accessed 5 Mar 2013]

CCAFS & GCF (2018) Sectoral guidance for GCF investments in agriculture. In press. CCAFS: Netherlands.

Challinor AJ et al (2016) Current warming will reduce yields unless maize breeding and seed systems adapt immediately. Nat Clim Chang (June). Available at: http://www.nature.com/doifinder/10.1038/nclimate3061 6, 954–958

Chavez E et al (2015) An end-to-end assessment of extreme weather impacts on food security. Nat Clim Chang 5(11):997–1001 Available at: http://www.scopus.com/inward/record.url?eid=2-s2.0-84945295462&partnerID=40&md5=05582dc08e918c2caf26b887d45ece3a

Giller KE et al (2009) Conservation agriculture and smallholder farming in Africa: The heretics' view. Field Crop Res 114(1):23–34

Gledhill R, Herweijer C, Hamza-Goodacre D (2012) MRV and data management for CSA mitigation. Climate-smart agriculture in Sub-Saharan Africa project

Graefe S et al (2016) Climate-smart livestock systems: an assessment of carbon stocks and GHG emissions in Nicaragua. PLoS One 1–19. https://doi.org/10.1371/journal.pone.0167949

Lin, B.B. (2007) Agroforestry management as an adaptive strategy against potential microclimate extremes in coffee agriculture. Agric For Meteorol, 144(1–2), 85–94. Available at: http://linkinghub.elsevier.com/retrieve/pii/S0168192307000548 [Accessed 25 Oct 2012]

Minang PA et al (2014) Climate-smart landscapes: multifunctionality in practice. Available at: https://books.google.com/books?id=rii-BQAAQBAJ&pgis=1

Paustian K et al (2016) Climate-smart soils. Nature 532(7597):49–57 Available at: http://www.nature.com/doifinder/10.1038/nature17174

Sala S, Rossi F, David S (eds) (2016) Supporting agricultural extension towards climate-smart agriculture: an overview of existing tools. Global Alliance for Climate Smart Agriculture (GASCA)/FAO, Italy Available at: http://www.fao.org/3/a-bl361e.pdf

Thierfelder C et al (2017) How climate-smart is conservation agriculture (CA)? – its potential to deliver on adaptation, mitigation and productivity on smallholder farms in southern Africa. Food Secur 9:537–560. https://doi.org/10.1007/s12571-017-0665-3

Thornton PK, Herrero M (2010) Potential for reduced methane and carbcon dioxide emissions from livestock and pasture management in the tropics. Pro Natl Acad Sci U S A, 107(46), 19667–19672. Available at: http://www.pubmedcentral.nih.gov/articlerender.fcgi?artid=2993410&tool=pmcentrez&rendertype=abstract [Accessed 12 Nov 2012]

World Business Council for Sustainable Development (WBCSD) (2015) WBCSD climate-smart agriculture action plan. Mid-term report. Available at: http://docs.wbcsd.org/2017/11/WBCSD_Climate_Smart_Agriculture-Action_Plan_2020-MidTermReport.pdf. (Accessed 14 Nov 2017)

Part I
Risks and Impacts of Climate Change

2

Climate Change in Africa: Past and Future

Evan Girvetz, Julian Ramirez-Villegas, Lieven Claessens, Christine Lamanna, Carlos Navarro-Racines, Andreea Nowak, Phil Thornton and Todd S. Rosenstock

2.1 Introduction

Farmers in Africa—like those across the globe—face rising temperatures and more extreme weather associated with climate change (Snyder 2016; IPCC 2012). Much of Africa's vulnerability to climate change lies in the fact that its agricultural systems remain largely rain-fed, with few technological inputs. The majority of Africa's farmers

E. Girvetz (✉)
International Center for Tropical Agriculture (CIAT), Nairobi, Kenya
e-mail: E.Girvetz@cgiar.org

J. Ramirez-Villegas · C. Navarro-Racines
International Center for Tropical Agriculture (CIAT), Cali-Palmira, Colombia

CGIAR Research Program on Climate Change, Agriculture and Food Security (CCAFS), Copenhagen, Denmark
e-mail: J.R.villegas@cgiar.org; C.E.navarro@cgiar.org

L. Claessens
CGIAR Research Program on Climate Change, Agriculture and Food Security (CCAFS), Copenhagen, Denmark

International Institute of Tropical Agriculture (IITA), Arusha, Tanzania
e-mail: l.claessens@cgiar.org

C. Lamanna · A. Nowak
World Agroforestry Centre (ICRAF), Nairobi, Kenya
e-mail: C.Lamanna@cgiar.org; A.Nowak@cgiar.org

P. Thornton
International Livestock Research Institute (ILRI), Nairobi, Kenya
e-mail: p.thornotn@cgiar.org

T. S. Rosenstock
World Agroforestry Centre (ICRAF), Kinshasa, Democratic Republic of the Congo
e-mail: T.Rosenstock@cgiar.org

work on a small-scale or subsistence level, with their opportunities limited by persistent poverty, lack of access to infrastructure and information, and challenges related to policy and governance. Climate change is expected to have major negative impacts on the livelihoods and food security of such farmers. Governments and development professionals must confront the challenge of helping them to adapt (Shackleton et al. 2015).

This paper offers a general overview of historical climate change in Africa, and in particular how it has already led to rising temperatures and increased rainfall variability. It then examines the models that provide projections—with varying levels of certainty—of what climate change will mean for farmers across eastern and southern Africa in the coming decades (ESA).

The paper also highlights the strengths and limitations of the available information regarding the effects of climate change. Adapting to climate change requires better projections of the specific climate hazards that will be faced at the national, regional and local levels (Challinor et al. 2007; Muller et al. 2011). In particular, implementation of climate-smart agriculture (CSA) projects has been constrained by the lack of information on the best responses in specific regions. There are serious gaps in observed historical weather data at the local level across the continent, and the continuing collection of such data still lags far behind where it should be. Strengthening the database of observed weather is critical to understanding the changes that have occurred already, to project future changes and their impacts, and to plan appropriately to address them. Once collected and analyzed, climate data must be communicated in ways that help development practitioners and decision-makers understand climate impacts in specific places. Good tools are available, but practitioners at the local level must have the access and training to use them.

Much work remains to be done. However, given that the impacts of climate change are already being felt on the ground, it is imperative that adaptation begins immediately. Even in places where projections are uncertain, steps can be taken right now to implement CSA practices and make farmers more resilient in the face of climate change.

2.2 Past and Present: Evidence Africa's Climate Has Already Changed

There is clear evidence that average temperatures have become warmer across the globe. In Africa these changes became apparent starting in about 1975, and since then temperatures have increased at a rate of about 0.03 °C per year (NOAA 2018; Hartmann et al. 2013). In those regions of Africa for which data are available, most have also recorded an increase in the incidence of extreme temperatures as well as longer heat waves (Seneviratne et al. 2012).

Historic variability can provide useful context for understanding climate change. Climate variability can be thought of as a bell curve, with weather in any given year most likely to cluster around the average (the top of the bell) and extremes of temperature or precipitation occurring less often (the flatter parts of the curve). Climate change can shift both the mean and the overall shape of the bell curve, often flattening it out because of the rising frequency of extremes (Kirtman et al. 2013).

Records are constantly being broken: in Africa 19 of the past 20 years have been hotter than any previous year on record. The new normal for temperature is hotter than ever experienced in the recorded past.

Historic precipitation patterns show that much of Africa is drying (Hartmann et al. 2013, Fig. 2.1). West Africa and parts of southern Africa, particularly Zambia

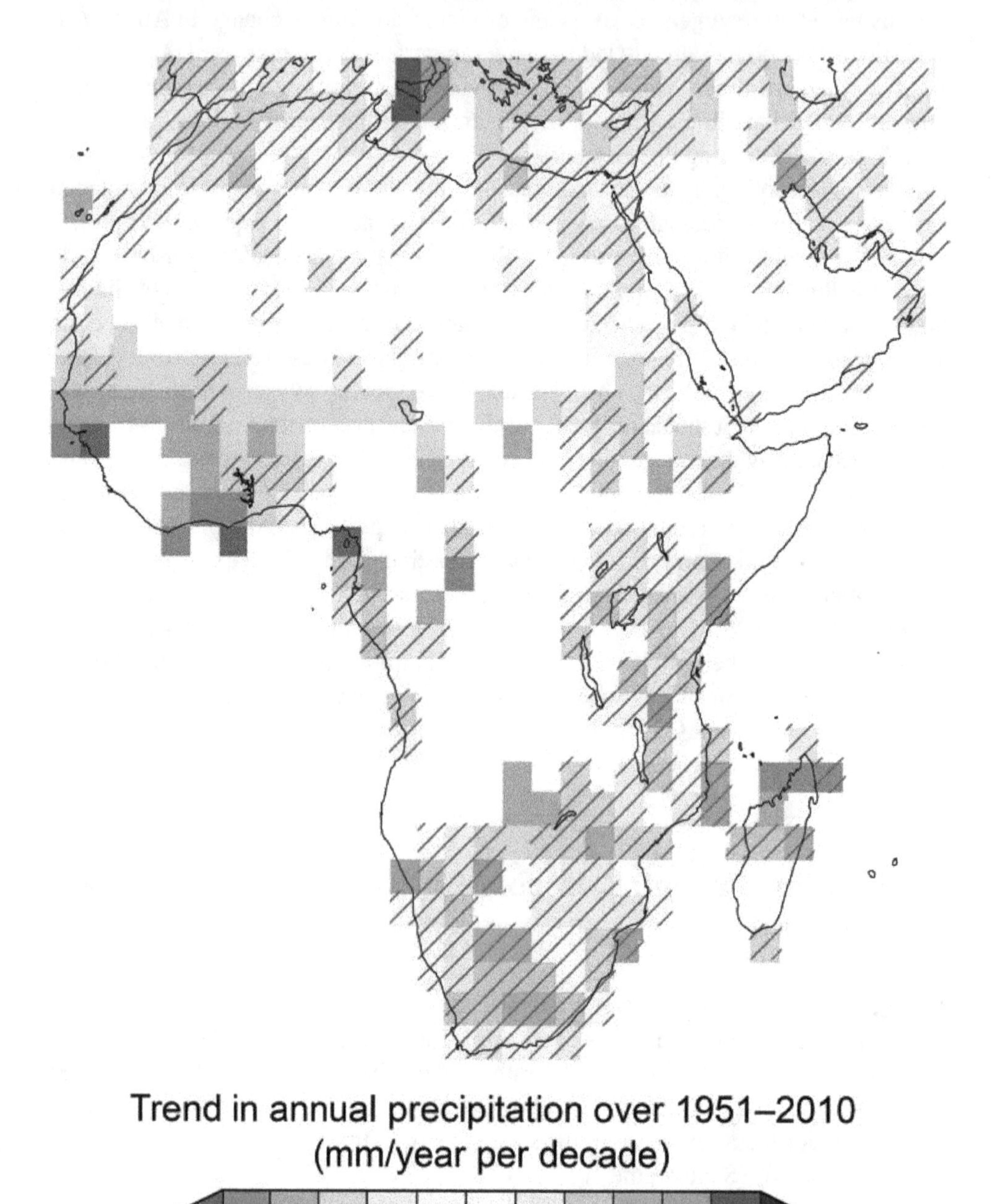

Fig. 2.1 Historical changes in precipitation from 1951 to 2010 (From Niang et al. 2014). The map has been derived from a linear trend. Areas with insufficient data are marked as white, solid colors indicate statistically significant trends at 10% level, and diagonal lines indicate areas where trends are not statistically significant

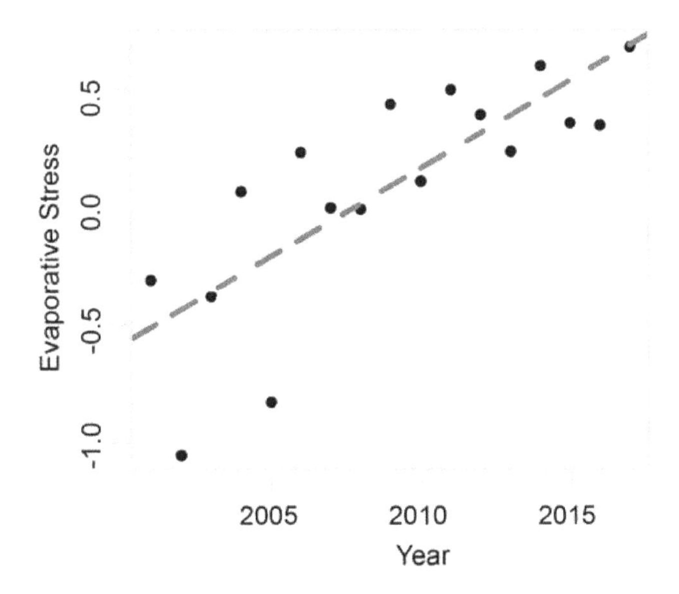

Fig. 2.2 Historic time series for evaporative stress in Zambia during 2001–2017, showing a highly significant ($p < 0.001$) increase during this time period. (From https://climateserv.servirglobal.net, accessed 27 January 2018)

and Zimbabwe, show rapid and statistically significant decreases in precipitation. By contrast, South Africa and limited parts of East and North Africa have experienced increased rainfall. At the same time, increased temperatures are leading to higher rates of evapotranspiration, which produces drier soil conditions (Girvetz and Zganjar 2014). Evaporative stress consistently increased in Zambia between 2001 and 2017 (Fig. 2.2). Even in the face of increasing precipitation, it is possible for the aridity of soils to increase. In southern Africa from 1961 to 2000, an increasing frequency of dry spells was accompanied by an increase in the intensity of daily rainfall, which has implications for runoff (New et al. 2006).

2.3 Future: Climate Model Projections for Africa

General circulation models (GCMs) provide the most straightforward and scientifically accepted way to project future climate conditions. However, climate-change simulations performed with GCMs are only possible at coarse resolutions (typically 50–100 km grid cells) that are not detailed enough to assess regional and national impacts. Agricultural livelihoods, soils and local climatic conditions vary vastly at much smaller spatial scales. Spatial downscaling techniques can and should be used to bring these coarse scale maps down to a finer resolution.

Despite their limitations, GCMs are the most commonly used tool to analyze changes in climates at a variety of spatial scales. The latest GCMs available—the Coupled Model Intercomparison Project Phase 5 (CMIP5)—suggest that temperature increases for Africa with the current emissions trajectory (i.e. RCP 8.5) is 1.7 °C by the 2030s, 2.7 °C by the 2050s, and 4.5 °C by the 2080s (Fig. 2.3). Even under the lowest greenhouse gas emissions scenario, by 2030 the climate average is

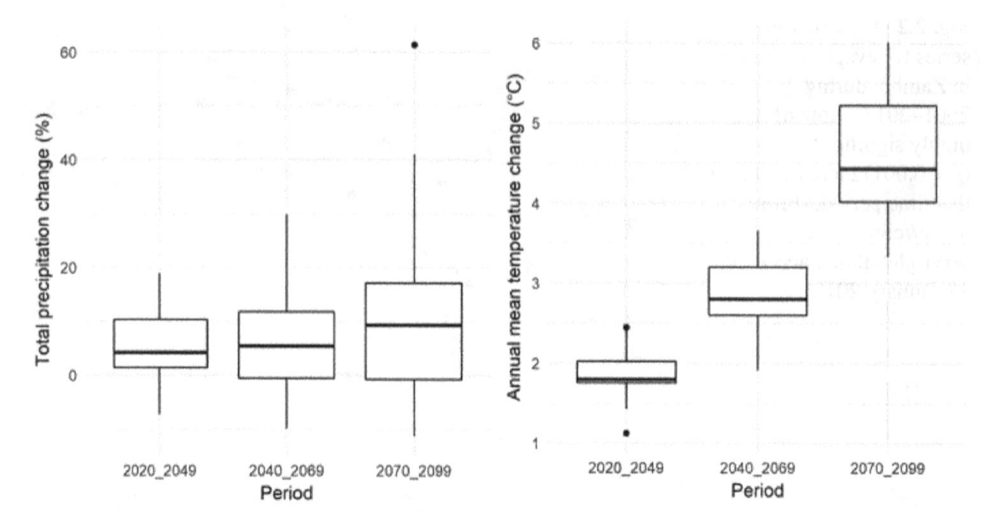

Fig. 2.3 Projected changes in annual mean temperature (in °C) and total annual precipitation (in percentage) for the African continent as projected by 33 general circulation models (GCMs) of the CMIP5 model ensemble under RCP 8.5 and three different time periods. Thick black horizontal lines represent the median, boxes show the interquartile range, and whiskers represent the 5–95th percentiles of the data

projected to be completely different from what has ever been experienced historically (Girvetz et al. 2009, climatewizard.ciat.cgiar.org).

Future precipitation is much more difficult to model (Sillmann et al. 2013; Ramirez-Villegas et al. 2013). The median of the CMIP5 models indicates that by 2050, under the higher emission scenario (RCP 8.5, Fig. 2.4), annual precipitation will increase across much of eastern and central Africa, while decreasing across parts of southern, western and northern Africa (Fig. 2.4). Increases of over 200 mm and more than 25% annually are shown in some places, as well as decreases of over 100 mm and more than 20% in other places. Not all climate models agree on the magnitude or even direction of change. However, there are some places with high agreement among the climate models: over 80% of the climate models agree on decreased precipitation in the future for some parts of northern and southern Africa (Niang et al. 2014).

Precipitation is also projected to change differently in different months, with alterations to the onset, length and cessation of the growing season. For example, in Tanzania precipitation is projected to increase during the middle of the wet season (November–May) and to decrease at the wet season's beginning (September–October) and end (May–June) (see http://climatewizard.ciat.cgiar.org/SBSTA/Africa_2050/). Overall precipitation is projected to increase, but within a shorter time frame, indicating both shortening of the rainy season and an increased frequency of extreme precipitation events.

Even in areas experiencing increased precipitation, crop production systems can be affected by worsening water stress. Depending on the timing of rainfall, the amount of the temperature increase, and the changes in cloud cover (and hence in

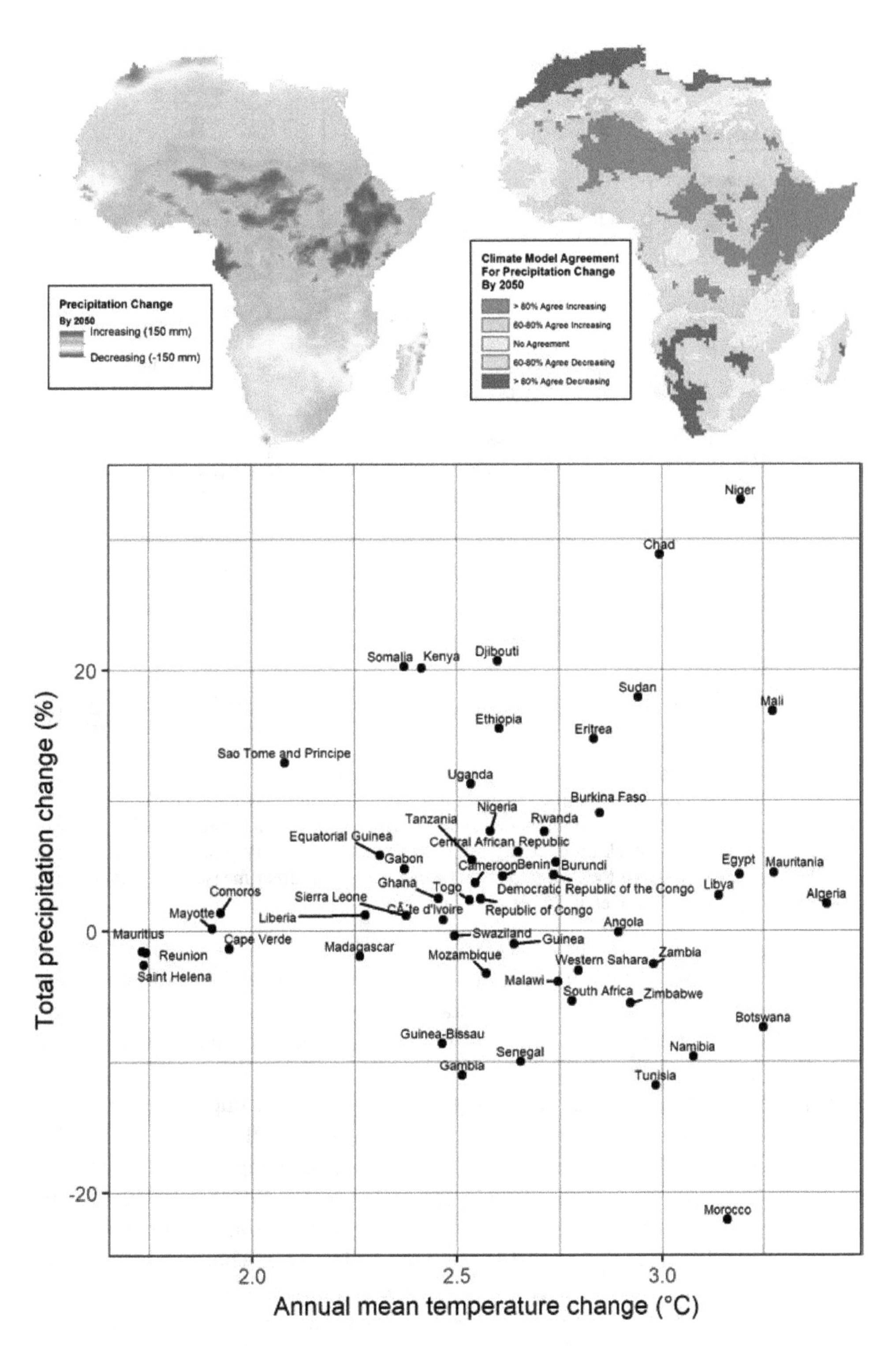

Fig. 2.4 Projected changes in total annual precipitation by 2050 (top left), climate model agreement (top right), and average change in precipitation and temperature, by country (bottom)

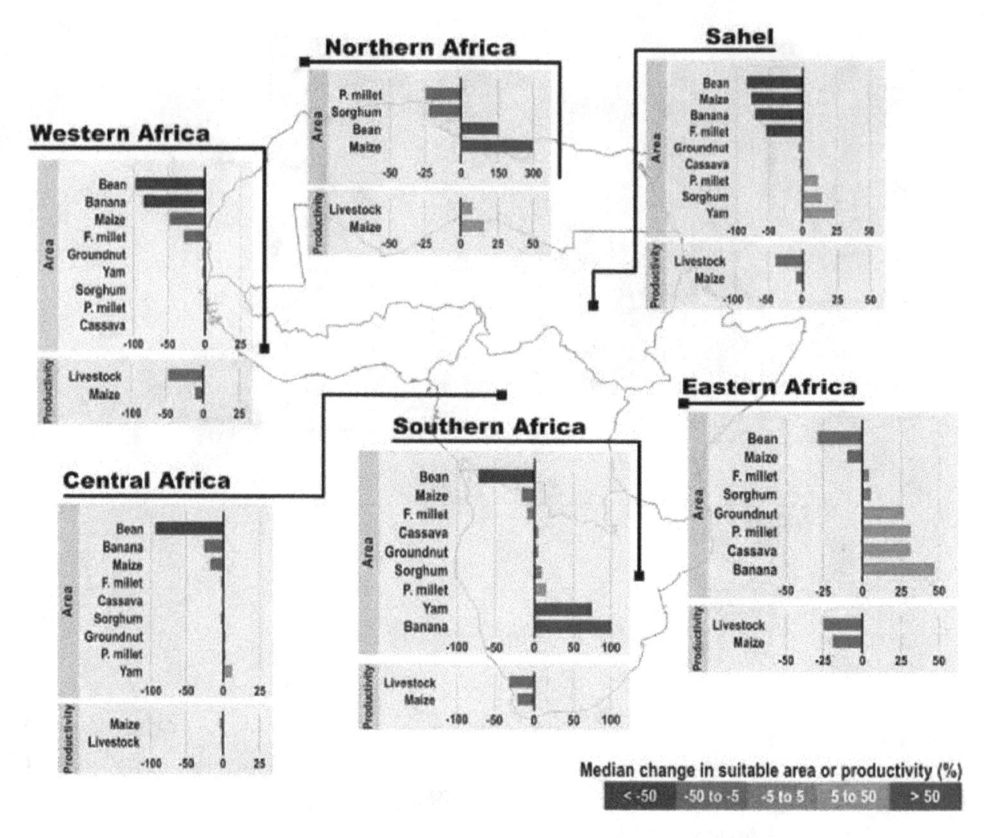

Fig. 2.5 Projected median changes in climatically suitable area and productivity by 2050s and RCP8.5, relative to a historical period (1970–2000). Median values given are based on ensemble simulations of niche and productivity models, and therefore should be interpreted in light of associated uncertainties. Livestock productivity refers to annual net primary productivity (ANPP) of rangelands (a proxy for livestock productivity), rather than to a direct measure of meat or milk productivity. (Source: Dinesh et al. 2015)

incoming shortwave radiation), many places are likely to have less available water both in streams and in the soil, because warmer temperatures will cause more water to evaporate directly from the soil or to be transpired through plants (Girvetz and Zganjar 2014).

Changes to temperature and precipitation have immediate implications for food production and security across the continent (Niang et al. 2014; Porter et al. 2014; Muller et al. 2011; Rosenzweig et al. 2014). Current growing areas of maize and beans are projected to experience yield reductions of 12–40% by the 2050s. The climate suitability of most major crops is also projected to shift as climate warms (Rippke et al. 2016; Zabel et al. 2014). Ramirez-Villegas and Thornton (2015) have shown that two of Africa's staple crops—maize and beans—are projected to have severe decreases in suitability across much of the continent (Fig. 2.5). Increasing atmospheric levels of carbon dioxide is likely to affect the nutrient content of plants, resulting in serious protein and micro-nutrient cold spots in parts of sub-Saharan Africa (Medek et al. 2017; Myers et al. 2014, 2015a, b). This poses a serious concern for food security and nutrition. While adaptation in the short- and mid-term may help some areas to continue growing these crops, by the end of the century it is estimated that over 30% of the area where maize is grown and over 60% of the area where beans are grown would need to grow entirely different crops (Rippke et al. 2016).

Some crops are much more resilient than maize and beans to changes in climate. In southern Africa, Ramirez-Villegas and Thornton (2015) showed that the suitable area for beans decreased greatly and maize decreased slightly, whereas other crops—banana, yam and pearl millet—increased in range. Similar results were found in East Africa, where maize and beans are projected to experience major decreases in suitable area, whereas cassava, groundnut, pearl millet and banana are projected to increase. These more resilient crops could be promoted as replacement options for areas that require adaptation.

Under climate change, pressures from pests, weeds and diseases are also expected to increase. In the highland regions of East Africa, warming trends could lead to the expansion of crop pests—such as the coffee berry borer—into previously cold-limited areas (Jaramillo et al. 2011). Threats to banana production could come from range expansion of burrowing nematodes (Nicholls et al. 2008) and black leaf streak disease. Striga weed, a major cause of cereal yield reduction in sub-Saharan Africa, could become a more widespread problem because of changes in temperature, rainfall and seasonality (Niang et al. 2014). By contrast, climate change may reduce the range of major cassava pests including whitefly, cassava mealybug, cassava brown streak virus and cassava mosaic geminivirus (Jarvis et al. 2012). However, certain areas of current cassava production—including Southeast Africa and Madagascar—may see an increase in whiteflies, mites and mealybugs (Bellotti et al. 2012). In the case of livestock, changes in temperature and rainfall could increase the suitability of the main tick vector of East Coast fever across much of Southern Africa (Olwoch et al. 2008).

2.4 Implications for Development

2.4.1 Adapting African Agriculture to Climate Change

African agriculture must adapt in order to ensure food and nutritional security. Management adjustments and crop breeding will be critical in the short- and mid-term, whereas at longer timescales planned transformations will likely be necessary (Rippke et al. 2016; Rickards and Howden 2012). Farmers and agricultural service providers—input suppliers, extension, financial services, safety net programs, etc.—will need to become resilient to new climate variability.

Although there is uncertainty in future climate projections, we have a great deal of solid information regarding how climate is already changing and the types of impacts farmers will need to address into the future. It is certain that temperatures are rising and will continue to rise at a rapid pace. Although climate models often do not agree on precipitation changes, there is considerable agreement on the trends in some locations. Moreover, precipitation is becoming more extreme in many places, often resulting in too much rain, too little rain, or rain that falls at the wrong time. A location might experience more overall rain during the growing season, but if it falls intensively in the beginning or middle of the season, the end of the season might be too dry, especially as hotter temperatures dry out soils. A single location might experience increased flooding during the middle of the rainy season and increased aridity later. This situation creates a need for crop varieties that can withstand waterlogging, help prevent erosion from heavy rains, and reach maturity during a shorter growing season.

Recent studies show that the types of management practices beneficial for adaptation and increased productivity are highly varied (Challinor et al. 2014; Lamanna et al. 2016). For instance, a recent review and meta-analysis of field studies in Uganda and Tanzania found more than 20 practices in each country that could improve adaptation and productivity, each with varying effectiveness depending on the farming system and site in question. The use of fertilizers (both organic and inorganic) and water saving techniques generally have the largest positive effects on crop productivity (Lamanna et al. 2015). Similar findings have been reported elsewhere in Africa (Rosenstock et al. this volume).

Improving the available crop varieties is a key mid-term strategy to increase productivity, improve production stability and adapt to projected climate changes. For example, although climate change will hurt bean production across Africa (Rippke et al. 2016; Ramirez-Villegas and Thornton 2015), heat-tolerant bean varieties could greatly reduce the impact (CIAT 2015). Current work on inter-specific crosses between common and tepary bean show promise for creating breeding lines that maintain yield under heat stress (CIAT 2015). Similarly, drought-tolerant varieties of maize could be an option for adaptation to reduced or inconsistent rainfall (Cairns et al. 2013; Rippke et al. 2016).

In the long-term, planned transformations will be required for some areas. Rippke et al. (2016) report that some 3–5% of the arable land of sub-Saharan Africa may require a transformation out of crop-based systems to either livestock-based systems or to an entirely new land use.

2.4.2 Collecting and Using Climate Data

Historical data and climate projections clearly establish the need to act quickly to help African farmers adapt to a changing climate. Too often, however, CSA interventions are being promoted without a proper understanding of the climate risks for the specific areas involved. In some cases, reliable information on tightly focused geographical areas simply has not been collected. And even when good information is available from climate models and impact studies, often this information is not presented in ways that are accessible and comprehensible for those doing the CSA planning and implementation.

Although historic climate trends can be identified across the African continent, there is a general lack of high-resolution data, including for key biomes and agricultural areas (white areas in Fig. 2.1 showing insufficient data). Some national meteorological agencies have made efforts to improve the information available by, for example, combining weather-station information with satellite imagery to create high-resolution gridded historical time-series climate datasets (Dinku et al. 2016). Overall, though, there remains a lack of precise information for decision-making. Weather-station record keeping has declined over the past decades due to lack of maintenance and a failure to install new stations. This trend must be reversed. Increasing the available data on observed weather across the continent is critical to understanding the changes that have occurred already, to predict future changes, and to plan appropriately to address them.

Even as new data are collected, development practitioners and decision-makers should make use of the information and tools now available to help them understand the climate context. CMIP5 projections are freely available through the Climate Wizard, a web application that allows anyone to easily query and map downscaled future climate change projections for specific places globally (ClimateWizard.org). Similarly, the Servir ClimateServ allows for easy online analysis and querying of historic observed precipitation, vegetation greenness and moisture stress, as well as seasonal forecasts looking forward in the short-term for most of the globe (climateserv.servirglobal.net/). More training is needed to help those implementing CSA learn how to access and use these tools. Such training should include profiling of CSA opportunities, prioritization of investment portfolios, design and implementation of CSA projects, and assessing the results of CSA projects (Girvetz et al. 2017).

In conclusion, we would like to emphasize these key points:

- The climate has already changed, with temperatures continuing to rise and precipitation patterns changing, and more disruption is certain in coming years and decades.
- The collection of weather observations at local weather stations must improve, and should be incorporated with satellite data.
- Climate data and tools are available and accessible to practitioners. More effort, however, should be put into disseminating this information and ensuring that development practitioners understand how it can be used for CSA planning and implementation.
- Given the uncertainties surrounding exactly how climate change will affect specific places, the best CSA options are those that build resilience and help farmers cope with a wide range of climate risks, especially heat, drought, erosion and flooding.

Farmers are already suffering from the effects of climate change. Average temperatures are rising, rainfall is becoming less predictable, and extreme weather events are growing more common. The situation poses a real and ever-increasing threat to rural livelihoods and food security. Government, civil society and the private sector must work together urgently to collect and analyze climate information, make it accessible to decision-makers on the ground, and to ensure that CSA planning and implementation are carried out based on the best information available.

References

Belloti A, Herrera Campo BV, Hyman G (2012) Cassava production and pest management: present and potential threats in a changing environment. Trop Plant Biol 5(1):39–72

Cairns JE, Crossa J, Zaidi PH et al (2013) Identification of drought, heat, and combined drought and heat tolerant donors in maize. Crop Sci 53:1335–1346. https://doi.org/10.2135/cropsci2012.09.0545

Challinor A, Wheeler T, Garforth C et al (2007) Assessing the vulnerability of food crop systems in Africa to climate change. Clim Chang 83:381–399. https://doi.org/10.1007/s10584-007-9249-0

Challinor AJ, Watson J, Lobell DB et al (2014) A meta-analysis of crop yield under climate change and adaptation. Nat Clim Chang 4:287–291. https://doi.org/10.1038/nclimate2153

CIAT (2015) Developing beans that can beat the heat. International Center for Tropical Agriculture (CIAT), Cali

Collins M, Knutti R, Arblaster J et al (2013) Long-term climate change: projections, commitments and irreversibility. In: Stocker TF, Qin D, Plattner GK et al (eds) Climate change 2013: the physical science basis, Contribution of Working Group I to the Fifth Assessment Report of the Intergovernmental Panel on Climate Change. Cambridge University Press, Cambridge, UK

Dinesh D, Bett B, Boone R et al (2015) Impact of climate change on African agriculture: focus on pests and diseases. CGIAR Research Program on Climate Change, Agriculture and Food Security (CCAFS), Copenhagen

Dinku T, Cousin R, Corral J et al (2016) The ENACTS approach: transforming climate services in Africa one country at a time. World Policy Papers, pp 1–24

Girvetz EH, Zganjar C (2014) Dissecting indices of aridity for assessing the impacts of global climate change. Clim Chang 126:469–483

Girvetz EH, Zganjar C, Shafer S et al (2009) Applied climate-change analysis: the Climate Wizard tool. PLoS One 4(12):e8320

Girvetz EH, Corner-Dolloff C, Lamanna C et al (2017) CSA–plan: strategies to put CSA into practice. Agric Dev 30:12–16

Hartmann DL, Klein Tank AMG, Rusticucci M et al (2013) Observations: atmosphere and surface. In: Stocker TF, Qin D, Plattner G-K et al (eds) Climate change 2013: the physical science basis, Contribution of Working Group I to the Fifth Assessment Report of the Intergovernmental Panel on Climate Change. Cambridge University Press, Cambridge, UK

IPCC (2012) IPCC 2012: summary for policymakers. In: Field CB, Barros V, Stocker TF et al (eds) Managing the risks of extreme events and disasters to advance climate change adaptation, A Special Report of Working Groups I and II of the Intergovernmental Panel on Climate Change. Cambridge University Press, Cambridge, UK

Jaramillo J, Muchugu E, Vega FE et al (2011) Some like it hot: the influence and implications of climate change on coffee berry borer (*Hypothenemus hampei*) and coffee production in East Africa. PLoS One 6(9):e24528

Jarvis A, Ramirez-Villegas J, Herrera Campo BV et al (2012) Is cassava the answer to African climate change adaptation? Trop Plant Biol 5(1):9–29

Kirtman B, Power SB, Adedoyin JA et al (2013) Near-term climate change: projections and predictability. In: Stocker TF, Qin D, Plattner G-K et al (eds) Climate change 2013: the physical science basis, Contribution of Working Group I to the Fifth Assessment Report of the Intergovernmental Panel on Climate Change. Cambridge University Press, Cambridge, UK

Lamanna C, Namoi N, Kimaro A et al (2016) Evidence-based opportunities for out-scaling climate-smart agriculture in East Africa. CCAFS Working Paper no. 172. Copenhagen

Medek DE, Schwartz J, Myers SS (2017) Estimated effects of future atmospheric CO2 concentrations on protein intake and the risk of protein deficiency by country and region. Environ Health Perspect 125(8):087002

Müller C, Cramer W, Hare WL et al (2011) Climate change risks for African agriculture. Proc Natl Acad Sci. https://doi.org/10.1073/pnas.1015078108

Myers SS, Wessells KR, Kloog I et al (2015a) Effect of increased concentrations of atmospheric carbon dioxide on the global threat of zinc deficiency: a modelling study. Lancet Glob Health 3(10):e639–e645

Myers S, Zanobetti Z, Kloog I et al (2015b) Increasing CO2 threatens human nutrition. Nature 510:139–142

New M, Porter JR, Xie L et al (2006) Evidence of trends in daily climate extremes over southern and west Africa. J Geophys Res 111:D14102. https://doi.org/10.1029/2005JD006289

Niang I, Ruppel OC, Abdrabo MA et al (2014) Africa. In: Barros VR et al (eds) Food security and food production systems. Climate Change 2014: impacts, adaptation, and vulnerability. Part B: regional aspects. Contribution of adaptation and vulnerability, Working Group II Contribution to the IPCC 5th Assessment Report of the Intergovernmental Panel on Climate Change. Cambridge University Press, Cambridge, UK, pp 1199–1265

Nicholls T, Norgrove L, Masters G (2008) Innovative solutions to new invaders: managing agricultural pests, diseases and weeds under climate change. In: Proceedings of agriculture in a changing climate: the new international research frontier, vol 3. ATSE Crawford Fund fourteenth annual development conference, pp 9–14

NOAA (2018) Climate at a glance: global time series. NOAA National Centers for Environmental information. Available at: http://www.ncdc.noaa.gov/cag/

Olwoch J, Reyers B, Engelbrecht F et al (2012) Climate change and the tick-borne disease, Theileriosis (East Coast fever) in sub-Saharan Africa. J Arid Environ 72(2):108–120

Ramirez-Villegas J, Thornton PK (2015) Climate change impacts on African crop production, CCAFS Working Paper no. 119. CGIAR Research Program on Climate Change, Agriculture and Food Security (CCAFS), Copenhagen Available at: www.ccafs.cgiar.org

Ramirez-Villegas J, Challinor AJ, Thornton PK et al (2013) Implications of regional improvement in global climate models for agricultural impact research. Environ Res Lett 8:24018

Rickards L, Howden SM (2012) Transformational adaptation: agriculture and climate change. Crop Pasture Sci 63:240–250. https://doi.org/10.1071/CP11172

Rippke U, Ramirez-Villegas J, Jarvis A et al (2016) Timescales of transformational climate change adaptation in sub-Saharan African agriculture. Nat Clim Chang 6(6):605–609

Seneviratne SI, Nicholls N, Easterling D et al (2012) Changes in climate extremes and their impacts on the natural physical environment. In: Field CB, Barros V, Stocker TF et al (eds) Managing the risks of extreme events and disasters to advance climate change adaptation. A special report of working groups I and II of the intergovernmental panel on climate change. pp 109–230

Shackleton S, Ziervogel G, Sallu SM et al (2015) Why is socially-just climate change adaptation in sub-Saharan Africa so challenging? A review of barriers identified from empirical cases. Wiley Interdiscip Rev Clim Chang 6(3):321–344

Sillmann J, Kharin VV, Zhang X et al (2013) Climate extremes indices in the CMIP5 multimodel ensemble: part 1. Model evaluation in the present climate. J Geophys Res Atmos 118:1716–1733. https://doi.org/10.1002/jgrd.50203

Zabel F, Putzenlechner B, Mauser W (2014) Global agricultural land resources—a high resolution suitability evaluation and its perspectives until 2100 under climate change conditions. PLoS One 9:e107522. https://doi.org/10.1371/journal.pone.0107522

3

Transmission of Infectious Diseases in Livestock: Impact of Climate Change

Bernard Bett, Johanna Lindahl, and Grace Delia

3.1 Background

The global-average surface temperature has risen steadily since the nineteenth century due to an increase in the concentration of heat-trapping gases such as carbon dioxide and methane in the atmosphere. These changes have had important consequences on rainfall patterns, the intensity of droughts, and the viability of ecosystems (Martin et al. 2008) among other changes. Taken together, these changes have substantial effects on the transmission patterns of infectious diseases.

A few studies have been done to identify processes through which climate change influences infectious disease occurrence. While more work needs to be done to fully characterise these processes, the existing knowledge suggests two broad categories of impact, often classified as 'direct' and 'indirect'. Direct impacts are realised when a rise in temperature, precipitation intensity, flooding, humidity, etc. increase pathogens' or vectors' metabolic processes, reproductive rates, and (or) population densities, resulting in enhanced vector–pathogen–host contact and, therefore, the risk of disease (Bett et al. 2017). These changes operate within defined biological limits. This is because an increase in temperature or flooding beyond a given threshold leads to the desiccation of these arthropods or the flushing of vector breeding sites, and hence a decline in disease transmission risk. Direct effects are often associated with diseases caused by pathogens that spend part of their life cycles outside

B. Bett (✉) · G. Delia
International Livestock Research Institute, Nairobi, Kenya
e-mail: b.bett@cgiar.org; d.grace@cgiar.org

J. Lindahl
International Livestock Research Institute, Nairobi, Kenya

Uppsala University, Uppsala, Sweden

Swedish University of Agricultural Sciences, Uppsala, Sweden
e-mail: J.Lindahl@cgiar.org

a mammalian host. These include vector-borne diseases, helminthoses and fungal infections.

Indirect effects are less apparent and would include changes in disease transmission patterns associated with climate-induced ecological, socio-cultural or behavioural disruptions. In pastoral areas, for example, prolonged droughts trigger more frequent and long-distance movements which enhance contact between distinct populations of animals. This would also include movement into previously uninhabited areas potentiating exposure to new disease agents. Different indirect effects are reviewed in Lindahl and Grace (2015).

We use two well-studied vector-borne diseases—Rift Valley fever, which often occurs in epidemics in East Africa, and tick-borne diseases, which are endemic in many parts of the world—to demonstrate the impacts of climate change on livestock diseases. Our review focuses on the direct effects given that indirect effects are not well studied and are also difficult to quantify.

3.2 Case Studies

3.2.1 *Rift Valley Fever*

Rift Valley fever (RVF) is a mosquito-borne viral zoonosis mainly affecting sheep, goats, cattle, buffaloes and camels. People become infected following a bite from an infected mosquito, or after close contact with acutely infected animals or infected tissues. In people, the disease manifests as a mild influenza-like syndrome in a majority of cases (more than 80%) or a severe disease with haemorrhagic fever, encephalitis, or retinitis in a few cases (Njenga et al. 2009). In livestock, the disease manifests as increased abortion and perinatal mortality rates.

3.2.1.1 Drivers

RVF outbreaks have been reported in some countries in East and southern Africa including Kenya, South Africa, Tanzania and Uganda following periods of above-normal precipitation. The disease has also been reported in other countries including the Comoros archipelago, Madagascar, Mauritania, Saudi Arabia, Senegal and Sudan (Madani et al. 2003) and Yemen (Abdo-Salem et al. 2006). In South Africa, recent RVF outbreaks observed in 2008–2011 were associated with relentless and widespread strong seasonal rainfall and high soil saturation (Williams et al. 2016). Areas affected by these outbreaks are shown in Fig. 3.1. In East Africa, major outbreaks are often associated with the warm phase of the El Niño/Southern Oscillation (ENSO) phenomenon, although there have been a few incidences (e.g. in mid-1989) when an elevated RVF activity was not ENSO-driven. There have also been localised outbreaks in Uganda associated with seasonal rainfall and flooding. Figure 3.2

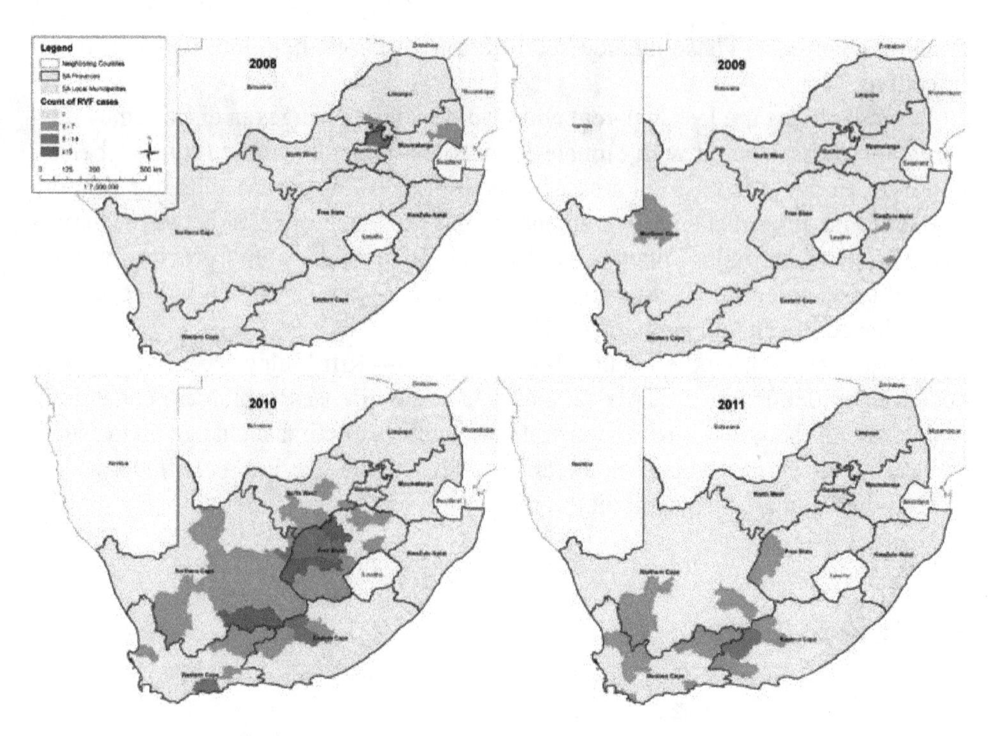

Fig. 3.1 The spatial distribution of laboratory-confirmed human cases in South Africa in local administrative municipalities 2008–2011 (Archer et al. 2013). The spatiotemporal distribution of the RVF cases in humans paralleled those of livestock, which were triggered by heavy rainfall

gives an RVF risk map based on data that were collected during the 2006/2007 outbreak. Not all El Niño events lead to RVF outbreaks; El Niño events recorded in Kenya in 1964, 1969, 1972–1973, 1981 and 1991–1995, for example, did not lead to RVF outbreaks.

In West Africa, RVF outbreaks occurred in 1998, 2003, 2010 and 2012 following an interlude between a dry period, lasting for about a week, and a period of heavy precipitation (Caminade et al. 2014). The 2009–2010 outbreak, which affected small ruminants, camels and people was associated with a fourfold increase in rainfall in a desert region in northern Mauritania (Faye et al. 2014). Similar outbreaks occurred in Senegal in 2013–2014, exacerbated by extensive livestock movements that aided the dissemination of the virus (Sow et al. 2016).

3.2.1.2 Climate Change and RVF

A few studies have been done to evaluate the expected impacts of climate change on RVF transmission. These suggest that climate change is likely to expand RVF's geographical range due to expansion of the vector niches (Mweya et al. 2017; Taylor et al. 2016). There are also indications that the average rainfall in eastern Africa, including the Horn of Africa, is expected to increase, while that for southern Africa is likely to decline with climate change (Conway 2009). ENSO-related precipitation

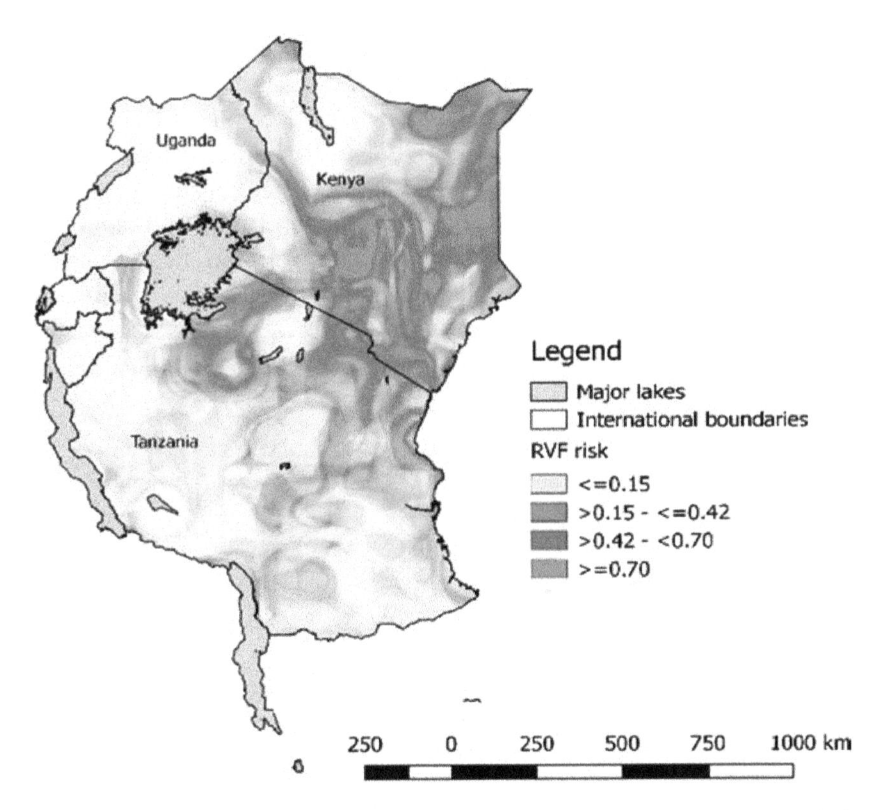

Fig. 3.2 Risk map for RVF in East Africa based on reported cases in livestock during the 2006/2007 outbreak and rainfall distribution (Bett et al. 2017). Probability estimates given on the map indicate the chance that a given area could experience an RVF outbreak based on the environmental conditions observed in December 2006

variability is also predicted to intensify (Intergovernmental Panel on Climate Change 2014) and this might have stronger implications for RVF, given that the intense precipitation which drives an RVF epidemic follows El Niño and La Niña events in East and South Africa, respectively. Further evidence for an increased precipitation under intermediate warming scenarios in parts of equatorial East Africa is provided by Hulme et al. (2001). They suggest that the region will experience a 5–20% increase in rainfall between December and February and 5–10% reduction in rainfall between June and August by 2050.

Climate change may indirectly increase RVF risk through land use change associated with the development of irrigation schemes and dams. The increasing frequency of droughts and erratic rainfall in arid and semi-arid areas would necessitate the construction of dams and irrigation schemes to support water supply and food production. Previously, outbreaks of RVF have been reported following flood irrigation in the Orange River region and Western Cape province, South Africa (Williams et al. 2016), and the construction of dams, i.e. Aswan High Dam in Egypt in 1977 and the Senegal Dam on the Senegal–Mauritania border in 1987/1988 (Martin et al. 2008). A recent study conducted in Kenya confirmed that irrigated areas in arid and semi-arid areas support endemic transmission of RVF (Mbotha et al. 2017) but more work is needed to isolate the virus from such cases to confirm observations made from serological studies. Drought-resistant livestock species, including goats, which are thought to play a critical role in the epidemiology of RVF and other zoonotic diseases, are increasingly being raised in arid and semi-arid areas as one of the adaptation measures for climate change and variability. These changes are likely to increase the risk of infectious diseases that would compromise health and livelihoods of a large population of pastoralists.

3.2.2 Ticks and Tick-Borne Diseases (TBDs)

Ticks are important vectors of a wide range of pathogens that cause many diseases in livestock such as anaplasmosis, babesiosis, cowdriosis, coxiellosis (Q fever), Crimean–Congo haemorrhagic fever, ehrlichiosis and theileriosis. East Coast fever (ECF)—the disease with the greatest economic importance in dairy animals—is caused by *Theileria parva* and transmitted by *Rhipicephalus appendiculatus.* The disease causes high mortality, especially in highly productive, susceptible breeds where mortality can reach 100%. Other losses associated with the disease include poor weight gain, fertility losses, reduced growth and productivity, paralysis and increased susceptibility to other diseases. Its geographical range stretches from South Sudan to South Africa and up to Democratic Republic of the Congo (DRC) (Olwoch et al. 2008).

There are few studies in East and southern Africa that have looked at the effects of climate change on ticks and TBDs. Olwoch et al. (2008) applied a simple climate envelope model to investigate the effects of climate change on the distribution of *R. appendiculatus* and ECF in sub-Saharan Africa, based on climate anomalies for 2020s versus 1990s. They predicted a reduction in the range of the tick in the western arid regions in Angola, southern DRC and Namibia, given that these areas were already hot and dry and further increases in temperature would make them unsuitable under the future climate scenarios used. On the contrary, the study established that some areas in Botswana, eastern DRC, the Northern and Eastern Cape provinces of South Africa, and Zambia would become more suitable in the 2020s, because of increased rainfall and a rise in the minimum temperatures.

From a global perspective, a rise in temperature has the potential to expand the geographical range of about 50% of tick species, with 70% of these involving economically important tick species (Cumming and van Vuuren 2006). This mainly represents the northern expansion of the northern limits of ticks as has been observed in Sweden and Russia among other places.

3.2.3 *Inferences from the Case Studies*

These two cases show that climate change will cause local shifts in geographical ranges of most vector-borne diseases in both dry/hot and cool/wet areas due to at least two distinct processes. In the hot/dry areas, scenarios of higher rainfall and humidity would promote higher survival rates of vectors, while in the cool/wet areas, increasing temperatures would allow overwintering of these vectors. The key determinants of vectors' population dynamics include temperature, humidity and water availability, especially for mosquitoes. Although we point to potential shifts in disease risk, we believe climate change would affect transmission patterns of infectious diseases in multiple ways, including lowering the effectiveness of existing intervention strategies. No studies have been done to verify this issue but given that the rate of development of most arthropods would increase with temperature and lead to changing population dynamics, the frequency of application of some of the vector control measures such as acaricides might need to be reviewed. High temperatures also reduce the hosts' immune responses (Dittmar et al. 2014) and studies need to be done to determine whether this has implications on the effectiveness of the available vaccines which confer protection by priming the hosts' immune system.

3.3 Mitigations and Adaptations

Projections from simulation models suggest that global warming will continue to worsen if the current levels of greenhouse gas emissions are not reduced. It is, therefore, expected that the incidence and impacts of climate-sensitive diseases—including RVF and TBDs—will increase, particularly among the most vulnerable populations in developing countries. These diseases, though, can be mitigated by established control measures including quarantine, import bans, the identification and removal of suspicious animals and premises, surveillance and reporting, vaccination, disinfection, and compensation (Grace and McDermott 2012). However, the effectiveness of some of these measures in the face of climate change has not been determined. Moreover, their deployment is inadequate as the animal health systems in most of these countries have deteriorated.

Vector control and vaccination are often used to control RVF and TBDs. Vector control is however not a reliable measure for controlling RVF in livestock (Gachohi et al. 2017). This is because floods that trigger RVF epidemics maintain high mosquito population densities and insecticide-induced mortality rates would be much lower compared to the rates of development and emergence of new adults. Conversely, acaricides have been used successfully for many years to control TBDs but recent observations indicate that tick resistance to acaricides is threatening to limit the effectiveness of this measure. Alternative ways of managing TBDs are therefore being developed, such as the use of tick vaccines (specifically for

Boophilus spp.), immunisation of animals through infection-and-treatment methods (ITM), breeding of TBD-resistant animals, and the strategic use of acaricides to balance the need to eliminate ticks versus the need to raise the endemic stability of TBDs in the livestock populations. There are ongoing studies to develop new vaccines to replace ITM.

RVF can be reliably controlled using livestock vaccination but its episodic occurrence, a predilection for remote, pastoral areas and lack of forward planning and pre-allocation of emergency funds in most animal health institutions cause a lot of delays in response. An assessment of emergency vaccination programmes that were implemented following the 2006/2007 outbreak in Kenya (Gachohi et al. 2012) as well as those deployed during the recent RVF scare in 2016/2017 showed that livestock vaccination was implemented late and at very low levels to attain sustainable herd immunity. It has now been realised that the administration of livestock vaccination as part of emergency response measures during periods of heightened RVF risk does not provide beneficial outcomes. They fall short of achieving critical levels of coverage that are required for the establishment of protective immunity. Research is underway to determine alternative vaccination strategies for RVF that might involve periodic vaccination in the high-risk areas in place of reactive or emergency vaccinations. In this case, reactive vaccinations can be used strategically to complement periodic vaccination following warnings for El Niño in East Africa or La Niña in South Africa.

Animal health programmes need to be underpinned by efficient surveillance systems which promptly detect and report disease occurrence patterns for action, and guide the prioritisation of interventions to geographical regions or periods where/when interventions can yield desirable outcomes. There have been multiple uncoordinated efforts towards improving disease surveillance and the development of risk maps and contingency plans in the target areas to help in rationalising interventions. New surveillance systems based on citizen science methods and cloud computing offer great opportunities for identifying the distribution of these infectious diseases; they might also provide clues on how to deploy measures for multiple diseases at the same time. In addition, these systems can be programmed to provide input data for real-time disease forecasting, enabling decision-makers to plan more effectively for impending disease risks. This would require analysing such surveillance data with climate and land use/land cover data as predictors to generate dynamic risk maps.

3.4 Conclusions and Implications for Development

Climate change is expected to increase the risk of many vector-borne diseases, including those of RVF and TBDs. It is also likely to reduce the effectiveness of some of the control measures—such as vector control efforts—and hence decision makers need to be sensitized more on how to make the best use of the existing interventions, as more research is implemented to determine optimal control options. A key trend of recent decades has been the greater integration of human and veterinary medicine. One World One Health is a growing movement built around the premise that the health of humans, animals and the environment are inextricably linked, and that disease is best managed in broad and interdisciplinary collaborations. Such a multidisciplinary approach can improve targeting of interventions.

References

Abdo-Salem S, Gerbier G, Bonnet P et al (2006) Descriptive and spatial epidemiology of Rift Valley fever outbreak in Yemen 2000–2001. Ann N Y Acad Sci 1081:240–242. https://doi.org/10.1196/annals.1373.028

Archer BN, Thomas J, Weyer J et al (2013) Epidemiologic investigations into outbreaks of Rift Valley fever in humans, South Africa, 2008–2011. Emerg Infect Dis 19:1918–1925. https://doi.org/10.3201/eid1912.121527

Bett B, Kiunga P, Gachohi J et al (2017) Effects of climate change on the occurrence and distribution of livestock diseases. Prev Vet Med. https://doi.org/10.1016/j.prevetmed.2016.11.019

Caminade C, Ndione JA, Diallo M et al (2014) Rift Valley fever outbreaks in Mauritania and related environmental conditions. Int J Environ Res Public Health 11:903–918. https://doi.org/10.3390/ijerph110100903

Conway G (2009) The science of climate change in Africa: impacts and adaptation. Imperial College, London

Cumming G, van Vuuren D (2006) Will climate change affect ectoparasite species ranges? Glob Ecol Biogeogr 15:486–497

Dittmar J, Janssen H, Kuske A et al (2014) Heat and immunity: an experimental heat wave alters immune functions in three-spined sticklebacks (*Gasterosteus aculeatus*). J Anim Ecol 83:744–757. https://doi.org/10.1111/1365-2656.12175

Faye O, Ba H, Ba Y et al (2014) Reemergence of Rift Valley fever, Mauritania, 2010. Emerg Infect Dis 20:300–303. https://doi.org/10.3201/eid2002.130996

Gachohi JM, Bett B, Njogu G et al (2012) The 2006–2007 Rift Valley fever outbreak in Kenya: sources of early warning messages and response measures implemented by the Department of Veterinary Services. Rev Sci Tech 31:877–887

Gachohi JM, Njenga MK, Kitala P et al (2017) Correction: modelling vaccination strategies against Rift Valley fever in livestock in Kenya. PLoS Negl Trop Dis 11:e0005316. https://doi.org/10.1371/journal.pntd.0005316

Grace D, McDermott J (2012) Livestock epidemic. In: Wisner B, Gaillard JC, Kelman I (eds) Handbook of hazards and disaster risk reduction and management. Routledge, London

Hulme M, Doherty R, Ngara T et al (2001) African climate change: 1900–2100. Clim Res 17:145–168

Intergovernmental Panel on Climate Change (2014) Synthesis report. In: Pachauri RK, Meyer LA (eds) Contribution of working groups I, II and III to the fifth assessment report of the IPCC. IPCC, Geneva

Lindahl JF, Grace D (2015) The consequences of human actions on risks for infectious diseases: a review. Infect Ecol Epidemiol 5:30048. https://doi.org/10.3402/iee.v5.30048

Madani TA, Al-Mazrou YY, Al-Jeffri MH et al (2003) Rift Valley fever epidemic in Saudi Arabia: epidemiological, clinical, and laboratory characteristics. Clin Infect Dis 37:1084–1092. https://doi.org/10.1086/378747

Martin V, Chevalier V, Ceccato P et al (2008) The impact of climate change on the epidemiology and control of Rift Valley fever vector-borne diseases Rift Valley fever and climate change. Rev Sci Technol Off Int Epiz 27:413–426

Mbotha D, Bett B, Kairu-Wanyoike S et al (2017) Inter-epidemic Rift Valley fever virus seroconversions in an irrigation scheme in Bura, south-east Kenya. Transbound Emerg Dis. https://doi.org/10.1111/tbed.12674

Mweya CN, Mboera LEG, Kimera SI (2017) Climate influence on emerging risk areas for Rift Valley fever epidemics in Tanzania. Am J Trop Med Hyg 97:109–114. https://doi.org/10.4269/ajtmh.16-0444

Njenga MK, Paweska J, Wanjala R et al (2009) Using a field quantitative real-time PCR test to rapidly identify highly viremic Rift Valley fever cases. J Clin Microbiol 47:1166–1171. https://doi.org/10.1128/JCM.01905-08

Olwoch JM, Reyers B, Engelbrecht FA et al (2008) Climate change and the tick-borne disease, Theileriosis (East Coast fever) in sub-Saharan Africa. J Arid Environ 72:108–120. https://doi.org/10.1016/j.jaridenv.2007.04.003

Sow A, Faye O, Ba Y et al (2016) Widespread Rift Valley fever emergence in Senegal in 2013–2014. Open Forum Infect Dis 3:ofw149. https://doi.org/10.1093/ofid/ofw149

Taylor D, Hagenlocher M, Jones AE et al (2016) Environmental change and Rift Valley fever in eastern Africa: projecting beyond healthy futures. Geospat Health 11:387. https://doi.org/10.4081/gh.2016.387

Williams R, Malherbe J, Weepener H et al (2016) Anomalous high rainfall and soil saturation as combined risk indicator of Rift Valley fever outbreaks, South Africa, 2008–2011. Emerg Infect Dis 22. https://doi.org/10.3201/eid2212.151352

4

Future Suitability of Large Scale Crops and Climate Change

Roland Hunter and Olivier Crespo

4.1 Introduction

The *planalto* midlands is a plateau that extends across central Angola, including the majority of the provinces of Huíla, Benguela, Cuanza Sul, Bié, Huambo and Malanje (see Fig. 4.1). The plateau ranges in altitude from 800 to 1600 m above sea level and extends eastwards from the escarpment above the semiarid coastal region towards the central highlands of the country. The Köppen-Geiger classification defines the climate of the interior plateau as "temperate with dry winters and warm/hot summers" (Köppen-Geiger abbreviations Cwa and Cwb, respectively), while the lowlands between the coast and plateau are classified as arid steppe (BSh). Collectively, the region represented by the arid lowlands of Huíla, Benguela, Cuanza Sul and the comparatively temperate highlands above supports a diverse and productive agriculture sector and is a major producer of economically important staple and cash crops.

There is a risk that climate change will undermine the potential contributions of these crops toward national objectives for sustainable development and food security. However, stakeholders are unable to plan for or respond to the risks posed by climate change to agricultural productivity, food security and socioeconomic development, due to the absence of more detailed information to assess the scope and scale of climate-change impacts.

This study assessed the likely impact of climate change on the future suitability of Angola's *planalto* region on two staple crops commonly grown in the region,

R. Hunter (✉)
African Climate and Development Initiative, University of Cape Town,
Cape Town, South Africa
e-mail: roland.hunter@uct.ac.za

O. Crespo
Climate System Analysis Group, University of Cape Town, Cape Town, South Africa
e-mail: olivier@csag.uct.ac.za

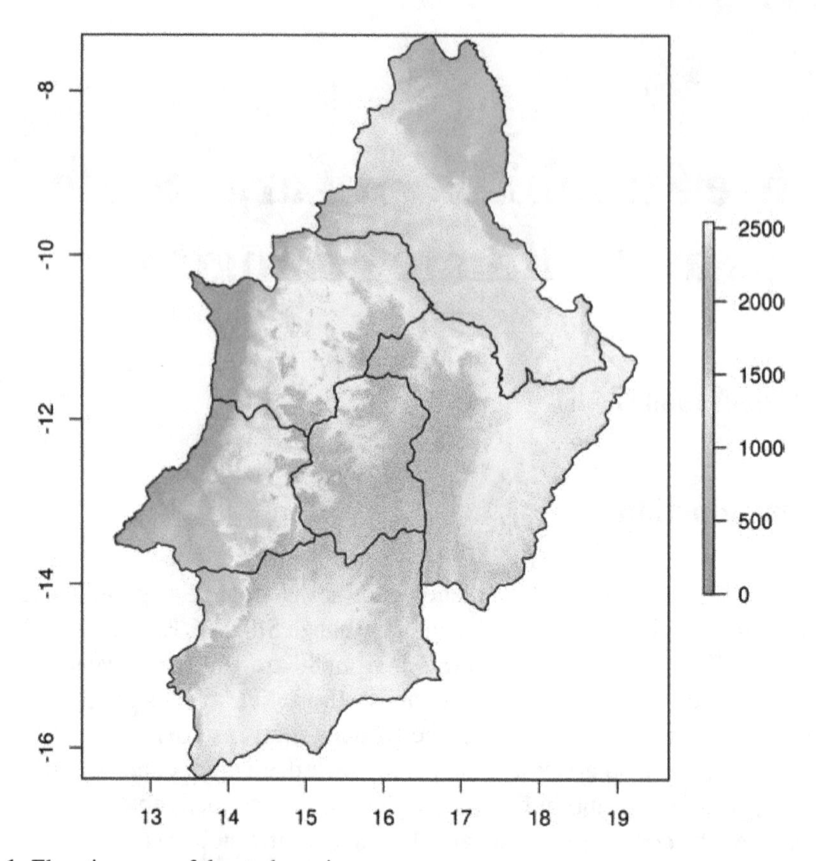

Fig. 4.1 Elevation map of the study region

namely cassava (*Manihot esculentum*) and maize (*Zea mays*), which are respectively the first and second most important staple crops by area of cultivation and total production. Using a model-based approach, this study assessed the impacts of climate change on the spatial extent of areas classified as climatically suitable for maize and cassava, between the "historical baseline" period (i.e., the present) and a future date (2050). The goal of these analyses is to improve decision-making and spatial planning regarding which crops, cultivars and farming practices should be promoted as part of a strategy for climate-resilient agricultural and socio-economic development in the *planalto*.

4.2 Materials and Methods

4.2.1 Sources of Climate Data

Baseline climate data for the study area was derived from Worldclim historical data, which provides average monthly climate data for minimum, mean and maximum temperature and for precipitation for the period 1960–1990 at a spatial resolution of about 1 km^2 (a resolution grid of 30 arc-sec) (Hijmans et al. 2005). Interpolations of observed data for the period 1960–1990 are henceforth referred to as the "historical baseline" period.

The future effects of climate change in the study area were computed based on analysis of 29 general circulation models (GCMs) downloaded from the AgMERRA dataset (Ruane et al. 2015). Future climate changes in 2050 for monthly mean temperature (Tmean), monthly minimum temperature (Tmin), and monthly mean precipitation (Precip) were computed assuming the scenario of RCP 8.5 (high emission pathway).

4.2.2 *Analysis of Crop Suitability*

The influence of future climate change predictions on crop suitability was assessed using the Ecocrop suitability model developed by the Food and Agriculture Organisation (Ecocrop 2010), based on the methodologies described in Ramirez-Villegas et al. (2013). The Ecocrop model calculates the relative suitability of a crop in response to a range of climate variables such as temperature, rainfall; and growing period, thereby generating a suitability index score ranging from 0 (totally unsuitable) to 1 (optimal/excellent suitability) as an output. It should be noted that this study did not undertake any additional ground-truthing or calibration of the range of climate parameters preferred for either crop, and therefore the default EcoCrop parameters were assumed. Suitability index scores were calculated for the range of climate variables reported for the historical baseline period (WorldClim data) and future (GCM predictions for 2050).

4.3 Results and Discussion

4.3.1 *Projected Climate Changes*

By 2050 a clear trend of warming is projected across the entire study region throughout all months of the year, with predictions of increases of Tmin and Tmean of approximately 1–2.5 °C. The mean and minimum monthly temperature (Tmean and Tmin) is predicted to increase by 1.5–2 °C in the eastern and southern interior of the country (including large areas of Bié, Huambo, Huíla and Malanje provinces), and increases of about 1 °C predicted for the coastal, central and northern regions of the country. Figure 4.2 depicts the spatial distribution of Tmin and Tmean (left and centre, respectively), with anomalies between the two time periods indicated by red shading (bottom row).

With respect to predicted effects of climate change on rainfall, it is projected that the onset of the rainy season (typically September–October) in 2050 will be charac-

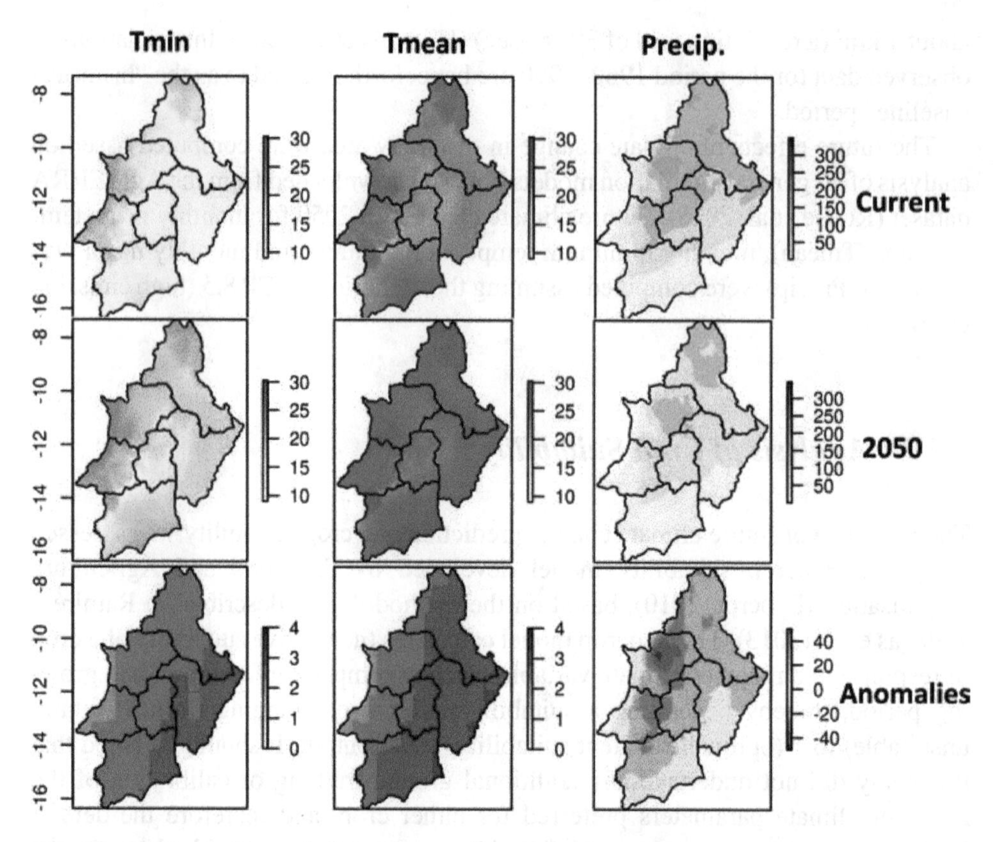

Fig. 4.2 Predicted effects of climate change in study region on average monthly minimum temperature (Tmin) (left), average monthly mean temperature (Tmean) (centre), and mean monthly precipitation (Precip) (right) by the year 2050 for the month of October. Top row depicts baseline (current) climate, centre row depicts predicted future (2050) climate and bottom row indicates anomalies between the two time periods

terised by reduced mean monthly precipitation across the entire region compared to the baseline. Anomalies in monthly rainfall (indicated by red shading in the centre of Fig. 4.2, bottom right) are particularly acute in the northern, central and western extents of the study area at the onset of the rainy season. The trend of reduced rainfall at the onset of the rainy season is projected to continue for the month of November in the central and southern extent of the study area (including the entire extent of Huíla province and majority of Bié and Huambo), whereas the majority of Cuanza Sul and Malanje provinces are projected to benefit from increased rainfall in November by 2050. The majority of the study area is projected to benefit from increased rainfall by 2050 during the midsummer months from December to February, with the exception of the southernmost extent of Huíla province. In March and April, the last months of the traditional maize-growing season, rainfall across the study area is projected to follow two distinct trends: (i) reduced rainfall in the southern and eastern areas, particularly Huíla, Bié and the south-eastern extent of Huambo; and (ii) increased rainfall in the central, western and northern areas, particularly Malanje, northern Huambo, and the highland interior of Cuanza Sul and Benguela provinces. No major changes to rainfall are projected for the dry winter months of May to August.

The predicted spatial and temporal shifts in temperature and precipitation are likely to result in diverse effects on crop productivity between different crops and regions. Increased temperatures are expected to increase crop water demand, which may lead to increased crop stress or reduced productivity. In certain areas, however, the increased temperatures may increase productivity and extend the length of growing season for some crops, particularly where supplementary irrigation is available or the duration or volume of rainfall received increases (such as in the centre and north of the study region). Climate change is projected to impact the distribution, timing and volume of rainfall, most notably showing a delayed onset of rainfall season or reducing the mean precipitation received during the growing season.

These projected climate changes are likely to result in long-term changes to the timing of various agricultural activities such as field preparation and sowing of seed. In the southern and eastern parts of the study area, notably Huíla and south-east Bié, climate change is expected to reduce precipitation across all months of the growing season, which will reduce the productivity of traditional agricultural approaches and force farmers to adopt new practices and crops. Drought-sensitive crops are likely to be increasingly unreliable or unproductive in the latter areas. In contrast, the central and northern extent of the study region is expected to benefit from increased rainfall during the middle and late summer months, which may extend the growing season or improve the yield potential of certain crops.

4.3.2 *Effects of Climate Change on Distribution of Crop Suitability*

Changes in the total spatial extent of suitable area were calculated for both crops for the period from the present to 2050. Figure 4.3 provides an example of the approach used to depict spatial distribution of crop suitability, where the relative proportion of each colour-shaded area indicates the spatial extent of each corresponding category of crop suitability. Modelled distribution of suitability for cassava and maize is depicted in Figs. 4.4 and 4.5, respectively, where current distribution of crop suitability is depicted on the left, projected future distribution of suitability is depicted on the right, and the anomalies (i.e. changes) between the two periods are depicted in the centre.

4.3.2.1 Cassava

Cassava is an important contributor to the diet and livelihoods of Angola's rural farmers and urban consumers, and is a particularly efficient crop in terms of calories generated per input cost. In addition, cassava is considered to be relatively tolerant of low rainfall conditions, and is increasingly promoted as a climate-resilient crop

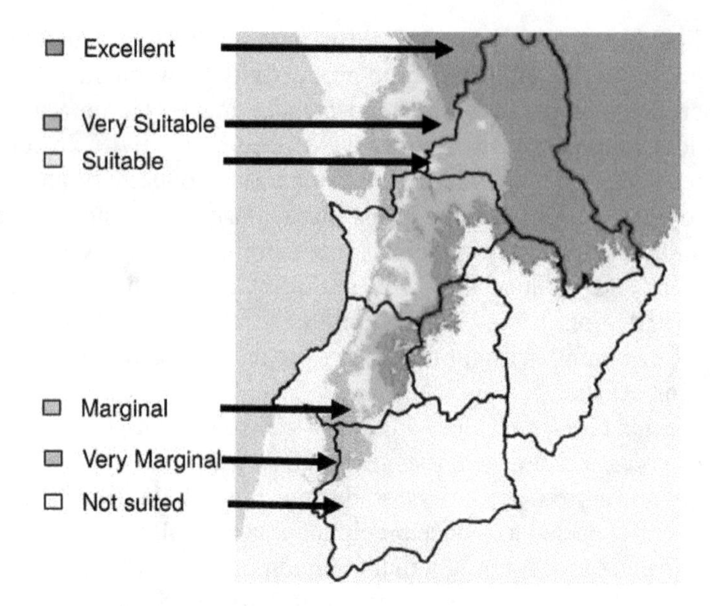

Fig. 4.3 Example demonstration of spatial variability in crop suitability index scores, where the relative proportion of each colour-shaded area indicates the spatial extent of each corresponding category of crop suitability

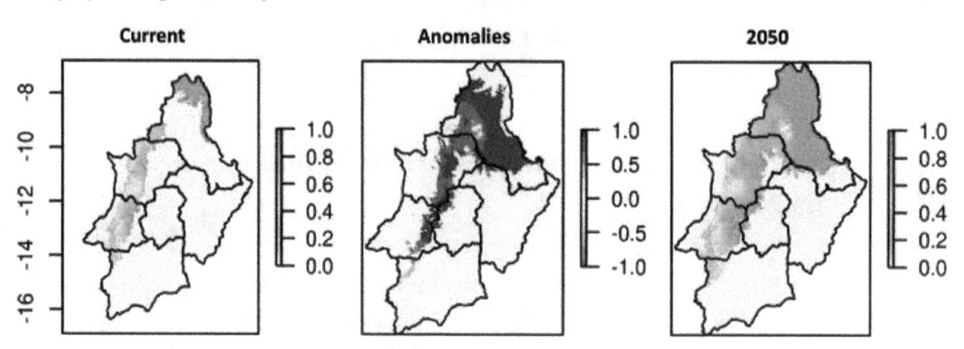

Fig. 4.4 Changes to spatial distribution of areas suitable for production of cassava (Manihot esculentum) in the 'historical' (left) and 'mid-century 2050' (right) scenarios as a result of climate change. Changes between the two-time periods are depicted in the centre

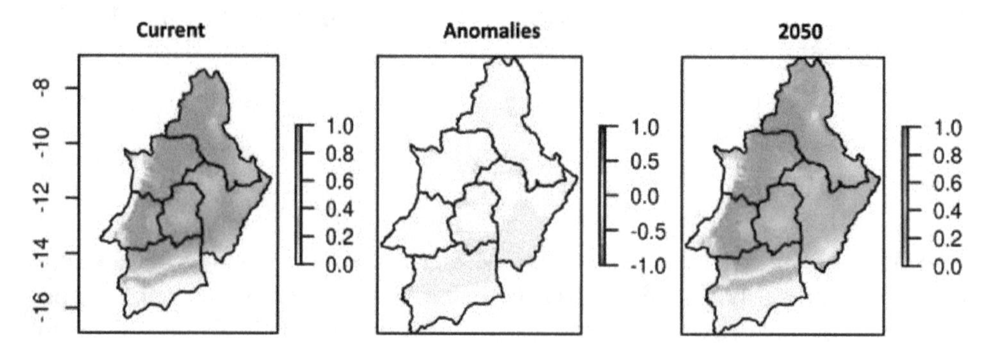

Fig. 4.5 Changes to spatial distribution of areas suitable for production of maize (*Zea mays*) in the 'historical' (left) and 'mid-century 2050' (right) scenarios as a result of climate change. Changes between the two-time periods are depicted in the centre

which has the potential to contribute meaningfully to pro-poor economic development (Theodory et al. 2014).

Previous studies of climate-change effects on cassava have suggested that productivity of cassava will not be negatively impacted by climate change and may enjoy slight increases in certain areas of Africa, particularly in Angola (Liu et al. 2008). Jarvis et al. (2010) also reported that climate change will result in a net increase in the range of suitable areas for cassava production (although noting that, despite a net increase in suitable area, other areas are predicted to decline in suitability as a result of temperature increases). The results of this study support the findings of the latter studies, suggesting that the main effect of climate change on cassava production will be to increase the spatial extent and relatively suitability (i.e., localised suitability index score) of existing cassava production zones. EcoCrop analyses predict that the extent of areas suitable for cassava production will increase in the interior highlands above the coastal escarpment, stretching northward from the border of Huambo and Benguela, through Cuanza Sul and northwards into Malanje. This expected improvement in the region's suitability for cassava may be attributable to the projected increase in Tmean from 20 to 21.5 °C, where the optimum temperature range for cassava is 20–29 °C. The trend towards increased suitability for cassava in the latter areas is projected to remain consistent from October through the rest of the summer months.

The suitable range for cassava production is limited by the arid low-lying southern interior of Huíla and western lowlands of coastal Benguela and Cuanza Sul, which are considered to be poorly suited for cassava production in both the baseline and future scenarios. Potential opportunities and adaptation options for such arid regions may include: (i) promotion of sweet potato as a perennial starch-rich alternative to cassava; (ii) adoption of relatively drought-tolerant cereals such as sorghum and millet; and (iii) promotion of increased crop diversification, including combinations of sweet potato, cassava, legumes and drought-tolerant cereals.

It should be emphasised that the potential benefits of cassava as a climate-resilient subsistence crop are unlikely to be realised without addressing existing structural barriers in the cassava value chain. Market accessibility for cassava farmers in remote areas is hindered by the short shelf life of unprocessed cassava. Therefore, it is recommended that efforts to promote the cultivation of cassava should be supported by simultaneous investments in capacity-building for improved post-harvest storage, processing and value-adding.

4.3.2.2 Maize

Maize is an important staple crop across Southern Africa and is broadly considered to be prone to climate risk such as drought, irregular rainfall and heat stress. Increased temperatures and an increased frequency of severe drought events pose major concerns to cereal production in sub-Saharan Africa, as do expected increased incidence of diseases, pests and parasitic plants (ADB 2015). Past studies have suggested that projected temperature increases could reduce the productivity of major cereal crops, including maize, by 20–30% by 2050 (Liu et al. 2008; Schlenker and Lobell 2010).

In agreement with past studies, EcoCrop analyses predict that climate change will result in minor but widespread decreases in the crop suitability index score for maize in the *planalto* region by 2050, particularly at the onset of the rainy season in October. In Huíla province, the absolute spatial extent of areas which are suitable for maize production are predicted to decrease considerably as a result of climate change, to the extent that the entire province is likely to become poorly suited to the crop by 2050. Of the remaining five provinces, the maize-suitable production areas are predicted to remain unchanged in absolute spatial extent but will undergo decreases in crop suitability index score.

In addition to the effects of increasing temperature, a major challenge that will affect maize farmers in Angola is the projected delay in onset of the rainy season as a result of climate change. The trend towards reduced suitability for maize production is likely to be attributable to the predicted decreases in rainfall over the growing season; EcoCrop's parameters specify a minimum seasonal rainfall of 400 mm and optimum rainfall of at least 600 mm. Analysis of GCMs suggest that onset of rainfall is likely to shift from October/November to December/January by 2050. For households practicing rain-fed maize cultivation, the delayed rains will increase the duration of the "lean" season, when households are reliant on the previous season's harvest. Households therefore will need to adopt new strategies to ensure that food reserves (and adequate seed for planting) can last through this longer lean season. It is unclear whether a long-term shift in the onset of Angola's rainy season will result in a delay to the planting season, or whether changing rainfall patterns will shorten the effective growing season. It is recommended that Angola urgently promote the development and adoption of locally adapted, improved maize cultivars that are more tolerant to heat and drought stress and that can grow to maturity within the confines of a shortened or variable growing season, as well as the promotion of comparatively drought-resilient cereals such as millet and sorghum.

4.4 Implications for Development

The approach and results presented in this chapter demonstrate the use of downscaled climate projections and crop suitability models as a useful but broad-level means of assessing the possible effects of climate change on the temporal and spatial distribution of crop suitability. This is particularly important in countries such as Angola where agronomic data and climate measurements are not readily available. In this case study of six provinces in the *planalto* region of Angola, the diverse impacts of climate change on the crops analysed cannot be easily generalised across the entire study area and indicate the need for detailed local-level studies and strategies for intervention.

The semi-arid regions in the south of Huíla and in the western lowlands of Benguela and Cuanza Sul have climates that are at the limit of the suitable range of the crops analysed. As a result, the spatial range of suitability for heat- and drought-sensitive crops such as maize is projected to be reduced in the low-lying, coastal and southerly parts of the study area by 2050. Climate change will also reduce the duration—or delay the onset—of the growing season for rain-fed crops such as maize across most of the study region. The negative effects of climate change on staple crops such as maize has the potential to undermine the wellbeing of rural households and jeopardise long-term objectives for economic development in climate-vulnerable regions such as Angola. In the affected regions, the primary options for adaptation include the promotion of both climate-resilient cultivars of maize and of alternative crops such as cassava, millet or sorghum.

However, despite the apparent threats posed by the declining productivity of cer tain crops in response to climate change, this study also indicates that climate change may create new opportunities for agricultural development through promotion of climate-resilient staples and alternative crops. In addition to crop-specific considerations, adaptation options for Angola's agriculture sector may include promotion of rural finance, food processing, development of irrigation infrastructure, increased access to extension services, development of early-warning systems and development of rural transport infrastructure.

These analyses provide a demonstration of the applications of crop suitability models for the identification of potential climate vulnerabilities related to food security, as well as identification of potential climate-resilient subsistence crops to be promoted as a strategy to adapt to changing climate conditions. Modelled approaches such as those applied in this study can be further strengthened through the inclusion of measures for calibration and incorporating field-level measurements and local crop performance data.

References

ADB (African Development Bank) (2015) Cereal crops: rice, maize, millet, sorghum, wheat. United Nations: Economic Commission for Africa

Ecocrop (2010) Ecocrop database. Food and Agriculture Organization of the United Nations, Rome http://ecocrop.fao.org/

Hijmans RJ, Cameron SE, Parra JL et al (2005) Very high resolution interpolated climate surfaces for global land areas. Int J Climatol 25:1965–1978 http://www.worldclim.org/version1

Jarvis A, Ramirez J, Anderson B et al (2010) Scenarios of climate change within the context of agriculture. In: Reynolds MP (ed) Climate change and crop production. CAB International, Wallingford, pp 9–37

Liu J, Fritz S, Wesenbeeck CFA et al (2008) A spatially explicit assessment of current and future hotspots of hunger in Sub-Saharan Africa in the context of global change. Glob Planet Chang 64:225–235

Ramirez-Villegas J, Jarvis A, Laderach P (2013) Empirical approaches for assessing impacts of climate change on agriculture: the EcoCrop model and a case study with grain sorghum. Agric For Meteorol 170(15):67–78

Ruane AC, Goldberg R, Chryssanthacopoulos J (2015) AgMIP climate forcing datasets for agricultural modeling: merged products for gap-filling and historical climate series estimation. Agric For Meteorol 200:233–248 https://data.giss.nasa.gov/impacts/agmipcf/agmerra/

Schlenker W, Lobell DB (2010) Robust negative impacts of climate change on African agriculture. Environ Res Lett 5:014010

Theodory M, Honi B, Sewando P (2014) Consumer preference for cassava products versus different technologies. Int J Innov Sci Res 2(1):143–151

5

Crops and Climate Change: Impact of Management and Soils

Patricia Masikati, Katrien Descheemaeker, and Olivier Crespo

5.1 Introduction

More than 50% of agricultural land in Africa is degraded and yields of the main staple crops have been at the lower end of the global range for decades (UNCCD 2014; Folberth et al. 2013). To meet the demands of a growing population, agricultural land has expanded into forests. This, coupled with unsustainable agricultural practices has led to increased land degradation (Lisk 2009; GGCA 2012). Africa is one of the most vulnerable continents because of its highly sensitive social and ecological systems and its limited institutional and economic capacity to respond appropriately to these emerging threats (Lisk 2009; GGCA 2012; Perez et al. 2015). Although climate change affects a number of development sectors, the risk to agriculture stands out since the sector represents a significant part of the economies of many African countries (Vermeulen et al. 2012). There is no doubt that climate change will amplify drivers of land degradation and pose increased threats on smallholders' livelihoods of which the majority are women (GGCA 2012; UNCCD 2014).

Degradation of agricultural land is causing annual yield reductions of 0.5–1% suggesting productivity loss of at least 20% in the next 40 years. In addition, climate

P. Masikati (✉)
World Agroforestry Center (ICRAF), Lusaka, Zambia
e-mail: P.Masikati@cgiar.org

K. Descheemaeker
Plant Production Systems, Wageningen University, Wageningen, the Netherlands

O. Crespo
Climate System Analysis Group, University of Cape Town, Rondebosch, South Africa

change impacts are projected to reduce yields by up to 25% (Ioras et al. 2014; Asseng et al. 2015; Rurinda et al. 2015). Agriculture-based livelihood systems that are already vulnerable to food insecurity will face immediate risk if such yield reductions would occur. Although there has been progress made to understand the impact of climate change and variability on different crops in Africa, there is limited knowledge on how crop-soil systems respond to climate change. Characteristics of different soils vary; for example, clay soils with high organic matter have low thermal conductivity as well as high water holding capacity (Makinen et al. 2017). In contrast, sandy soils, which are predominant in smallholder farming systems, have high thermal conductivity and low water holding capacity (Moyo 2001; Nyamangara et al. 2001). However, the levels of fertility of sandy soils within and across farms greatly depend on the soil-fertility management practices used (Tittonell et al. 2007; Zingore et al. 2011). Soil-climate combination also plays a key role. The magnitude of crop responses to climate is highly sensitive to the soil type (Folberth et al. 2016; Makinen et al. 2017). Farmers in Nkayi, Zimbabwe, have already experienced this; during years of above-average rainfall, farming on clay soils generated a better harvest than on sandy soils, while the reverse is also true.

Empirical and quantitative information regarding the dependency of yield responses to agro-climatic variables on soil type is needed for designing effective climate-smart adaptation methods and enhancing the resilience of smallholder farming systems in the region (Piikki et al. 2015; Folberth et al. 2016; Makinen et al. 2017). Crop models are important tools that can be used to unravel the importance of soil type on crop responses to climate change and variability. However, model choice is also important as different model configurations, operation time steps, physiobiological processes, and others determine the model outputs (Asseng et al. 2015). Here we use the Decision Support System For Agrotechnology Transfer (DSSAT) model and the Agricultural Production Systems Simulator (APSIM) model (McCown et al. 1996; Jones et al. 2003; Hoogenboom et al. 2010; Holzworth et al. 2015). The two models simulate the dynamics of phenological development, biomass growth and partitioning, water and nitrogen cycling in an atmosphere-crop-soil system driven by daily weather variables that include rainfall, maximum and minimum temperatures and solar radiation (Hoogenboom et al. 2010; Holzworth et al. 2015). We use the two models to (1) assess the sensitivity of maize and groundnuts to individual climatic factors such as rainfall, temperature and CO_2 concentration, under three soil types differentiated by levels of organic carbon and plant available soil water (2) simulate the combined impacts of future climate (2040–2070) on the two crops across the three soil types. Both soil fertility and climate are important issues in smallholder farming systems and will have different impacts on plant production and crop yields under future climate change. Production may increase or decrease depending on plant response to the interactions between climate and soil type, hence the importance to assess these impacts to inform adaptation decision-making.

5.2 Methods

5.2.1 *Study Site*

Nkayi district is located in the northwestern part of Zimbabwe. Soils in the area are predominantly sandy. Soil organic carbon varies between 0.4% and 0.8% in the top layers while the water holding capacity varies from 52 to 102 mm (Moyo 2001). Maize is the major staple cereal while groundnuts are generally considered women crops that can improve both household income and nutrition. Groundnut is a multi-purpose crop that can be used to improve soil fertility as well as the quality of live-stock feed, especially during the dry season. Nkayi has a short growing season and limited water availability. Yields are not necessarily linked with higher nitrogen input due to interactions between nitrogen-induced growth and its effects on water use and water availability at different growth stages, especially during grain filling.

Historical changes in climate in the district show increasing temperature trends and recent projections show increases of approximately 1–2 °C in the near future, 2–3 °C in the mid-century, 2–5 °C by end of century (Masikati et al. 2015). Projections (medium confidence) show that rainfall change direction and amplitude are uncertain, yet averages would remain within or close to baseline variability. Seasonality seems to remain unchanged with possible rainfall reduction at the beginning of the rainy season (Masikati et al. 2015). These projected changes will have different impacts on plant production and crop yields as production may increase or decrease depending on the interactions between crops, climate (CO_2, temperature and rainfall) and soil type.

5.2.2 *Climate Data*

The best available historical weather record was gap-filled with AgMERRA data to create a 30 yearlong daily climate data set for Nkayi district (Ruane et al. 2014). To assess sensitivity of maize and groundnuts to different climatic factors; temperature (minimum and maximum, CO_2 and rainfall) we used increments as shown in Table 5.1. To assess the second objective, we use two climate scenarios generated under two Representative Concentration Pathways (RCP), RCP4.5 and RCP8.5 for mid-century (2040–2070) (Ruane and McDermid 2017) (Fig. 5.1a, b) The two

Table 5.1 Factors and levels considered for sensitivity analyses

Sensitivity analyses	
CO_2 (ppm)	360, 450, 540, 630, 720 (with 0 and 60 kg N/ha)
Temperature (°C)	−2, 0, 2, 4, 6, 8
Rain (% change)	0.25, 0.5, 0.75, 1, 1.25, 1.5, 1.75, 2
Fertilizer response (kg/ha)	0, 30, 60, 90, 120, 150, 180

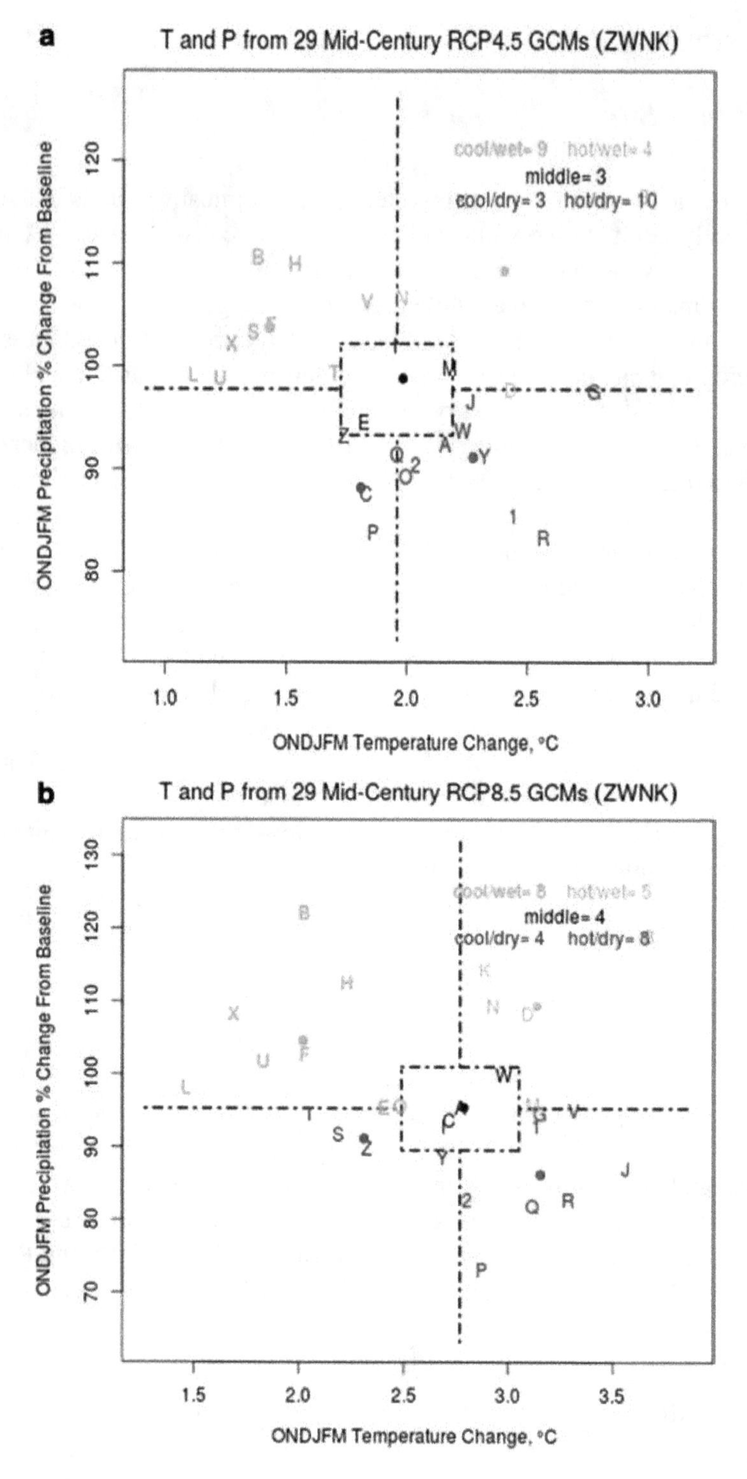

Fig. 5.1 (**a**) Mid-century temperature and precipitation changes in Nkayi, Zimbabwe, from 29 General Circulation Models (GCMs) under RCP4.5 and RCP 8.5 (**b**) Sub-setting of (GCMs) represented by different letters with different colored dots showing average changes in precipitation and temperature as predicted by GCMs showing projected cool/wet, cool/dry, hot/wet, hot/dry and middle conditions

selected were the hot/wet and hot/dry scenarios in view of likely increases of temperature while rainfall change direction is uncertain.

5.2.3 Crop Model Setup and Sensitivity Tests

DSSAT and APSIM models were used to test the effects of climate change on crop production (McCown et al. 1996; Jones et al. 2003; Hoogenboom et al. 2010; Holzworth et al. 2015). Both DSSAT and APSIM are models that have been developed to simulate biophysical processes in crop farming systems in relation to economic and ecological outcomes of management practices in current or future farming systems (Hoogenboom et al. 2010; Holzworth et al. 2015; Steduto et al. 2009). For Nkayi the models have been calibrated (Masikati et al. 2014, 2015) and can be used with confidence in conducting ex-ante climate impact assessments on crop production systems.

For this study we assess the impacts of single climate factors (CO_2, temperature and rainfall) at varying levels (Table 5.1) and also the combined effects on maize and groundnut. Model simulations were done on three soil types which differed in soil physical and chemical characteristics (Table 5.2): poor, average and better soils representing about 29%, 59% and 12% of farms in the district, respectively. Current farmer management practices were used and these are defined in Table 5.3. Outputs from the models, which were considered for the current analyses, include grain and stover crop yields. Planting was set to be done automatically after the model detected that the set soil moisture conditions were met. For this study the sowing window was set between 1 November and 31 December, and planting was done when at least 15 mm of rain was received in three consecutive days.

5.3 Results

5.3.1 Maize Response to CO_2, Temperature, Rainfall and Fertilizer

Maize sensitivity to CO_2 was evaluated at different concentrations with 60 kg N/ha and without nitrogen fertilizer. Without fertilizer, only the APSIM model simulates slight yield increases in response to increasing CO_2 concentrations on better soils. However, when fertilizer is added both models simulate increases of maize yields with increasing CO_2 across all soil types. Maize sensitivity to CO_2 levels differed between the two models and across soil types (Fig. 5.2). Both maize grain and stover show incremental yields up to the maximum level evaluated here, 720 parts per million (ppm), which is more than double the current CO_2 levels.

Table 5.2 Soil initial conditions used for evaluation of APSIM and DSSAT crop models. Soil samples were collected from experimental sites in December 2008 from the Nkayi district

	Soil layers (cm)														
	Low carbon soil					Medium carbon soil					High carbon soil				
	PAWC to rooting depth (65 mm)					PAWC to rooting depth (72 mm)					PAWC to rooting depth (92 mm)				
Parameter	0–15	15–30	30–45	45–60	60–75	0–15	15–30	30–45	45–60	60–90	0–15	15–30	30–60	60–90	90–120
Organic carbon (%)	0.33	0.27	0.21	0.19	0.09	0.49	0.47	0.43	0.32	0.28	0.89	0.86	0.76	0.57	0.36
NO_3-N[a] (ppm)	1.70	1.21	1.10	0.11	0.11	2.13	2.00	1.71	0.43	0.43	2.95	2.86	2.84	0.69	0.55
LL 15[a] (mm/mm)	0.15	0.18	0.23	0.24	0.27	0.15	0.18	0.23	0.24	0.27	0.15	0.18	0.23	0.24	0.27
DUL[a] (mm/mm)	0.28	0.30	0.30	0.30	0.32	0.28	0.30	0.30	0.30	0.32	0.28	0.30	0.30	0.30	0.32
SAT[a] (mm/mm)	0.38	0.40	0.40	0.40	0.42	0.38	0.40	0.40	0.40	0.42	0.38	0.40	0.40	0.40	0.42
Bulk density (g cm^{-3})	1.43	1.42	1.55	1.55	1.61	1.43	1.42	1.55	1.55	1.61	1.43	1.42	1.55	1.55	1.61

Source: Masikati (2011)

[a]*NO_3-N* nitrate-nitrogen, *LL 15* crop lower limit, *DUL* drained upper limit, *SAT* saturation, *PAWC* plant available water capacity

Table 5.3 Treatments used to assess the sensitivity of maize and groundnuts crops to different climate factors in Nkayi, Zimbabwe

Crop	Treatment
Maize	Maize production under farmer practice (low-input system), average fertilizer application: 3 kg/ha[a] and average manure application: 300 kg/ha[a]
Groundnuts	Groundnut production under farmer practice, use of low yielding recycled seed with no fertilizer

[a]ICRISAT (2008) and Masikati (2011)

In response to temperature the two models show divergent results on both maize grain and stover on poor soil. The APSIM model showed a slight increase for grain yields and a decrease for stover yields, however, the DSSAT model showed the opposite. Temperature increase of up to 2 °C would see a slight decrease of grain yields on poor soils, while the same temperature increase would substantially reduce grain yields on average and better soils with higher impact on the latter. Both models show almost no effects of temperature increases on maize stover across soil types.

In response to rainfall, a 25% reduction in rainfall, show yield reductions across all soil types, however impacts are higher on average and better soils. For example at 25% rainfall reduction yield losses on poor soils simulated by the APSIM model are 68 kg/ha, while on the average and better soils are 138 and 487 kg/ha, respectively. Simulated average grain yields for current rainfall are 434, 759 and 2110 kg/ha for poor, average and better soils, respectively. Conversely rainfall that was higher than the defined baseline was not beneficial to maize grown on poor soils. On better soils, maize yield increases were simulated only up to about 25% rainfall increases but after that there is a yield plateau.

On all soil types maize yields show positive response to increases in fertilizer application rates. Maize yields reach a plateau at about 60–70 kg N/ha for all soil types, however, from 30 kg N/ha, the rate of yield increases on better soils is low compared to the other two soil types. Although increases are simulated across soils with increasing rates of fertilizer higher yield gains were simulated for poor than the other two soils at application rate of 30 kg N/ha. Grain yields gains with application of 30 kg N/ha from base yields simulated by the APSIM model were 1314, 1190 and 466 kg/ha for poor, average and better soils, respectively. The average base yields were 434, 759 and 2110 kg/ha. Above 60 kg N/ha, there is a yield plateau, meaning that the water environment at Nkayi becomes the limiting factor to achieving higher average yield.

5.3.2 *Groundnuts Response to CO_2, Temperature, Rainfall and Fertilizer*

Groundnuts show high response to CO_2 concentrations on all soil types (Fig. 5.3). The two models show similar trends although yields from the APSIM model are higher than those simulated by the DSSAT model. Both grain and stover yields

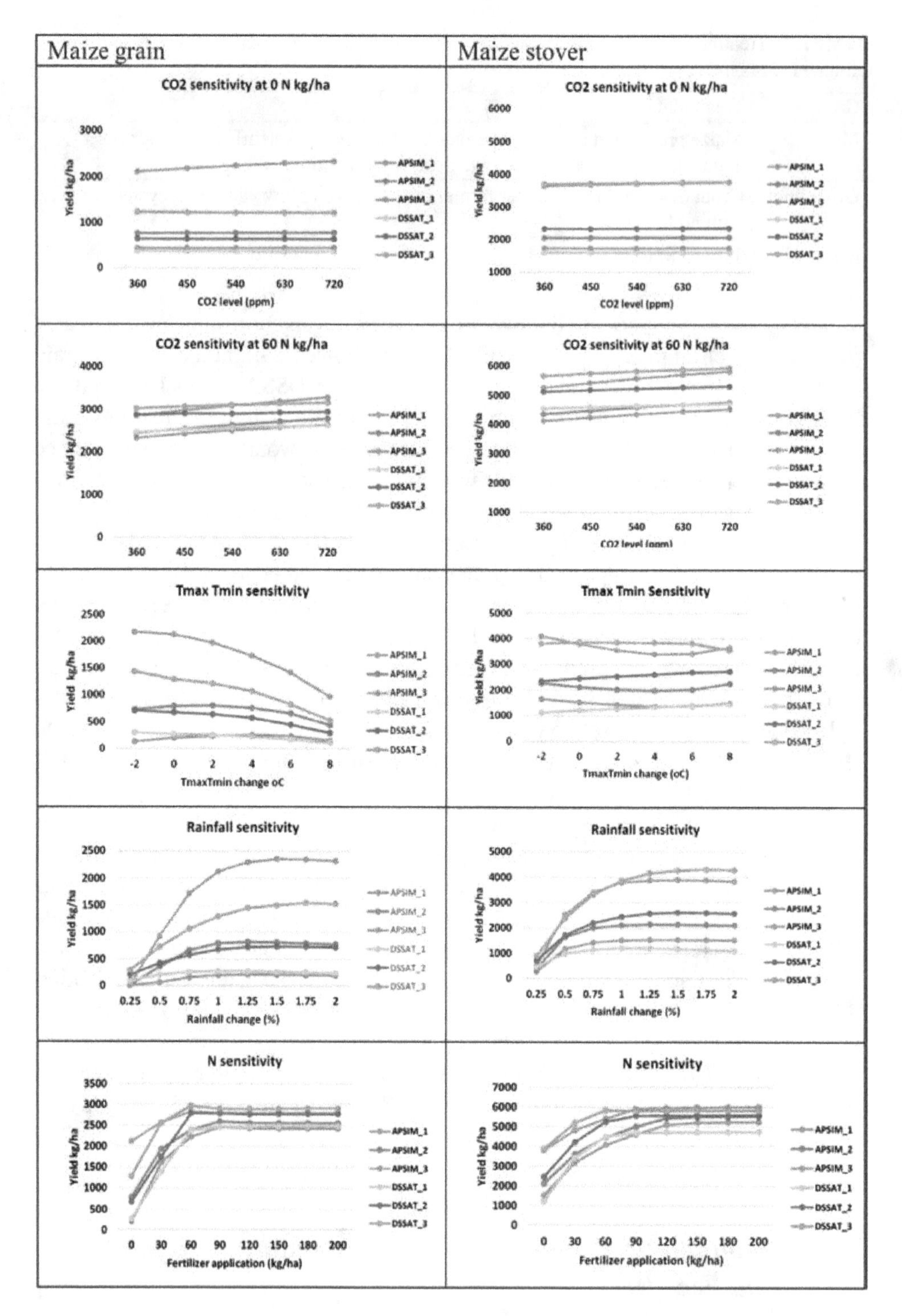

Fig. 5.2 Sensitivity of maize grain and stover to temperature, CO_2, rainfall change and fertilizer application rates on different soil types, in Nkayi Zimbabwe. APSIM_1, APSIM_2 and APSIM_3 show simulations by the APSIM model for the three soil types: 1 = poor; 2 = average; 3 = better. The same applies for the DSSAT model

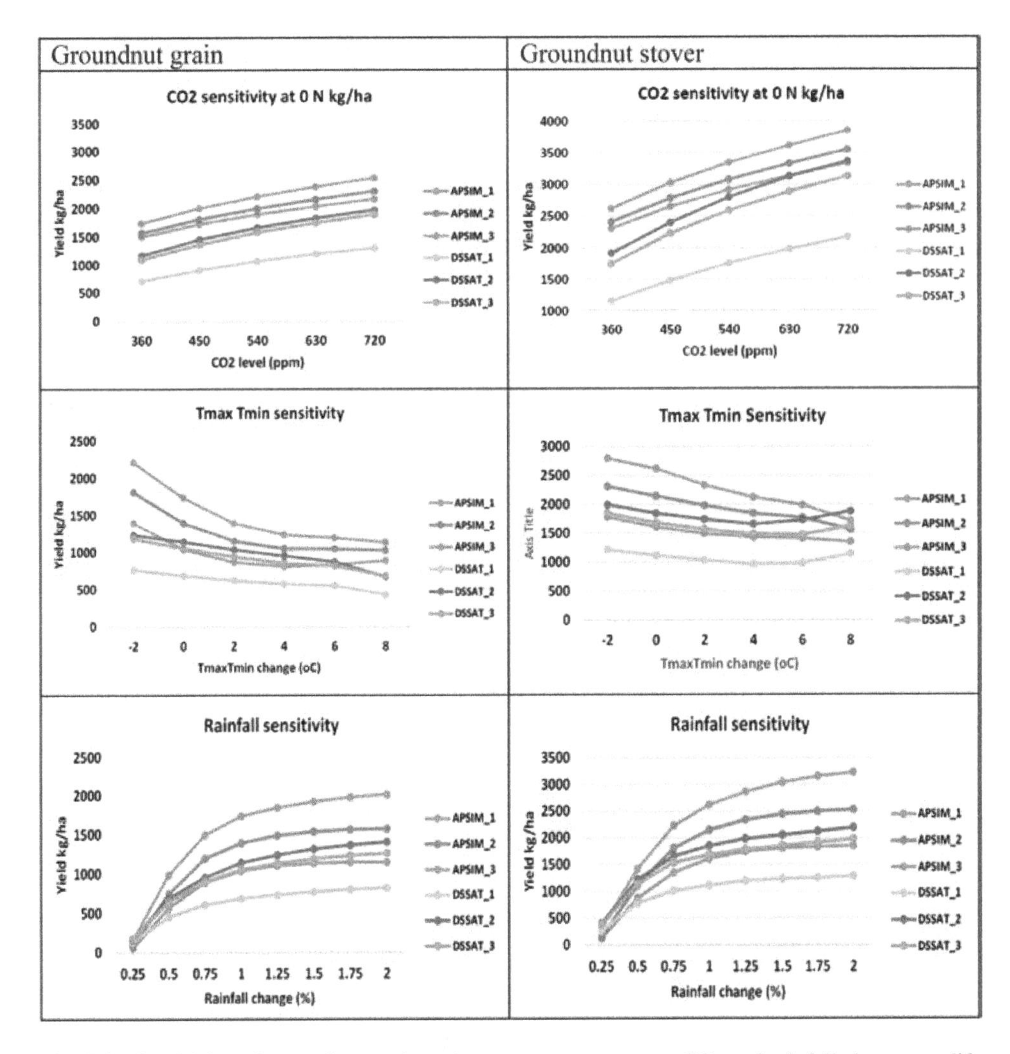

Fig. 5.3 Sensitivity of groundnut grain and stover to temperature, CO_2 and rainfall change on different soil types, in Nkayi Zimbabwe. APSIM_1, APSIM_2 and APSIM_3 show simulations by the APSIM model for the three soil types: 1 = poor; 2 = average; 3 = better. The same applies for the DSSAT model

increase up to the highest CO_2 level evaluated here. The APSIM model simulated higher increases on the better soils while the DSSAT model simulated similarly large increases on the average soil.

In response to temperature changes, both models show negative effects of increased temperature and positive effects of decreased temperatures. Higher yield reductions were simulated for better soils than for the other two soils types. Both models show high yield reductions with temperature increases of about 2 °C, however at higher temperature increases, for example at +6 °C, the DSSAT model simulates slight stover yield increases across all soil types.

Reduction in rainfall by about 50% shows substantial reductions in both grain and stover yields. Both stover and grain yields continue to increase as rainfall increases, however, the increases are very small.

5.3.3 *Combined Effects of Climate Factors on Maize*

We evaluated the combined effects of climate elements on maize grain and stover yield. Hot/wet and hot/dry climate scenarios for both RCP4.5 and 8.5 were used. The effects were simulated by both the APSIM and DSSAT model on three soil types. The two models show divergent effects of combined climate elements on both maize grain and stover. The APSIM model shows more stover yield reductions while the DSSAT model shows more grain reductions across soils and climate scenarios. Yield reductions are more pronounced on better soils than on the other two soil types, while the hot/wet climate scenario shows more positive effects than the hot/dry scenario. Climate effects are more distinct for grain than for stover and this is more pronounced for RCP8.5 than RCP4.5. Generally, the hot/dry conditions show substantial reductions with probability of 35, 40 and 70% of getting reduced grain yields as simulated by the APSIM model while the DSSAT model shows 85, 75 and 85% on poor, average and better soils respectively under RCP4.5 hot/dry conditions. Generally, maize production will decrease under future climate scenarios though the degree of impact differs among soil types.

5.3.4 *Combined Effects of Climate Factors on Groundnuts*

In contrast to maize, groundnuts mostly showed positive effects with yield increases of more than 50% for stover in some instances. The APSIM model generally simulated positive grain yields under RCP4.5 and reductions at a probability of 19, 30 and 50% for poor, average and better soils, respectively under RCP8.5 hot/dry conditions. Stover yields showed positive yield increases for both RCP4.5 and RCP8.5 under hot/wet conditions. However, reductions were simulated for hot/dry conditions mainly for RCP8.5. The DSSAT model shows more negative effects on grain yields than stover yields across soil types and climate scenarios. Grain yield reductions are more pronounced for average and better soils under hot/dry climate scenarios for both RCP4.5 and 8.5. Stover yields on average soil are mostly affected showing about 40% probability of getting negative yields. Although groundnuts seem to be benefitting on average, however there are years when yield changes are negative.

5.4 Discussion

5.4.1 Maize and Groundnut Response to CO_2, Temperature, Rainfall and Fertilizer

Temperature increases in areas such as Zimbabwe where crops are grown near thresholds can be detrimental to rain-fed crop production. Increased temperatures can negatively affect crop yields by accelerating crop phonological stages hence less time for biomass accumulation (Asseng et al. 2015). In this study, increased temperatures show negative effects on both maize and groundnut yields across soil types with higher yield reductions simulated on the better soils. Although simulated yield reductions were higher on better soils, average grain and stover yields were always higher than those for poor soils.

Responses to CO_2 can vary by crop species (Asseng et al. 2015). In our study, maize showed minimal increases of about 5% for both the APSIM and DSSAT model with fertilizer, however, groundnuts showed average increases of about 23%. Asseng et al. (2015) reported that C4 (e.g., maize, sorghum, millet) and C3 (e.g., wheat, groundnuts, potatoes) plants when CO_2 is increased to 500–550 (ppm), grain yield can be increased by 10–20% and by <13% for C3 and C4, respectively. Responses to CO_2 also depend on soil water and nutrient availability with highest responses being reported under soil water limiting conditions (Kang et al. 2002). However low soil fertility can reduce the possible positive effects of elevated CO_2 on yields (Yang et al. 2006). This was also simulated in the current study where there were minimal to no benefits at all with increases in CO_2 across all soil types when no fertilizer was added. However positive responses were simulated with application of 60 kg N/ha with higher increases simulated on better soils. Both stover and grain yields increased as CO_2 concentrations increased up to the 720 ppm level. Increases of CO_2 in the atmosphere is one of the most certain aspects of climate over the coming decades and leguminous crops such as groundnuts have the potential to benefit from this. Leguminous crops fix the atmospheric nitrogen, release high-quality organic matter in the soil and allow sequestration of carbon in soil. If used as feed (provided the quality is not affected), leguminous crops could reduce methane emissions from livestock. These multiple benefits provide both mitigation and adaptation benefits to farmers.

Rainfall variability can have both positive and negative impacts on agriculture depending on the environment. Reduced rainfall by about 25% can be detrimental to crop yields while increases by similar magnitude would not be as beneficial in low input systems and more importantly on poor soils. Rainfall distribution also plays an important role, as lack of rainfall at crop critical growth stages such as anthesis can substantially reduce grain yield.

Smallholder farming systems are low input systems with an average nitrogen application rate of 3 kg/ha and zero fertilizer application for legumes such as groundnuts. General fertilizer recommendations for different soil types are up to 110, 110–140 and 140–180 kg N/ha for better, average and poor soils, respectively, which is beyond what most farmers can afford (FAO 2006; Vanlauwe and Giller 2006). However, in our current study both models simulate yield plateau at 60 kg N/ha for better soils and at around 70 kg N/ha for poor soils. The biophysical and socio-economic situation needs to be considered for establishing recommendations and these should be location-specific and dynamic because soil changes depending on how it is managed.

5.4.2 Combined Effects of Climate Factors on Maize and Groundnuts

Temperatures are projected to increase in Nkayi district, however, rainfall is likely to change by −15% to +10%. Average annual rainfall for Nkayi is about 650 mm per year and the projected reductions and increases can lead to 552 and 715 mm per year, respectively. Variability will be high especially under hot/dry conditions as shown by the variations in yields reductions (Figs. 5.4 and 5.5). Yield variability is higher for maize grain than stover while for groundnuts high variability is only simulated by the DSSAT model for both grain and stover. Increased temperature effects supersede the other factors and will be mostly detrimental to maize while high response to CO_2 exhibited by groundnuts negate the negative effects of increased temperature. Increased temperature reduces crop yields by accelerating crop phonological stages, hence, reducing the time for biomass accumulation. Another adverse effect of high temperature is heat and/or water stress, which at the critical crop growth stages, such as anthesis or grain filling, also reduces crop yields. It will be important to assess which effect will be more prominent and this information could be used when developing adaptation strategies. Important is also to assess the particular times when crops are water stressed so that farmers can adjust water and nutrient management, planting and sowing dates, plant densities and cultivar choice. Climate-smart agricultural practices such as agroforestry that make use of water more efficiently and have the potential to induce microclimatic conditions can be recommended in areas affected by heat stress (Mbow et al. 2014).

5.5 Implications for Development

Crop models are important tools that can be used to understand disaggregated effects of climate elements on crop production. However, models do differ in the way they are constructed and in their responses to different effects of climatic factors on crop production. We used two crop models DSSAT and APSIM and both models generally agreed on the effects of different climatic factors on maize and groundnuts. It is only the magnitude of the effects that vary, for example, reductions on maize grain yields are more pronounced in the APSIM model while the DSSAT

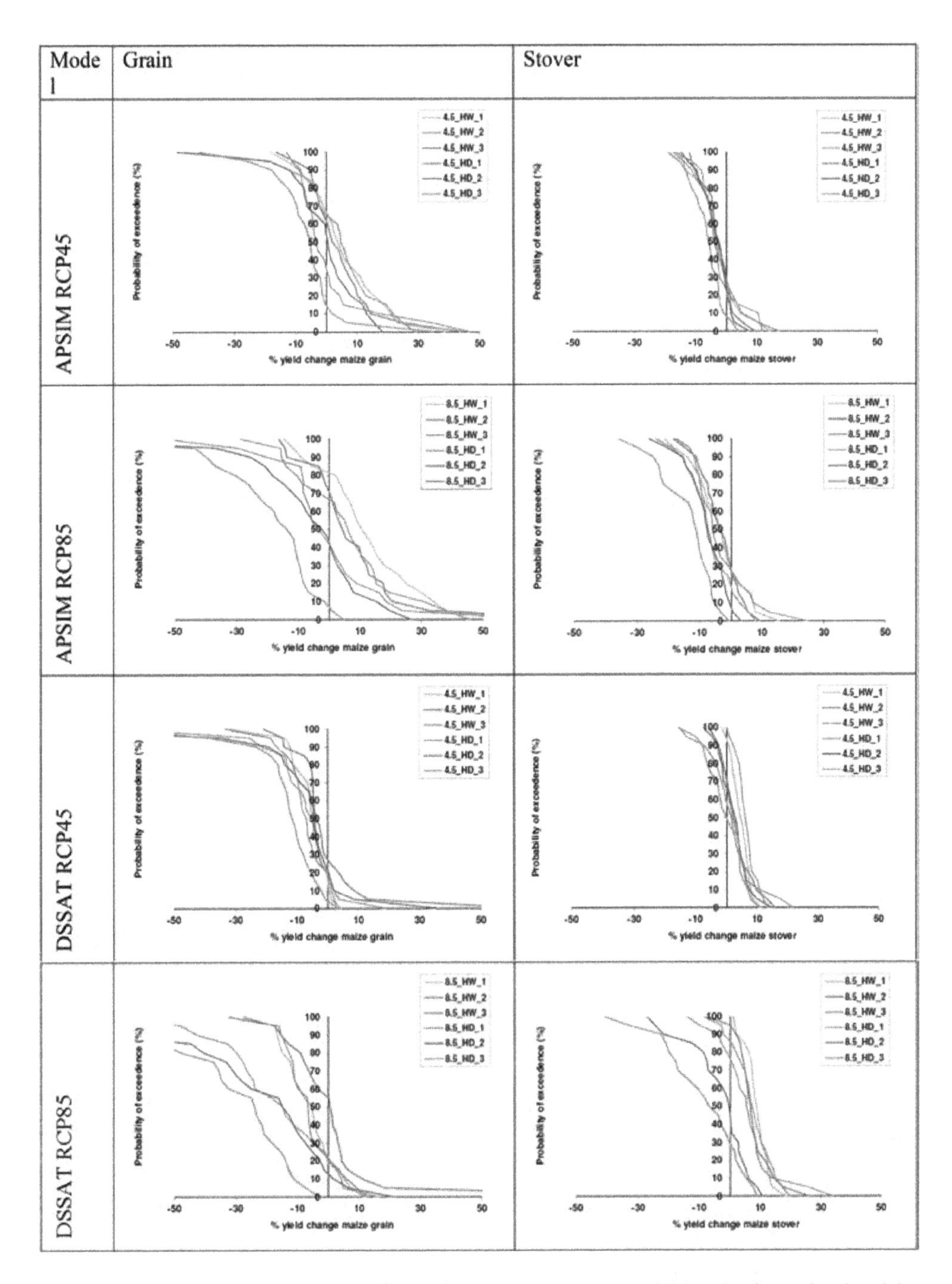

Fig. 5.4 Probability of exceedance for maize grain and stover yield reductions simulated by APSIM and DSSAT models for RCP4.5 and 8.5 for hot/wet and hot/dry climate scenarios. 8.5_HW_1, 8.5_HW_2, 8.5_HW_3 = RCP8.5, hot/wet for poor, average and better soils, respectively. HD represent hot/dry; HW represent hot/wet

model shows more pronounced reduction of maize stover yields. Both models show yield benefits under elevated CO_2 concentration for groundnuts negating the effects of increased temperatures when evaluating the combined effects of the climatic factors. However, yield increases for both groundnut grain and stover are more

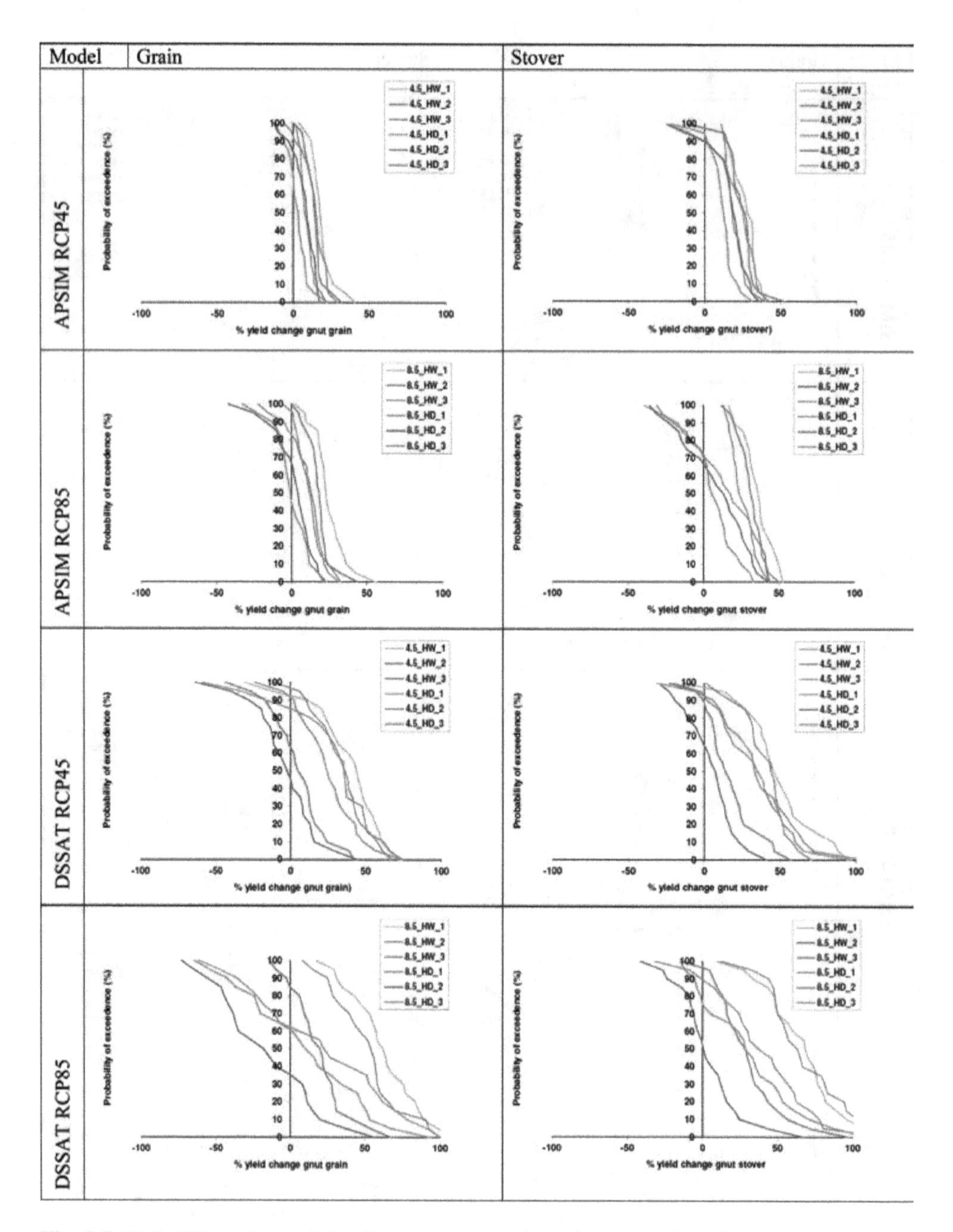

Fig. 5.5 Probability of exceedance for groundnut grain and stover yield reductions simulated by APSIM and DSSAT models for RCP4.5 and 8.5 for hot/wet and hot/dry climate scenarios. 8.5_HW_1, 8.5_HW_2, 8.5_HW_3 = RCP8.5, hot/wet for poor, average and better soils, respectively. HD represents hot/dry, HW represent hot/wet

pronounced in the DSSAT model than in the APSIM model. Soils play an important role in determining outputs of crop-climate interactions; they can buffer or aggravate climatic impacts. Better soils exhibited higher responses to positive influences such as increased rainfall and CO_2 concentrations compared to poor soils. Better soils would be more important in future farming systems.

Acknowledgements In this paper we use the Agricultural Model Intercomperison Project approach (AgMIP, www.agmip.org), using the case study of the crop-livestock Intensification Project (CLIP). UKAID and USDA funded the research. We are grateful to communities in Nkayi district, Matabeleland North and Zimbabwe national stakeholders for their contributions. The research contributes to CGIAR research program CCAFS.

References

Asseng S, Zhu Y, Wang E et al (2015) Crop modeling for climate change impact and adaptation. In: Crop physiology: applications for genetic improvement and agronomy, 2nd edn. Elsevier, London, Waltham, San Diego, pp 505–546. https://doi.org/10.1016/B978-0-12-417104-6.00020-0

FAO (2006) Fertilizer use by crop in Zimbabwe. Land and plant nutrition management service. Land and water development division. Food and Agriculture Organization of the United Nations, Rome

Folberth C, Yang H, Gaiser T et al (2013) Modeling maize yield responses to improvement in nutrient, water and cultivar inputs in sub-Saharan Africa. Agric Syst 119:22–34

Folberth C, Skalsky R, Moltchanova E et al (2016) Uncertainty in soil data can outweigh climate impact signals in global crop yield simulations. Nat Commun 7:11872. https://doi.org/10.1038/ncomms11872

GGCA (2012) Gender climate change and food security. Policy brief. Global Gender and Climate Alliance, United Nations Development Programme, New York

Holzworth D, Huth NI, Fainges J et al (2015) APSIM next generation: the final frontier. In: Weber T, McPhee MJ, Anderssen RS (eds) MODSIM2015, 21st international congress on modelling and simulation. Modelling and Simulation Society of Australia and New Zealand, December 2015, p 490–496. Available from: www.mssanz.org.au/modsim2015/B4/walmsley.pdf

Hoogenboom G, Jones JW, Wilkens PW et al (2010) Decision support system for agrotechnology transfer (DSSAT) version 4.5 [CD-ROM]. University of Hawaii, Honolulu

ICRISAT (2008) Crop-livestock water productivity project household survey (Nkayi District, Zimbabwe). International Crops Research Institute for the Semi-Arid Tropics, Harare

Ioras F, Bandara I, Kemp C (2014) Introduction to climate change and land degradation. In: Arraiza MP, Santamarta JC, Ioras F et al (eds) Climate change and restoration of degraded land. Colegio de Ingenieros de Montes, Madrid, pp 15–48

Jones JW, Hoogenboom G, Porter CH et al (2003) DSSAT cropping system model. Eur J Agron 18:235–265

Kang SZ, Zhang FC, Hu XT et al (2002) Benefits of CO_2 enrichment on crop plants are modified by soil water. Plant Soil 238(1):69–77. https://doi.org/10.1023/A:1014244413067

Lisk F (2009) The current climate change situation in Africa. In: Besada H, Sewankambo NK (eds) Climate change in Africa: adaptation, mitigation and governance challenge. The Centre for International Governance Innovation, Waterloo, pp 8–15

Makinen H, Kaseva J, Virkajarvi P et al (2017) Shifts in soil-climate combination deserve attention. Agric For Meteorol 234–235:236–246

Masikati P (2011) Improving the water productivity of integrated crop-livestock systems in the semi-arid tropics of Zimbabwe: ex-ante analysis using simulation modeling. Dissertation, Centre for Development Research (ZEF), University of Bonn

Masikati P, Manschadi A, van Rooyen A et al (2014) Maize–mucuna rotation: an alternative technology to improve water productivity in smallholder farming systems. Agric Syst 123:62–70

Masikati P, Homann-KeeTui S, Descheemaeker K et al (2015) Crop-livestock intensification in the face of climate change: exploring opportunities to reduce risk and increase resilience in Southern Africa using an integrated multi-modeling approach. In: Rosenzweig C, Hillel D (eds) Handbook of climate change and agroecosystems: the Agricultural Model Intercomparison and Improvement Project (AgMIP) integrated crop and economic assessments, ICP Series on

Climate Change Impacts, Adaptation, and Mitigation, vol 3. Imperial College Press, London, pp 90–112

Mbow C, Smith P, Skole D et al (2014) Achieving mitigation to climate change through sustainable agroforestry practices in Africa. Curr Opin Environ Sustain 6:8–14

McCown RL, Hammer GL, Hargreaves JNG et al (1996) APSIM: a novel software system for model development, model testing, and simulation in agricultural research. Agric Syst 50:255–271

Moyo M (2001) Representative soil profiles of ICRISAT research sites Chemistry and Soil Research Institute, Soils Report No A666. Agriculture Research Extensions (AREX), Harare, p 97

Nyamangara J, Gotosa J, Mpofu SE (2001) Cattle manure effects on structural Stability and water retention capacity of a granitic sandy soil in Zimbabwe. Soil Tillage Res 62(3–4):157–162

Perez C, Jones EM, Kristjanson P et al (2015) How resilient are farming households and communities to a changing climate in Africa? A gender-based perspective. Glob Environ Chang 34:95–107

Piikki K, Winowiecki L, Vagen TG et al (2015) The importance of soil fertility constraints in modeling sustainability under progressive climate change in Tanzania. Procedia Environ Sci 29:199–211

Ruane AC, McDermid SP (2017) Selection of a representative subset of global climate models that captures the profile of regional changes for integrated climate impacts assessment. Earth Perspect 4:1. https://doi.org/10.1186/s40322-017-0036-4

Ruane AC, Goldberg R, Chryssanthacopoulos J (2014) Climate forcing datasets for agricultural modeling: merged products for gap-filling and historical climate series estimation. Agric For Meteorol 200:233–248. https://doi.org/10.1016/j.agrformet.2014.09.016

Rurinda J, van Wijk MT, Mapfumo P et al (2015) Climate change and maize yield in southern Africa: what can farm management do? Glob Chang Biol 21(12):4588–4601

Steduto P, Hsiao TC, Raes D et al (2009) AquaCrop–the FAO crop model to simulate yield response to water: I. Concepts and underlying principles. Agron J 101(3):426–437

Tittonell P, Zingore S, van Wijk MT et al (2007) Nutrient use efficiencies and crop responses to N, P and manure applications in Zimbabwean soils: exploring management strategies across soil fertility gradients. Field Crop Res 100:348–368

UNCCD (2014) Land-based adaptation and resilience: powered by nature. 2nd edn 2014. Secretariat of the United Nations Convention to Combat Desertification. Secretariat of the United Nations Convention to Combat Desertification, Bonn. Available from: http://www.eld-initiative.org/fileadmin/pdf/Land_Based_Adaptation_ENG_Sall_web.pdf

Vanlauwe B, Giller KE (2006) Popular myths around soil fertility management in sub-Saharan Africa. Agric Ecosyst Environ 116:34–46

Vermeulen SJ, Aggarwal PK, Ainslie A et al (2012) Options for support to agriculture and food security under climate change. Environ Sci Policy 15:136–144

Yang L, Huang J, Yang H et al (2006) The impact of free-air CO_2 enrichment (FACE) and N supply on yield formation of rice crops with large panicle. Field Crops Res 98:141–150

Zingore S, Tittonell P, Corbeels M et al (2011) Nutr Cycl Agroecosyst 90:87–103

Part II
Delivery of Quality Crop Germplasm: Opportunities and Challenges

6

Climate-Smart Crop Varieties: Development and Distribution by Private Seed Sector

Biswanath Das, Francois Van Deventer, Andries Wessels, Given Mudenda, John Key, and Dusan Ristanovic

6.1 Introduction

CC poses a significant risk to crop production across sub-Saharan Africa (SSA), with ESA particularly vulnerable to the projected changes. Temperature increases are estimated to rise at a rate above the global average during the twenty-first century and it is predicted that by 2050 will significantly change the cropping duration for key staple crops (Cairns et al. 2013; Schlenker and Lobell 2010; Niang et al. 2014; James and Washington 2013; Challinor et al. 2016). Meanwhile, precipitation is projected to increase in parts of eastern Africa but decrease significantly in southern Africa. The combined heat and drought stress in parts of ESA is projected to reduce yields of staple cereals by as much as 30% within two decades (Niang et al. 2014; Lobell et al. 2008).

Smallholder, subsistence farmers constitute over 70% of the population in ESA and account for over 75% of agricultural output (AGRA 2017). They are the group most vulnerable to CC and require urgent, scalable access to CS crop varieties with adaptive characteristics that can tolerate future climes. These include; tolerance to combined heat and drought stress, waterlogging and lodging stress, post-harvest storability, maintenance of nutritive value in warmer climes, and adaptation to new and shifting incidences of pests and diseases. To deliver CS crop varieties in CC

B. Das (✉) · G. Mudenda
Syngenta, Lusaka, Zambia
e-mail: biswanath.das@syngenta.com

F. Van Deventer · A. Wessels
Syngenta, Midrand, South Africa

J. Key
Syngenta, Basel, Switzerland

D. Ristanovic
Plant Breeding Consultant, Lusaka, Zambia

affected areas of ESA will largely depend on increasing the rate of genetic gain (genetic improvement through artificial selection) for CS traits and the establishment of scalable, competitive seed delivery systems that ensure improved varieties reach smallholder farmers in the shortest time (Atlin et al. 2017).

Smallholder farmers' adoption of improved crop varieties in SSA is amongst the lowest in the world (estimated to be 20% by the Alliance for a Green Revolution in Africa (AGRA) 2017), yet the formal seed sector has grown significantly following deregulation of the seed industry regionally in the early 1990s. The emerging private seed sector provides a unique and timely opportunity to promote the development and dissemination of improved, CS crop varieties through certified, scalable seed systems that can potentially impact millions of livelihoods in SSA. In this chapter, the specific roles and constraints for the private sector in ESA in developing and disseminating improved, CS crop varieties are discussed, with particular emphasis on maize (*Zeae maydis*), the staple food crop and primary source of daily calorie intake in the region.

6.2 The Emerging Private Seed Sector in ESA

In most of ESA, the plant breeding and seed industries were dominated by public institutions until the mid-1990s, when the seed sector was deregulated. Since then, dozens of private, local seed companies have been established, and several global multinational seed corporations have entered the ESA seed market. The primary focus of seed companies in ESA is maize, the driver of the global seed industry by virtue of acreage and potential for hybridization. The effect of deregulating the seed sector in ESA is highlighted in Fig. 6.1a, b, which show maize variety releases in Zambia and Kenya respectively. Both countries have emerged as leading centers for the seed industry in SSA and serve as important bellwethers of regional trends. In both cases, deregulation of the seed industry has led to a marked increase in the total number of seed companies and, subsequently, maize variety releases. However, the majority of these variety releases have been licensed from existing public breeding pipelines, and it is estimated that less than 25% of seed companies in the region (estimated to be 80 in total) have invested in proprietary germplasm improvement (Langyintuo et al. 2008).

Variety releases of other important staple crops in ESA have not emulated maize, in large part due to low commercialisation opportunities for the private sector. Total variety releases for maize, sorghum (*Sorghum bicolour*), common bean (*Phaseolus vulgaris* L.) and cassava (*Manihot esculentum*) are shown in Fig. 6.2a, b for Zambia and Kenya respectively. Even though variety releases of these crops have doubled since deregulation, the cumulative number of releases (for sorghum, common bean and cassava) is still less than 30% that of maize, and dominated by the public sector (over 80% of releases). In Zambia, cassava is an important secondary staple crop, yet only seven varieties (all publically bred) have been released since 1970, the latest in 2001. These crops are important components of food and

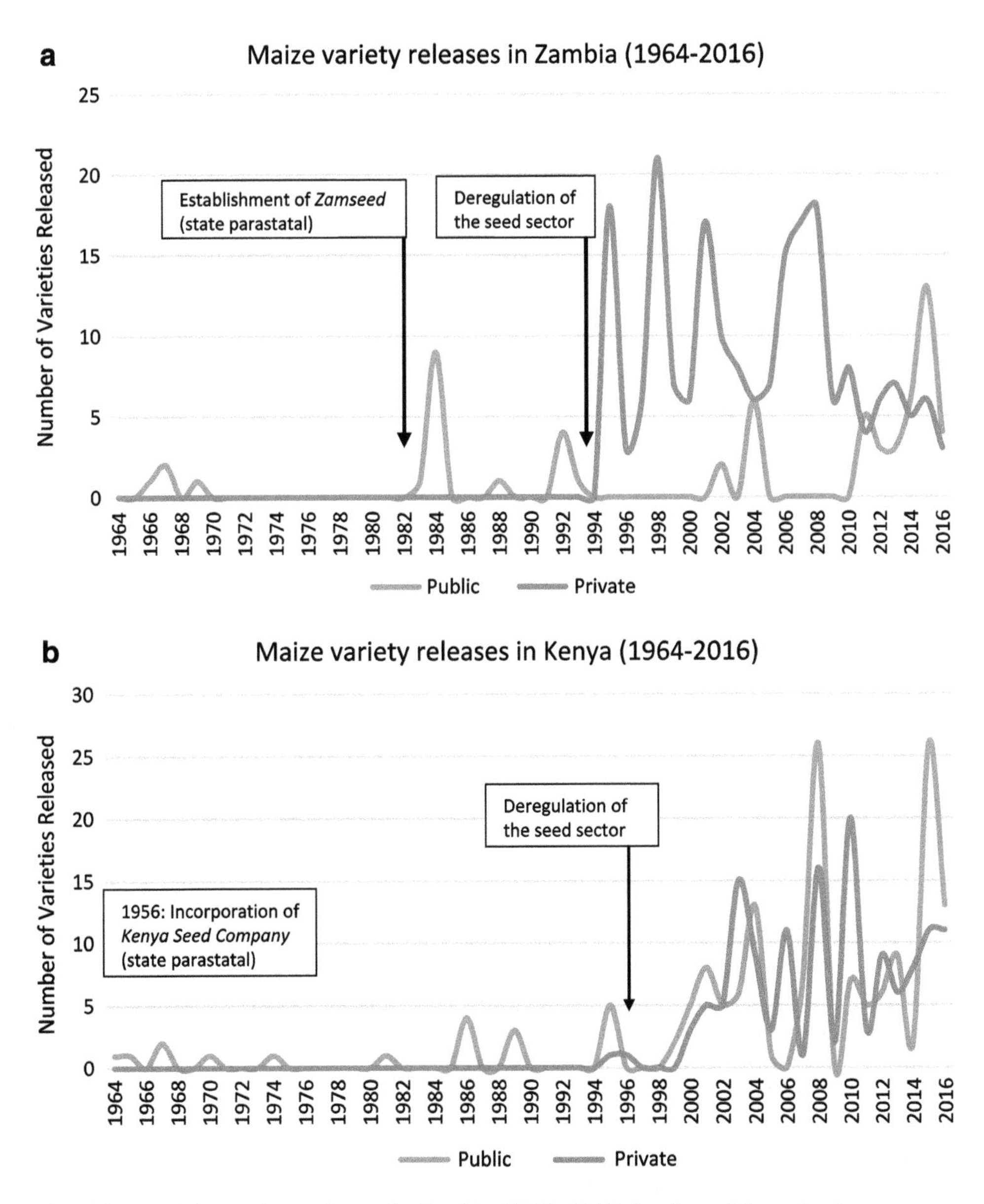

Fig. 6.1 (**a**) Maize variety releases in Zambia (1964–2016) by the public and private sectors. (Source: SCCI 2017). (**b**) Maize variety releases in Kenya (1964–2016) by the public and private sectors (KEPHIS 2017)

nutritional security in ESA, where they will play a critical role in diversified, CS agricultural systems. Market incentives are urgently required to better integrate these open pollinated and vegetatively propagated crops into scalable, certified seed systems in the region.

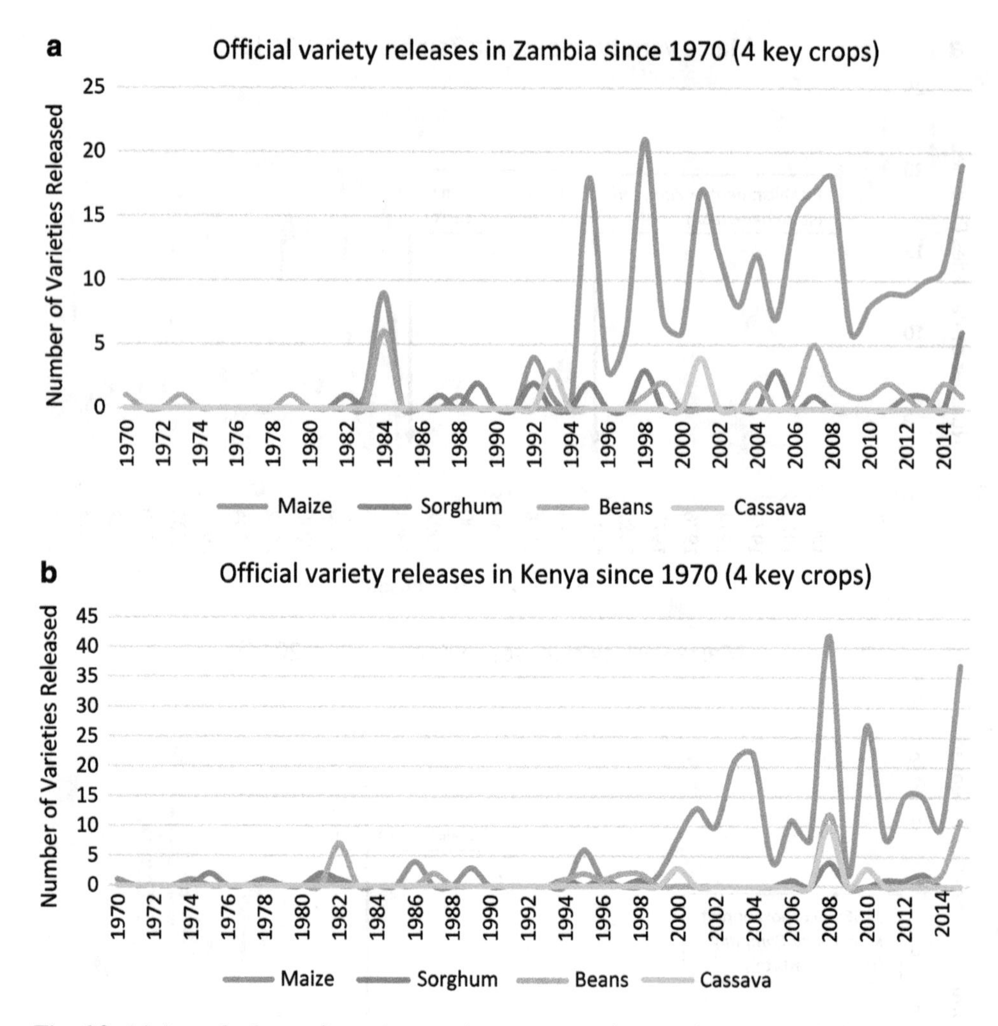

Fig. 6.2 (**a**) Annual releases for maize, sorghum, common bean and cassava varieties in Zambia since 1970. (Source: SCCI 2017). (**b**) Annual releases for maize, sorghum, common bean and cassava varieties in Kenya since 1970 (KEPHIS 2017)

6.3 Low Rates of Variety Turnover and Agricultural Research and Development Investment in ESA

Despite the growth of the seed industry in ESA since the 1990s, rates of variety turnover remain slow, and investment into agricultural research and development is extremely low. A handful of established varieties also continue to dominate markets in most countries (Abate et al. 2017). In Kenya, H614D (a variety released in 1986 by the state parastatal) accounts for over 40% of area cultivated to improved maize varieties while in Zambia, the three most widely grown maize varieties were released almost two decades ago, shortly after the deregulation of the seed sector (Smale and Olwande 2014; Smale et al. 2015). The average age of commercial

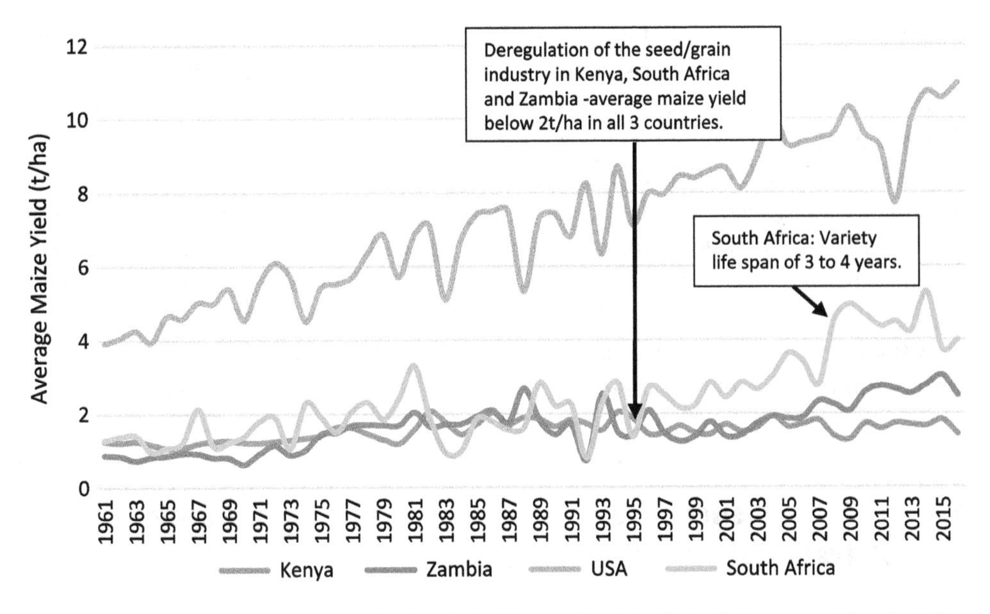

Fig. 6.3 Average maize yields (1961–2016) in Kenya, Zambia, United States and South Africa (FAO 2018)

maize varieties in ESA is estimated to be 13 years. South Africa is an exception; with the most competitive seed industry in the region, the life of the typical maize variety here spans 3–4 years, similar to the United States. Most smallholder farmers in ESA are therefore not cultivating the best available varieties for their environment, and in many cases are persevering with obsolete cultivars that were developed under climatic, agronomic and pest conditions distinct from current and future climes. This has contributed to modest yield gains for maize in many countries in ESA (Fig. 6.3).

Reasons for slow rates of maize variety turnover in ESA are several and complex. The majority of smallholder farmers in the region grow maize in unpredictable, rain-fed conditions, and are risk averse to investing in inputs and new technologies. Average yields throughout the region are low and genetic gains in yield through crop improvement (usually 1% per year in well managed breeding programmes) are frequently overshadowed by seasonal variations in on-farm climatic conditions and crop management. The incentive for smallholder farmers to invest in new agricultural technologies is further reduced by limited access to grain markets, poor storage and transport infrastructure, as well as counterfeit seed and fertilizer. Without strong demand for new varieties, seed companies are reluctant to withdraw established, well-known varieties and invest in launching and marketing new products.

In addition to low rates of variety turnover, investment in agricultural research and development is very limited in ESA. Low income countries (including most of those in ESA) account for less than 3% of global agricultural research and development expenditure, despite being some of the most vulnerable to CC (Pardey et al. 2016). Of this expenditure in ESA, 90% is by the public sector, which continues to dominate the development of new technologies, including crop varieties (Beinteman and Stads 2011). By comparison, private sector investment in agricultural research and development in member countries of the Organization for Economic Co-operation and Development (OECD) regularly accounts for over 70% of total expenditure (OECD 2018) and the role and costs of developing new agricultural technologies has been assumed by a vibrant private sector, driven by competition for market share. Private sector investment in agricultural research and development remains low in ESA in part due to small, fragmented markets and a lack of commercial incentive in the region. Given the projected impacts of CC in ESA, increased investment in crop improvement is vital, as are mechanisms to drive faster rates of variety turnover to ensure farmers have sustained access to the latest genetics.

6.4 Driving Genetic Gain for CS Traits Through Public-Private Partnerships (PPP)

Increasing rates of genetic gain will be fundamental to ensuring plant breeders are able to react quickly to changing dynamics caused by CC, many of which are difficult to predict (e.g., shifting incidence and severity of pests and disease). Driving genetic gain for CS traits will require access to appropriate germplasm, reliable phenotyping platforms for traits of interest, and adoption of modern breeding methods that reduce breeding cycle time. Given the current levels of investment in agricultural research and development in ESA, driving genetic gains for CS traits is unlikely to be achieved in the near term without the combined efforts of PPPs.

Effective PPPs will utilize the public sector's experience and capacity in the region whilst exploiting the emerging private sectors access to regional markets and expertise in commercial plant breeding, particularly in the case of regional or international companies. Public research institutions in ESA, for example, have developed germplasm adapted to local conditions and are strategically positioned to establish long term regional phenotyping networks for key CS traits, such as drought or emerging disease tolerance (e.g., the maize lethal necrosis (MLN) screening facility in Kenya, developed by the Kenya Agricultural and Livestock Research Organization (KALRO) and the International Maize and Wheat Improvement Center (CIMMYT)).

Conversely, the emerging private sector offers a sustainable route to market whilst assuming the costs and responsibility for seed production, quality, purity and distribution. Currently, most small and medium scale enterprise (SME) seed companies in ESA rely on this model to license and commercialise publically developed varieties, although significant bottlenecks persist in accessing foundation seed and legal services to enter mutually beneficial licensing agreements (Cramer, this volume).

The entry of multinational corporation (MNC) seed companies into the ESA seed market provides an additional opportunity to develop PPPs around technology

Table 6.1 Strength rating (low, medium or high) of selected drivers of genetic gain within the private and public sectors in ESA

Drivers of genetic gain in maize in ESA	Relative strengths	
	ESA public pipeline	MNC pipeline
Germplasm		
Locally adapted germplasm	High	Medium
Access to commercial, global germplasm	Medium	High
Phenotyping		
Establishment of regional phenotyping platforms for CS traits	High	Low
Phenotyping technology (high throughput precision screens, remote sensing, electronic data capture, etc.)	Medium	High
Access to modern breeding technology		
Double Haploids	Medium	High
Marker assisted selection, genomic selection	Medium	High
Data management systems	Low	High
Mechanisation of breeding programmes		
Seed inventory management, tracking and processing	Medium	High
Planting, harvesting, seed drying and storage	Low	High
Market orientated breeding programme		
Development of target product profiles	Medium	Medium
Cost of goods and production research	Low	High
Adoption of new technology through extension	High	Low

transfer and optimisation of breeding pipelines. MNCs have led the global development of applied breeding technology in genomics, phenomics and mechanisation, and can therefore complement ongoing public breeding efforts with modern technology to drive genetic gain. Technologies such as doubled haploids,[1] marker assisted selection,[2] precision phenotyping tools and data management platforms have transformed plant breeding in mature seed markets to develop products quickly in response to customer requirements. MNCs also have access to global sources of elite germplasm for a range of traits that will become more important in ESA as a result of CC (in terms of tolerance to drought, new pests and diseases). PPPs between public institutions and MNCs are likely to focus on germplasm exchange, the creation and release of joint products, the provision of technological services, and shared phenotyping platforms. The relative strengths of MNCs and public breeding pipelines in ESA in terms of driving genetic gain are shown in Table 6.1.

[1] Artificial doubling of haploids to develop homozygous lines in one generation rather than six generations as required by conventional breeding

[2] Use of genetic markers to drive selection for a trait of interest

6.5 Enhancing the Delivery of CS Maize Varieties: Harmonising Seed Laws and Promoting Adoption

In addition to increasing the rate of genetic gain for CS varieties, regional bottlenecks in releasing, disseminating and adopting new varieties in ESA must be addressed in order to incentivise the private sector to invest in crop improvement, to reduce product life cycles, and to ensure certified seed of CS varieties reach smallholder farmers. ESA presents an attractive maize seed market (currently 20% that of North America) and many countries share common agro-ecologies which eases regional scaling of competitive varieties (Fig. 6.4). The reality, however, is nearly twenty individual nation states with distinct laws, regulations and trade agreements, making ESA a fractured and challenging seed market.

For over 20 years regional, intergovernmental bodies such as the Common Market for Eastern and Southern Africa (COMESA) and the Southern African Development Community (SADC) have strongly recommended the harmonisation of seed laws governing variety release, the protection of plant breeder rights and cross border movement and sale of certified seed in ESA (personal communication). For example COMESA's Seed Trade Harmonization Regulations Programme (COMSHIP) calls for the harmonization of release processes across member countries and the development of a regional variety list, where varieties that have been released in two countries can be sold in similar agro-ecologies in all other COMESA member nations (COMESA 2014). However, actual adoption of these recommendations has been slow and most nations maintain separate release processes and laws. As a result, of the hundreds of improved maize varieties that have been released in

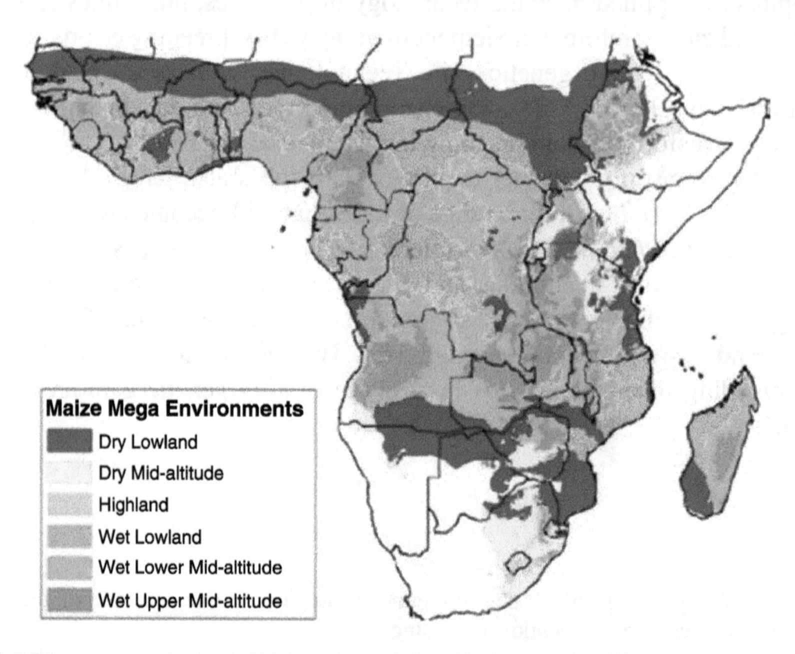

Fig. 6.4 Maize agro-ecologies in SSA. (Adapted from Hodson et al. 2002)

ESA since the 1990s, less than 5% have been successfully released and marketed in more than one country (Abate et al. 2017).

A formal variety release process remains essential in emerging seed markets such as ESA, to protect both farmers and the nascent seed industry from the entry of substandard products on to the market. However, the current regulatory environment in ESA is widely acknowledged to be costly and cumbersome for the seed industry (Bett 2017). Table 6.2 shows the current status of variety release processes in six ESA countries; the intercountry variations that exist throughout the variety release process are limiting market opportunities for seed companies and complicating both stock inventory and the consolidation of production bases.

Currently, very few variety release committees (VRCs) in ESA explicitly consider CS traits for variety release (Table 6.2). Given the extra investment and effort required to develop CS varieties, it is necessary to prioritise the release of varieties with these traits to secure private sector interest and commitment. The recent decision by the Kenya Plant Health Inspectorate Service (KEPHIS) to fast track the release of varieties tolerant to maize lethal necrosis (MLN) in Kenya is an example of engaging seed sector support to address an urgent challenge for smallholder farmers. The current outbreak of fall armyworm (*Spodoptera frugiperda*) throughout Africa provides another opportunity to prioritise a trait that is likely to become more relevant as temperatures increase in ESA as a result of CC.

Developing a brand around a CS trait such as drought tolerance represents a major commitment by a seed company to accept a certain degree of responsibility for varietal performance. To support the private sector to assume these risks, regulatory bodies in ESA need to provide a solid framework to protect intellectual property and clamp down on counterfeit seed that can damage farmer confidence in improved varieties. In recent years, there has been growing concern about the prevalence of counterfeit seed on sale in ESA and inadequate efforts by governments and regulatory authorities to address the problem (Mabaya et al. 2017; Bold et al. 2015). In Uganda for example, it is estimated that up to 50% of seed sold as certified seed is either fake or of substandard quality (AGRA 2011; Bold et al. 2015). Joining the International Union for the Protection of New Varieties of Plants (UPOV) and adopting global plant variety protection standards will increase private sector confidence in intellectual property protection and seed quality in ESA, though only Kenya, Tanzania and South Africa are currently members in ESA (Table 6.2).

Replacing old varieties with new, improved varieties will be a key pillar to driving agricultural productivity in ESA in the coming years, as it has in other parts of the world (Atlin et al. 2017). The benefits of cultivating improved, CS varieties need to be promoted (via extension services) to smallholder farmers who are operating in rain-fed, suboptimal environments at risk from CC. To drive uptake, the withdrawal of obsolete mega varieties should be encouraged and varieties with CS traits should be prioritised in farmer demonstrations and seed distribution programmes.

Table 6.2 Variety release regulations in six countries in ESA

		Zambia	Zimbabwe	Tanzania	Kenya	Malawi	South Africa
1	Number of seasons of Official National Performance Testing (NPT)	2	0 (preceded by 2 years of company trialing +1 year DUS)	1 (preceded by 1 years of company trialing +1 year DUS)	2	3	0
2	How many locations are NPT trials conducted in?	6	5 sites for company trialing	3 per agro-ecology	6–12	6–12	N/A
3	Is Farmer evaluation necessary for release?	No	No	Yes	No	No	No
4	What are the release criteria?	Must be DUS and have VCU	Must be DUS and competitive	Must be DUS and competitive	Superior yield to checks by 5–10% or special attribute(s)	Superior to checks with proven VCU	DUS and entry into national variety list
5	Are CS traits (e.g., drought) considered and tested during release?	Considered, but not tested	Considered, but not tested	Considered, but not tested	Yes, tested	No	Yes for Biotech traits
6	Is the country a member of UPOV*?	No	No	Yes	Yes	No	Yes
7	Can releases from neighbouring countries be sold?	No	No	No	No	No	Yes if Variety List is Open for Crop
8	Does certified seed have to be produced in-country?	No	Varies	No	No	Yes to qualify for govt. input programme	No
9	Average age of commercial hybrids	10 years	13 years	14 years	14 years	11 years	4 years
10	Maize hybrid adoption rate	65%	95%	20%	80%	15%	98%

Source: Abate et al. (2017)

* UPOV stands for International Union for the Protection of New Plant Varieties

6.6 Implications for development

The emerging private seed sector in ESA provides a significant opportunity to develop partnerships with established public plant breeding programmes, to accelerate the development of improved varieties with CS traits and their subsequent distribution through scalable, certified seed systems. Some 50% of yield gains in most global regions are commonly attributed to genetic gains made through plant breeding. Providing smallholder farmers in ESA with access to the latest, improved germplasm can therefore play a major role in adapting agricultural systems in ESA to CC. The promotion of an enabling regulatory environment for the release and adoption of improved varieties with CS traits will further stimulate private sector interest and investment. This is particularly applicable to the smallholder maize seed market, which is the primary basis for the growth of the emerging seed industry and the foundation of regional food security in ESA.

References

Abate T, Fisher M, Abdoulaye T et al (2017) Characteristics of maize cultivars in Africa: how modern are they and how many do smallholder farmers grow? Agric Food Secur 6:30

AGRA (2011) Country case studies on the PASS value chain strategy/approach and its impact/effect on smallholder farmer yields in Africa: Kenya, Tanzania, Uganda, East Africa synthesis report. Alliance for a Green Revolution in Africa (AGRA), Nairobi

AGRA (2017) Africa agriculture status report: the business of smallholder agriculture in Sub-Saharan Africa (issue 5). Alliance for a Green Revolution in Africa (AGRA), Nairobi

Atlin GN, Cairns J, Das B (2017) Rapid Breeding and varietal replacement are critical to adaptation of cropping systems in the developing world to climate change. Glob Food Sec 12:31–37

Beintema N, Stads GJ (2011) African agricultural R&D in the new millennium progress for some, challenges for many. International Food Policy Research Institute/Agricultural Science and Technology Indicators, Washington, DC/Rome

Bett W (2017) Harmonizing Seed Laws in COMESA. Opening address of Secretary of State for Agriculture, Republic of Kenya. COMESA national seed law harmonization conference, Nairobi, Jun 2017

Bold T, Kaizzi KC, Svensson J et al (2015) Low quality, low returns, low adoption: evidence from the market for fertilizer and hybrid seed in Uganda, IGC working paper F-43805-UGA-1. International Growth Centre, London

Cairns JE, Hellin J, Sonder K et al (2013) Adapting maize production to climate change in sub-Saharan Africa. Food Secur 5:345–360

Challinor AJ, Koehler A-K, Ramirez-Villegas J et al (2016) Current warming will reduce yields unless maize breeding and seed systems adapt immediately. Nat Clim Chang 6(10):954–958

COMESA (2014) COMESA seed harmonization implementation plant (COM-SHIP). In: Mukuka J (ed) Common market for Eastern and Southern Africa (COMESA). Alliance for Commodity Trade in Eastern and Southern Africa (ACTESA), Lusaka Available from: http://africalead-ftf.org/wp-content/uploads/2016/09/COMESA-Seed-Harmonisation-Implementation-Plan-COM-SHIP-_JULY-2014.pdf

FAO (2018) FAOSTAT. http://www.fao.org/faostat/en/#data/QC. Accessed 18 Jan 2018

Hodson DP, Martinez-Romero E, White JW et al (2002) Africa maize research atlas (v. 3.0). CD-ROM publication. International Maize and Wheat Improvement Center (CIMMYT), Mexico

James R, Washington R (2013) Changes in African temperature and precipitation associated with degrees of global warming. Climate Change 117(4):859–872

KEPHIS (2017) Kenya official variety register (2017). Kenya Plant Health Inspectorate Service, Nairobi

Langyintuo AS, Mwangi W, Diallo AO et al (2008) An analysis of the bottlenecks affecting the production & deployment of maize seed in Eastern and Southern Africa. International Maize and Wheat Improvement Center (CIMMYT), Harare

Lobell DB, Burke MB, Tebaldi C et al (2008) Prioritizing climate change adaptation needs for food security in 2030. Science 319:607–610

Mabaya E, Alberto ME, Tomo AA et al (2017) Mozambique brief 2017 – the African seed access index. Available from: http://tasai.org/reports/

Niang I, Ruppel OC, Abdrabo MA et al (2014) Africa. In: Barros VR, Field CB, Dokken DJ et al (eds) Climate change 2014: impacts, adaptation, and vulnerability. Part B: regional aspects, Contribution of working group II to the fifth assessment report of the intergovernmental panel on climate change. Cambridge University Press, Cambridge, UK/New York, pp 1199–1265

OECD (2018) OECD statistics. http://stats.oecd.org/Index.aspx?DataSetCode=GERD_FUNDS

Pardey PG, Chan-Kang C, Dehmer SP et al (2016) Agricultural R&D is on the move. Nature 537:301–303

SCCI (2017) Official variety registrar 2017. Seed Control and Certification Institute, Mount Makulu, Chilanga, Zambia

Schlenker W, Lobell DB (2010) Robust negative impacts of climate change on African agriculture. Environ Res Lett 5(1):014010 Available from: http://iopscience.iop.org/article/10.1088/1748-9326/5/1/014010/meta

Smale M, Olwande J (2014) Demand for maize hybrids and hybrid change on smallholder farms in Kenya. Agric Econ 45:1–12

Smale M, Simpungwe E, Birol E et al (2015) The changing structure of the maize seed industry in Zambia: prospects for orange maize. Agribusiness 31(1):132–146

7

Drought-Tolerant Maize Variety for Combating Risks of Climate Change

Berhanu T. Ertiro, Girum Azmach, Tolera Keno, Temesgen Chibsa, Beyene Abebe, Girma Demissie, Dagne Wegary, Legesse Wolde, Adefris Teklewold, and Mosisa Worku

7.1 Introduction

In Ethiopia, annual maize production is 7.8 million tonnes with an average yield of 3.6 tonnes per hectare (t ha^{-1}) in 2016—the highest of any cereal in the country (Food and Agriculture Organization Corporate Statistical Database (FAOSTAT) 2017). Currently, 66% of cereal-farming households in Ethiopia cultivate maize on 2.1 million hectares (ha), making it the second most widely cultivated cereal in the country after teff. It is estimated that each household owns around 1 ha of crop land, of which at least half is allocated for maize cultivation in major maize-growing areas. Subsistence maize farming accounts for more than 95% of the total maize area and production, with 75% of all maize produced being consumed by the farming household (Abate et al. 2017).

Ethiopia started growing hybrid maize relatively late, even by African standards (Harrison 1970, Tolessa et al. 1993). Early maize research in Ethiopia focused on the identification of locally adapted open-pollinated varieties (OPVs) to replace low yielding, tall and lodging susceptible landraces. The national hybrid maize breeding programme in Ethiopia was launched in the early 1980s, targeting four major maize growing agro-ecologies: mid-altitude sub-humid, highland sub-humid, low-moisture stress, and lowland sub-humid maize agro-ecologies. A top-cross hybrid variety, BH140, was released for the mid-altitude sub-humid agro-ecology in 1988 (Tolessa et al. 1993). A late maturing three-way cross hybrid, BH660, adapted to the mid-altitude moist and transitional highland maize agro-ecologies was released in

B. T. Ertiro (✉) · G. Azmach · T. Keno · T. Chibsa · B. Abebe · G. Demissie · L. Wolde
Ethiopian Institute of Agricultural Research, Addis Ababa, Ethiopia

D. Wegary · A. Teklewold
Centro Internacional de Mejoramiento de Maíz y Trigo (CIMMYT), Addis Ababa, Ethiopia

M. Worku
CIMMYT, Nairobi, Kenya

1993; and, later, another intermediate-maturing single-cross hybrid, BH540, adapted to the mid-altitude moist maize agro-ecology was released in 1995. The launch of the National Extension Intervention Program in 1993 by the Ethiopian Government, in partnership with Sasakawa Global 2000, played a key role in the popularisation and dissemination of these hybrids (Worku et al. 2012; Abate et al. 2015). In the period 2002–2010, the three hybrids accounted for over 90% of total maize seed sales (35,000 tonnes) by the Ethiopian Seed Enterprise (ESE)—the primary public seed supplier in Ethiopia. BH660 constituted over 55% of total hybrid seed sales (Worku et al. 2012).

7.2 Climate Change and Drought in Ethiopia

The average age of current maize varieties under production in Ethiopia is 11 and 18 years for hybrids and OPVs respectively (Abate et al. 2017). On average, 80% of maize varieties commonly grown in Ethiopia were developed using germplasm not improved for drought tolerance over 20 years ago (Abate et al. 2015).

As for many countries of sub-Saharan Africa (SSA), climate change projections for Ethiopia suggest an increase of maximum and minimum temperatures and a decreasing trend in precipitation (Deressa 2007). Since 1971, Ethiopia has experienced eight drought episodes of varying severity due to reduced rains in different parts of the country (Viste et al. 2012). These episodes lasted from a single year to 4 years in duration. In general, decline in precipitation has been observed in southern Ethiopia during both the February–May and June–September seasons, although a similar trend was not observed in the central and northern highlands. The study by Viste et al. (2012) found 2009 to be an exceptionally severe drought year, and the second driest year overall in the period 1971–2011, surpassed only by the historic drought of 1984. The study also revealed increasing frequency of spring (February–May) droughts in all parts of the country in recent years. Despite highly variable rainfall, Ethiopia relies on a rain-fed agriculture with irrigated areas accounting for only 1% of the total maize area (Abate et al. 2015). This makes Ethiopian agricultural system highly vulnerable to drought events as was clearly demonstrated recently during the 2016 drought, caused by El Niño, that severely affected maize production.

Breeding and disseminating drought- and heat-tolerant, climate-smart maize varieties can play a major role in mitigating the risks of droughts today as well as projected climate change in Ethiopia. Old varieties that are currently in commercial production were not selected for drought tolerance and are less likely to be adapted to future climates. CIMMYT, in collaboration with national agricultural research programmes in SSA, has been intensively developing drought-tolerant (DT) varieties that are also high-yielding under optimum conditions. These new DT varieties should replace old popular varieties to minimise the risk of climate change on productivity.

BH660 is the most popular maize hybrid in Ethiopia but is over 20 years old and was not developed for drought tolerance. Between 2010 and 2012, annual certified seed production of this variety peaked at 6000 tonnes—an amount sufficient to cover more than 240,000 ha. Replacing the dominant but ageing crop varieties with new climate-smart varieties is a critical step in reducing the risks of climate change in SSA.

7.3 Research Efforts to Develop New Hybrids

The Ethiopian maize breeding programme was initiated in 1952 by first collecting germplasm from various national and international sources (Tolessa et al. 1993). The programme later focused on using germplasm of East African origin, owing to agro-ecological similarity (Harrison 1970; Tolessa et al. 1993). BH660 was developed using early generation inbred lines (Harrison 1970; Tolessa et al. 1993; Ertiro et al. 2015) derived from Kitale Synthetic II and ECU573. The current breeding strategy is to exploit CIMMYT, International Institute of Tropical Agriculture, and locally developed inbred lines—separately or in combination—to develop new DT hybrids in Ethiopia.

Because of its wide cultivation and popularity, the replacement of BH660 with a DT variety was considered crucial to address the increased frequency of drought as a result of climate change in Ethiopia. The first approach was aimed at replacing only the female parent of the single cross (SC) seed parent of BH660 with CIMMYT DT inbred lines, and then replacing the third (male) parent with other ECU573-derived inbred lines. This, however, was not successful as none of the hybrids outyielded the commercial checks. The second approach focused on a complete replacement of the SC seed parent with a CIMMYT DT SC tester with excellent general combining ability for drought stress, while maintaining the original male parent. Over a period of 4 years (2006–2009), 9–12 hybrid combinations were tested in more than 30 optimum and random drought environments (Table 7.1). Among the tested hybrids, BH661 consistently outperformed the commercial checks (BH660 and BH670) in most trial sites. In head-to-head comparison, BH661 showed an average grain-yield advantage (GYA) of 10.2% and 12.9% over BH660 and BH670, respectively. The new hybrid also showed an average of 2% reduction in plant height, 6% reduction in ear placement and, crucially, 34% reduction in lodging over BH660 (Fig. 7.1).

7.3.1 Release and Adoption of BH661

Unlike the established practice of embarking on variety demonstration only following official release, breeders conducted popularisation, demonstration and pilot seed production concurrently with the variety verification trial of BH661. The variety verification trial is the final stage of variety evaluation in Ethiopia, whereby

Table 7.1 Head-to-head comparison of BH661 with BH660 and BH670 for grain yield in different sets of trials conducted at eight mid-altitude, transitional highland and true highland subhumid maize agro-ecologies of Ethiopia during 2006–2011

Location	Altitude (masl)	Annual rainfall (mm)	Number of trials	Mean grain yield			% GYA over	
				BH661	BH660	BH670	BH660	BH670
Bako	1650	1211	6	9.79	9.43	9.07	3.8%	7.9%
Hawassa	1708	945	5	10.79	9.39	8.39	14.8%	28.5%
Areka	1750	1401	5	6.82	6.50	6.57	5.0%	3.8%
Arsi Negelle	1940	900	6	7.83	8.19	8.09	−4.4%	−3.1%
Jimma	1725	1448	6	10.56	8.71	8.54	21.3%	23.6%
Adet	2203	1118	4	8.10	7.87	7.12	2.9%	13.8%
Finote Selam	1853	1125	3	8.84	7.81	9.09	13.2%	−2.8%
Haramaya	2015	820	2	11.61	8.60	9.45	35.0%	22.9%
Across				**9.09**	**8.25**	**8.06**	**10.2%**	**12.9%**

Source: Bako National Maize Research Center; altitude and annual rainfall (Worku et al. 2012)
masl metres above sea level, *mm* millimetres

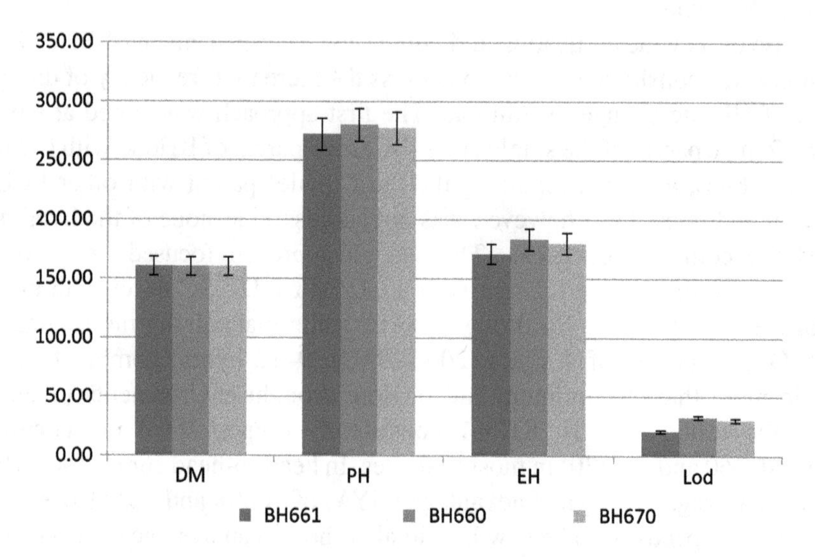

Fig. 7.1 Head-to-head comparison of BH660, BH661 and BH670 for secondary traits evaluated at Bako, Hawasa, Areka, Arsi Negele, Finote Selam and Jimma during 2006–2011; DM (days to maturity); PH (plant height); EH (ear height); Lod (lodging); error bars show ±5%

candidate varieties are compared with the current commercial checks on large plots (10 × 10 m) on research stations and farmers' fields. The plots are evaluated by an ad hoc Technical Variety Release Committee which incorporates farmers' perspectives. The promotion of BH661 began through farmers' participatory variety selection (PVS) with financial support from various CIMMYT-managed projects.

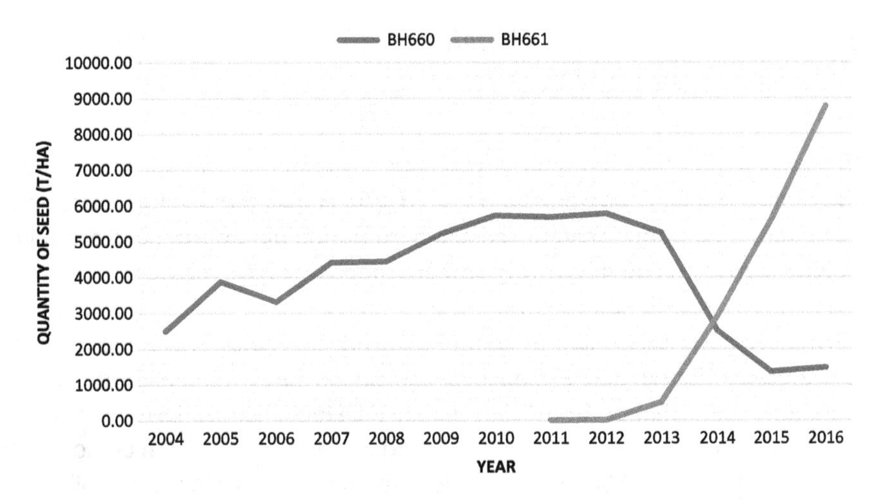

Fig. 7.2 Comparison of the amount of certified seed production of BH660 and BH661 from 2012 to 2016. (Source: compiled by the authors)

Farmers and other stakeholders who took part in the variety evaluation consistently ranked BH661 first for on-farm grain yield, maturity, prolificacy, standability, uniformity and biomass yield. In 2011, the National Variety Release Standing Committee officially approved the release of BH661 for commercial cultivation in the mid-altitude sub-humid and transition highland maize growing areas (Worku et al. 2012). Farmers who were impressed by the outstanding performance of BH661 during the verification and PVS trials started to demand its seed immediately, forcing seed companies to quickly scale-up certified seed production.

At the end of 2011, Bako National Maize Research Center supplied 400 kilogrames (kg) of breeder seed of each of the parental lines, and 450 kg of the SC parent to five certified seed producers—namely, Amhara Seed Enterprise (ASE), Bako Agricultural Research Center (BARC), ESE, Oromia Seed Enterprise (OSE) and South Seed Enterprise (SSE). The Center also produced and distributed 1.7 tonnes of certified seed to seed companies, agricultural offices, research centres and non-governmental organizations who were interested in popularising the new hybrid. In addition, the breeders of BH661 established demonstration plots, organised farmers' field days and intensively used public television stations to promote the hybrid in the two most widely spoken Ethiopian languages, Amharic and Afaan Oromo. By 2012, many institutions were actively promoting BH661, while four of them (BARC, ESE, OSE and SSE) had already begun producing basic seed. ESE produced 6.0 tonnes of certified seed, which was enough to cover 240 ha. By 2014, five companies—namely ASE, Avallo, ESE, OSE, and SSE—produced nearly 2900 tonnes of certified seed (Abate et al. 2015). This rose to almost 9000 tonnes by 2016 (Fig. 7.2). This concerted effort by the national maize research and extension programmes in Ethiopia, along with various national and international stakeholders, was instrumental in fast-tracking the dissemination and adoption of the new DT hybrid.

7.3.2 *How Were Farmers Convinced About the Superior Performance of BH661?*

In 2012, when the national maize research programme and other institutions embarked on large-scale popularisation of BH661 across the country, major maize growing areas—including western and southern Ethiopia—experienced drought, by coincidence during the main growing season. In most places, the onset of rains was delayed and ceased well before grain filling, which affected maize yields especially for late-maturing varieties such as BH660. During that season, the Bako National Maize Research Center had established several on-farm demonstration plots to showcase the performance of BH661 against BH660. Local farmers and other stakeholders from across the country converged at one of the on-farm demonstration sites in a village called Abakora to witness the superior performance of BH661, which was planted alongside BH660 on 400 m^2 plot. The participants clearly noticed the drought tolerance of BH661 and were convinced of its superiority relative to BH660.

In the same year, demonstration plots established by the SSE in the Hawassa area were also affected by drought, similarly convincing the Enterprise of the superior drought tolerance of BH661; thus fast-tracking large-scale seed production and marketing of BH661 in southern Ethiopia. From similar observations on demonstrations plots throughout the country in 2012, various key stakeholders were convinced that BH661 was a better option than BH660.

7.4 Outcome of the Replacement Activities

In 2016, nearly 9000 tonnes (Fig. 7.2) of certified seed of BH661 was produced and marketed by various seed producers in Ethiopia. From the current estimated 55% of maize area planted to improved seed, seed production of BH661 was sufficient to cover 360,000 ha (18% of the total maize area, or about 30% of maize area under improved seed). Improved seed in Ethiopia is produced and marketed by public seed companies (60% market share), local private seed companies (10% market share) and multinational seed companies with proprietary hybrids (30% market share). Public seed companies dominate the market due to incentives from the Government that include royalty-free licensing of public hybrids and access to state-owned land for seed production. As a result, hybrid seed price in Ethiopia is affordable (estimated at less than 1 United States Dollar per kg) to the small-scale farmer. Credit facilities are also available to the poorest farmers through farmers' cooperative unions for the purchase of seed and fertiliser, which is repaid after harvest.

Key to driving the adoption of BH661 was the willingness of seed companies to produce and market the new hybrid and withdraw BH660. Three key producibility features of BH661 were critical in driving the adoption of the hybrid by seed companies: disease- and drought-tolerance of the new SC seed parent which resulted in higher seed yields and, therefore, lower cost of goods; similar female and male planting splits to BH660, which meant that production systems did not need to be modified to accommodate production of BH661; and, having the same male parent as BH660, which meant both hybrids could be produced in proximal areas.

As of 2017, there remains high demand for BH661. Seed demand is recorded by development agents. Demand is compiled at district and zonal levels, which is later passed on to the regional Bureau of Agriculture and Natural Resources (BoANR) and finally to the Ministry of Agriculture and Natural Resources (MoANR). Seed produced by diverse seed growers, including public and private seed companies as well as farmers' cooperative unions, is reported to BoANR and MoANR. Finally, BoANR and MoANR are responsible to match demand and supply.

7.5 Conclusion and Implications for Development

The narrow genetic base of late-maturing germplasm adapted to East Africa has long hampered the development of DT varieties that could replace BH660—the dominant maize hybrid in Ethiopia for the last 25 years. Free access to DT maize germplasm from CIMMYT enabled breeders to develop new DT maize hybrid combinations that better yield under both moisture-stress and optimum conditions. The introduction of new DT varieties, like BH661, into stress-prone maize farming systems has contributed to improved productivity and food subsistence in Ethiopia. Better seed producibility parameters of the parents of BH661, compared to BH660, resulted in rapid adoption of the hybrid by the seed sector. The successful development and commercialisation of BH661 can serve as a valuable case study for breeders, seed companies, extension agents, regulatory and policy makers in how to aggressively replace ageing crop varieties with new climate-smart varieties. Success with BH661 was due to a higher grain yield than BH660 under DT conditions, the disease resistance and DT characteristics of its SC seed parent and the involvement of various stakeholders in popularising the variety. Nonetheless, an overreliance on a single new mega-variety presents risks and, therefore, the development and release of new climate-smart varieties should be a continuous process. Though the public-sector played a crucial role in the dissemination of BH661, this may not be sustainable in all instances and increased participation of the private-sector is likely to play a vital role in the dissemination of climate-smart varieties in SSA.

Acknowledgements Financial support from the Government of Ethiopia and various CIMMYT-managed projects (e.g. Drought Tolerant Maize for Africa and the Sustainable Intensification of Maize-Legume Cropping Systems for Food Security in Eastern and Southern Africa) are highly appreciated. All maize collaborating centres in Ethiopia are especially acknowledged for their data collection.

References

Abate T, Shiferaw B, Menkir A et al (2015) Factors that transformed maize productivity in Ethiopia. Food Secur 7:965–981

Abate T, Fisher M, Abdoulaye T et al (2017) Characteristics of maize cultivars in Africa: how modern are they and how many do smallholder farmers grow? Agric Food Secur 6:30

Deressa T (2007) Measuring the economic impact of climate change on Ethiopian agriculture: Ricardian approach, World Bank policy research paper no. 4342. World Bank, Washington, DC

Ertiro B, Ogugo V, Worku M et al (2015) Comparison of Kompetitive Allele Specific PCR (KASP) and genotyping by sequencing (GBS) for quality control analysis in maize. BMC Genomics 16:908

Food and Agriculture Organization Corporate Statistical Database (2017). Statistical databases and data sets of FAOSTAT. http://faostat.fao.org/default.aspx. Accessed 18 Dec 2017

Harrison MN (1970) Maize improvement in East Africa. In: Leakey CLA (ed) Crop improvement in East Africa. Commonwealth Agricultural Bureaux, Farnham Royal, pp 27–36

Tolessa B, Gobezayehu T, Worku M et al (1993) Genetic improvement of maize in Ethiopia. In: Tolesa B, Ranson JK (eds) Proceedings of the first national maize workshop of Ethiopia. May 5–7 1992. IAR/CIMMYT, Addis Ababa, pp 13–22

Viste E, Korecha D, Sorteberg A (2012) Recent drought and precipitation tendencies in Ethiopia. Theoretical and Applied Climatology 112(3–4):535–551

Worku M, Twumasi-Afriyie S, Wolde L et al (2012) Meeting the challenges of global climate change and food security through innovative maize research: proceedings of the third national maize workshop of Ethiopia. EIAR/CIMMYT, Addis Ababa

8

Delivery of Climate-Smart Varieties of Seeds: Issues and Challenges

Laura K. Cramer

8.1 Introduction

As anthropogenic climate change alters environments across eastern and southern Africa (ESA), farmers will need new crop varieties to counter predicted drops in yields and address new disease threats (Challinor et al. 2016; Spielman and Smale 2017). To help mitigate against potential declines in food security, farmers will need access to crop varieties that perform better under water stress; are shorter-maturing and therefore better adapted to changing rainfall patterns; and are more resistant to new pests and diseases. For one crop—common bean (*Phaseolus vulgaris* L.)—climate models predict that by the 2020s approximately 3.8 million ha of suitable area in Africa would benefit from a bean seed with improved drought tolerance (Buruchara et al. 2011). Research is also beginning to show that climate change is affecting the nutritional value of crops (Myers et al. 2017), suggesting that, to combat malnutrition, farmers must have access to varieties that maintain nutrient levels under higher levels of atmospheric carbon dioxide.

Agricultural researchers and plant breeders have been producing improved seeds in ESA for many decades, but adoption rates remain low (Table 8.1). Despite large investments in breeding, many varieties have never been commercially disseminated (Walker and Alwang 2015). In addition, the length of the breeding, dissemination and adoption (BDA) cycle for some varieties of crops can be as long as 30 years (Challinor et al. 2016). Given the rapid pace of climate change, this cycle is too long to meet the needs of farmers. The process needs to be shortened so that improved

L. K. Cramer (✉)
International Center for Tropical Agriculture (CIAT), Nairobi, Kenya

CGIAR Research Program on Climate Change, Agriculture and Food Security (CCAFS), Wageningen, The Netherlands
e-mail: l.cramer@cgiar.org

Table 8.1 Adoption of modern varieties of bean in Sub-Saharan Africa in 2009

Country	Area planted with modern varieties of beans (%)
Burundi	8.1
DR Congo	16.1
Ethiopia	43.7
Malawi	54.6
Mozambique	13.5
Rwanda	19.0
Tanzania	45.8
Uganda	31.0
Zambia	9.5

Source: Muthoni and Andrade (2015)

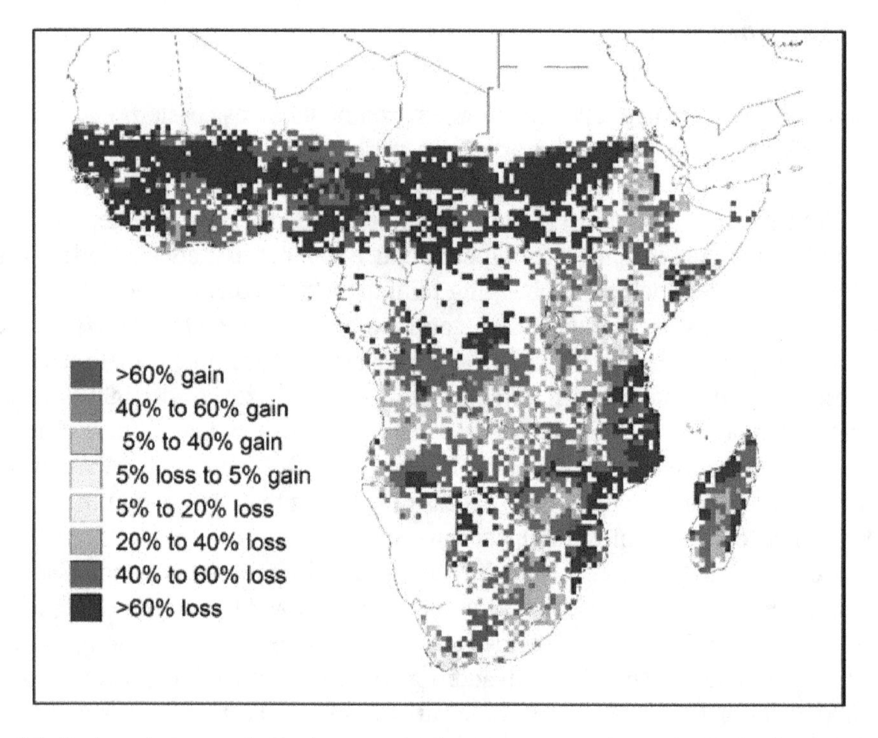

Fig. 8.1 Projected changes in dry bean production: percentage change in production by 2050s, assuming RCP8.5 (high-end emissions), in relation to the mean production of 1971–2000. (Source: Ramirez-Villegas and Thornton 2015)

varieties can reach farmers in a timelier manner (Atlin et al. 2017). Although large areas of Africa are expected to become climatically unsuitable for bean cultivation, certain areas of Kenya are expected to shift in the opposite direction, becoming more suitable (Fig. 8.1). These areas will become increasingly important for bean production as the climate continues to change. Given that beans are a key source of protein and micronutrients in the Kenyan diet, it is critical that new varieties are made available.

The research-for-development community has in recent years been paying increased attention to seed systems (Scoones and Thompson 2011). Initiatives such as the Alliance for a Green Revolution in Africa (AGRA) are working toward improving crop-seed systems, and approaches such as Integrated Seed Sector Development (ISSD) are being created to help in such efforts (Louwaars and de Boef 2012). But obstacles remain. In order for the private sector to play a key role in the dissemination of climate-adapted varieties, major bottlenecks in access to EGS need to be reduced. Such bottlenecks include complicated and disparate licensing agreements among the various regional genetics suppliers; lack of availability of sufficient breeder seed from licensors; and lack of financial resources, technical knowhow and infrastructure to maintain EGS. National governments, international institutions and universities need to work together with seed suppliers to harmonize regulations and streamline access to EGS for the benefit of small-scale farmers. This chapter will briefly explain the structure of seed systems and then address the bottleneck of EGS availability. It will do so through the presentation of two contrasting case studies involving access to EGS in Kenya. The chapter concludes with lessons from the case studies and implications for policy.

8.2 Brief Overview of Seed Systems and Related Interventions

Crop-seed systems in ESA involve a variety of sources and both formal and informal actors. According to McGuire and Sperling (2016), small-scale farmers obtain their seeds from local markets (51%), their own saved stocks (31%), friends and relatives (9%), agro-dealers (2%) and other sources. Supply systems vary depending on the type of crop, with hybridized row crops, cereals and legumes being of greater interest to the private seed sector compared to vegetatively propagated species such as sweet potato and cassava (see Parker et al. in this volume). A generic, much-simplified schematic of a seed system is presented in Fig. 8.2 for illustrative purposes, highlighting the position of EGS in the pathway.

National agricultural research services (NARSs), together with international agricultural research centers (IARCs) and other partners, have spent many decades breeding new crop varieties (see the top left portion of Fig. 8.1) (Walker and Alwang 2015). For many years non-governmental organizations (NGOs) and government extension services have distributed seeds directly to farmers or conducted seed and voucher fairs (Sperling and McGuire 2010). In recent years development partners have been providing funding for broader seed-sector development, including AGRA's support for capacity-building of agro-dealers (marketing and distribution column in Fig. 8.1) and local commercial seed companies

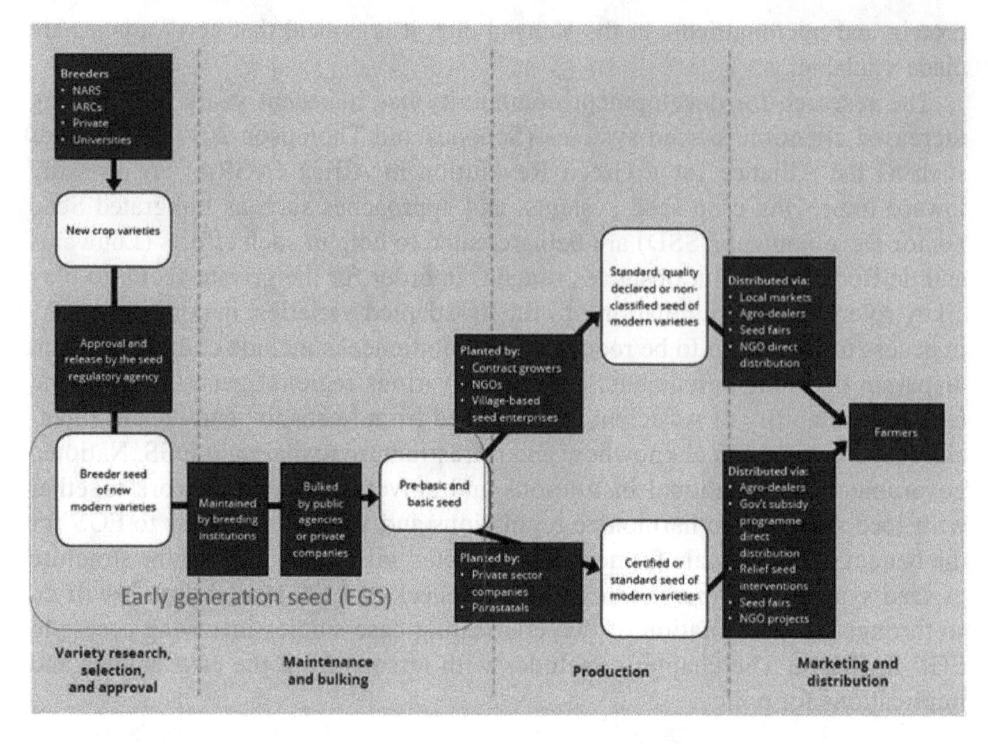

Fig. 8.2 Generic schematic diagram of a seed system. (Source: AgriExperience 2016)

(production column of Fig. 8.1) (McGuire and Sperling 2016). This moves the interventions from the top left along the supply chain toward the bottom center of the diagram. However, historically there has been a disconnect in the interventions: breeding activities from the top left often stopped at the release of a new variety, and seed distribution initiatives often promoted varieties that had been on the market for a long time. This caused a gap to develop around the issue of EGS for newly released varieties.

EGS—which encompasses breeder seed, pre-basic and basic seed—is the critical connection between breeding activities and the eventual production and distribution of varieties to farmers. The inattention paid to this part of the African food crop-seed sector as it has expanded in the last 20 years has created a major constraint within seed systems (EGS Africa Communiqué 2016). Recently a number of country case studies were published along with a communiqué setting forth principles for commercial and sustainable EGS supply (see Box 8.1). The following two case studies highlight the need for national governments to adopt clearer seed-sector policies abiding by these guiding principles, and for actors along the seed value chain to work in an integrated manner.

Box 8.1 Principles of Commercial and Sustainable EGS Supply (Source: Adapted from EGS Africa Communiqué 2016)

1. EGS interventions should be guided by the seed value chain, starting with farmer use of—and willingness to pay for—new, improved varieties.
2. EGS operations should include hybrid, self-pollinated and vegetatively propagated crops; cover formal, intermediary and informal seed systems; and engage public, private and community-based partners.
3. EGS supply should cater to different seed systems (not just the formal system).
4. Effective EGS supply should be part of enhanced seed-value-chain management and integration.
5. An assessment of the division of responsibilities for financing and supply of EGS should be undertaken, with specific consideration of public-private partnerships for open-pollinated and vegetatively propagated crops.
6. National agricultural research organizations (NAROs) are and should remain responsible for the production of breeder seed for improved varieties developed by publicly-financed crop-improvement programmes.
7. A gradual withdrawal of Consultative Group for International Agricultural Research (CGIAR) programmes in direct involvement with EGS production is recommended to move toward a more sustainable seed sector.
8. Research organizations and their breeders should partner in the maintenance of breeder seed so as to keep their priority focus on further crop improvement.
9. Several aspects of regional seed-trade harmonization efforts are relevant to EGS supply, including facilitating movement of EGS supply across borders.
10. Donors should transition from direct interventions in the seed value chain to strengthening public and/or private-sector capacity in EGS supply.

8.3 EGS Case Studies

The case studies presented in this chapter were developed through key informant interviews with a small, local seed company based in Kenya, correspondence with a university bean breeder, and a review of documents provided by the company and found in online searches. Names of some individuals and entities have been omitted to maintain the anonymity of those consulted. The case studies involve the same seed company interacting with two different universities. The examples are contrasting: one is a successful example of EGS sharing, and the other is an example of how the lack of EGS access can hinder commercialization of a new variety.

8.3.1 *Case Study 1: Successful Partnerships for Highland Bean Varieties*

The maize lethal necrosis (MLN) disease was first reported in Kenya in 2011 and caused significant losses to households (FSNWG 2013). According to de Groote et al. (2016), an estimated 500,000 mt of maize were lost in 2013, an amount valued at US$180 million and equivalent to 22% of average annual production. As the disease spread in the highlands of the Rift Valley, Kenya's agricultural extension service recommended that small-scale farmers plant crops other than maize to avoid infection and spread of the disease. Given that the main staple crops in the Kenyan diet are maize and beans, beans were a logical choice for farmers to plant. Common beans have not been grown historically in the highland areas because average temperatures were too low,[1] but in the last two decades rising average temperatures have made it possible to grow beans in these locations (MoALF 2016). Beans grown at high elevation, however, can be more vulnerable to diseases such as angular leaf spot, rust, common bacterial blight and root rot (Wagara and Kimani 2007).

Egerton University, in collaboration with the International Center for Tropical Agriculture (CIAT), developed and was starting to commercialize medium- to high-altitude bean varieties with disease resistance, and three varieties (Chelalang, Ciankui and Tasha) were deemed suitable for the situation arising from the MLN crisis. Between 2011 and 2014, the university partnered with several local seed companies. The involved parties signed contracts through which the university sold breeder seed (a type of EGS) to the seed companies, who received non-exclusive rights for multiplication, upscaling and commercialization. With funding received from AGRA in 2014, the university promoted and marketed the new varieties to help farmers learn about them. One variety in particular, Chelalang, has since been adopted by farmers in the highlands affected by MLN and is growing in popularity in the Rift Valley region. In 2017, Chelalang was included in an input-supply programme of the Kenyan government, the Kenya Cereal Enhancement Programme, through which 30,000 farmers each received enough seed to plant half an acre with that variety. A total of 300 mt of Chelalang seed was distributed prior to the long rainy season, stretching from March to May (pers. comm.).

With the spread of MLN disease creating an urgent need for action, the university and the seed companies quickly reached an agreement on the provision of the necessary breeder (EGS) seed, which was readily available in sufficient quantities. The cost for one ton of breeder seed was approximately US$2900. Egerton did not charge the companies a licensing fee, and under the agreement the companies pay a 5% royalty on gross sales to the university. The marketing efforts funded by development partners (received after the licensing agreements had been signed) helped raise awareness among farmers and stimulate demand. Using Fig. 8.1 as an analytical framework for assessment of this case study, the breeding had already

[1] Common bean is suited to optimal average daily temperatures of 20–25 °C, and below 15 °C seeds germinate poorly (De Ron et al. 2016).

been done in anticipation of future needs, the early generation breeder seed was in stock at the university, and the university quickly shared it through an agreement with local seed companies. The breeder was eager to see it reaching farmers and helped align the public good with commercial interests. Because they had previously worked together, there was an adequate level of trust between the university breeder and the seed companies that the contracts would be honored and the correct royalties paid. The companies were then able to bulk up the breeder seed and obtain all the necessary approvals from the Kenya Plant Health Inspectorate Service (KEPHIS) to get certified seed to market and thus available to the farmers who needed an alternative to maize. In this case, there was sufficient trust between the actors and smooth coordination for contract establishment, and EGS availability and access were not a hurdle.

8.3.2 Case Study 2: The EGS Hurdle for a Bean Variety High in Iron and Zinc

The second case study, by contrast, illustrates the difficulties that often arise. Two NGOs serving small-scale farmers were interested in procuring a bean variety high in iron and zinc that had been developed by a breeder at a local university using breeding lines provided by CIAT. This variety was developed through a project aimed at increasing the micronutrient quality of beans (CIAT 2005). The variety had been registered by the breeder in the Kenya seed catalogue of 2012, but it was not yet commercially available. One of the NGOs requested a sample of the seeds from the university breeder to use in demonstration plots, and the other NGO placed an advertisement in a national daily newspaper tendering for seed of a bean variety with the required characteristics. The first NGO also asked a seed company (one of the same companies involved in the project described in the first case study) if seeds of bean varieties with high iron and zinc bioavailability were available. The seed company, realizing a market need due to both the inquiry and the newspaper advertisement, made contact with the breeder at the university with the hope of paying for non-exclusive rights to commercialize the variety, which would allow the company to provide certified seed to the interested NGOs and to sell the variety in the open market.

The seed company, breeder and university office for intellectual property management went through several rounds of negotiation over the course of a year to try to reach an agreement but were unable to do so. The seed company was planning to start with production of 20 mt/year (increasing in subsequent years as demand rose). The draft contract proposed by the university stipulated that the seed company must purchase at least 2 mt of breeder seed every season to cover the costs of human resources, land, technical support, etc., in addition to paying an annual licensing fee. The company, however, was requesting a one-time purchase of only 50 kg of breeder seed. It planned to then bulk up this seed into the required amount of pre-basic and basic seed (see Fig. 8.1). The university also requested that the seed company carry out the production of the breeder seed on the university's behalf and then purchase it, with the cost of production subtracted from the price. This request indicated that an adequate amount of breeder seed was not available at the university. According to the university breeder, there was lack of a framework for producing certified breeder, pre-basic and basic seed for new varieties. It was assumed that breeders and/or their institutions would have the resources for this production and to support commercialization of new varieties, but in fact this is rarely the case (pers. comm.).

The agreement foundered because the seed company was unwilling to commit to buying such a large amount of seed every season and did not have the capacity to undertake the production of the breeder seed as the university was requesting. The upfront investment was considered too high for a variety that had not yet been commercialized. The total fixed cost of the stipulated 2 mt of breeder seed per season, along with the annual licensing fees and royalties as laid out in the proposed contract, would have amounted to a cost of 33% of the seed price at current market rates for a similar bean variety, squeezing already tight profit margins for the company when production, marketing and distribution costs were factored in. This made the agreement unattractive for the company. The university was unwilling to yield to the seed company's requests for a smaller amount of breeder seed, and the seed company was unable to meet the university's purchasing requirements. A lack of an urgent focusing event, apparent unavailability of EGS for the variety in question, and low motivation for commercialization all contributed to the failure to reach an agreement.

This case study illustrates how the lack of access to EGS has prevented a needed variety from reaching farmers. The university that holds the rights to the variety was seemingly not able to provide the breeder seed requested, and the seed company could not shoulder the burden to produce and purchase an amount that was uneconomical for its needs. There was no other bean variety released in Kenya with the nutritional properties specified. As a result, the bean variety that was developed to help reduce micronutrient malnutrition remains uncommercialized and unavailable to Kenyan farmers. The university breeder reported that he has released 24 bean varieties in Kenya but that only 10 have been commercialized (pers. comm.). The university has worked with the largest seed company in Kenya and provided it with exclusive rights to other bean varieties, but there appears to be a lack of trust and understanding between the university and the smaller seed company, leading to a failure to reach an agreement.

8.4 Possible Solutions and Implications for Development

Development partners such as the Bill and Melinda Gates Foundation (BMGF) and United States Agency for International Development (USAID) have started paying attention to the EGS hurdle and are studying the multiple causes of the constraints and how to overcome them (BMGF and USAID 2015). A host of recommendations by country and crop have been proposed to help address the relevant issues, including partial or full subsidization of EGS production costs for crops such as beans. In the Kenya country report released as part of the BMGF and USAID EGS studies, one of the main bottlenecks identified in the common bean EGS supply system is "inadequate supply of breeder seed from public sector breeders" which "precludes private sector involvement in EGS production and limits EGS production overall" (Context Network 2016, p. xvii). The second case presented here illustrates this constraint. The recommendation is for a public-private partnership (PPP) to reduce production burdens on the public sector and costs imposed on the private sector in an effort to improve the availability of EGS for common bean (Context Network 2016). Given the existing lack of trust between actors, there is a role for a neutral broker to help bring stakeholders together and facilitate the development of such a partnership.

The Pan-Africa Bean Research Alliance (PABRA), a long-running research consortium led by CIAT, comprises hundreds of actors and is active in 31 African countries (PABRA 2018). Along with its successes in breeding, variety release and dissemination, the alliance has learned lessons about breeding and seed production. As described by Buruchara et al. (2011, p. 241), "[p]ublic sector research has to commit to producing breeder seed—as an integral part of the variety development process. There is no sense in releasing a variety (or engaging in breeding) if that variety is not set on a course for multiplication."

Getting the actors represented in the top left of Fig. 8.1 to work in harmony with other stakeholders and building trust is key to overcoming the constraints of EGS supply. Clearer seed regulation policies within national governments and their agencies on the maintenance and supply of EGS would assist in overcoming such barriers. More commercially viable licensing options, particularly in cases where one supplier has a monopoly on a variety that has high value for cultivation, would also be useful. While there is a government-recommended royalty in Kenya (3%), this does not prevent the addition of other licensing fees and costs that drive up the expense to procure breeder seed and begin production (pers. comm.). Stronger commitment from those funding breeding programmes to take the new varieties through all the stages to commercialization (instead of stopping at release) would also help overcome the EGS hurdle.

The research-for-development community, national governments and private sector actors need to work together. New crop varieties are costly to produce, and leaving them uncommercialized is a waste of public research money (Rubyogo et al. 2010). According to Muthoni and Andrade (2015), funding for bean improvement alone at CIAT peaked at US$13.8 million in 1990 and recently stabilized at about US$5.5 million per year. The funders of breeding programmes should have specific goals in mind, including the ways in which any varieties that are developed will ultimately reach farmers, either through commercial processes or with public support for multiplication and distribution. Crops such as the common bean that have low multiplication rates and high transport costs require public research support and dissemination (BMGF and USAID 2015). The commercial sector, however, should not be overlooked. The company featured in these two case studies originally became interested in producing and selling common bean seed after the production manager realized that Kenyan farmers purchase seed at higher rates than is often assumed. They may be purchasing farmer-saved seed through the informal market, but there is also demand for higher-quality seed through the formal market (pers. comm.).

Easing the availability of EGS is now recognized as crucial to the crop-seed systems of ESA. Urgent action is required to remove the hurdles faced by private-sector companies who are ready and willing to begin commercializing new varieties and to build trust between the actors involved in breeding and commercialization. The experiences from the two cases presented here can be summarized into eight lessons:

1. Public subsidies to promote and market new varieties can encourage commercial interest.
2. Publicly funded agricultural input subsidies can kick-start the market and encourage commercial interest.
3. Access to breeder seed and multiplication rights must be at a cost low enough to attract commercial interest, and annual licensing fees should not be cost-prohibitive.
4. Universities and public breeders must invest in maintaining minimum supplies of breeder seed of promising varieties.
5. Non-exclusive rights are helpful in getting new varieties to market because they allow several small companies to sell the same variety.
6. Mandates for minimum seed production discourage commercial interest, especially for varieties that have not yet been marketed.
7. Public and private sectors ought to share the risks of initial seed multiplication and commercialization; commercial companies should not bear all of that risk.
8. Seed-sector actors should make use of focusing events (such as the MLN crisis) for public pressure to encourage successful collaboration.

Giving farmers more choices of crops and varieties through the availability of quality seeds will help enable them to better adapt to an altering environment. Because climate change will increase the spread of pests and diseases, alternative crops and resistant varieties are needed (Beebe et al. 2011). This is a critical component of climate-smart agriculture (CSA) because it allows farmers to increase their resilience through selection of appropriate varieties and boost their production by using higher-quality seeds and better-adapted varieties. If intensified production can be achieved, the third pillar of CSA—mitigation—may also be realized through reduced rates of expansion for agricultural land.

References

AgriExperience (2016) Reaching more farmers with high quality seed for drought tolerant crops, Vuna Research Report. Vuna, Pretoria

Atlin GN, Cairns JE, Das B (2017) Rapid breeding and varietal replacement are critical to adaptation of cropping systems in the devleoping world to climate change. Glob Food Sec 12:31–37. https://doi.org/10.1016/j.gfs.2017.01.008

Beebe S, Ramirez J, Jarvis A et al (2011) Genetic improvement of common beans and the challenges of climate change. In: Yadav S et al (eds) Crop adaptation to climate change. Wiley-Blackwell, Oxford. https://doi.org/10.1002/9780470960929.ch25

BMGF and USAID (2015) Early generation seed study. Bill and Melinda Gates Foundation, United States Agency for International Development, Seattle, WA, USA

Buruchara R, Chirwa R, Sperling L et al (2011) Development and delivery of bean varieties in Africa: the Pan-Africa Bean Research Alliance (PABRA) model. Afr Crop Sci J 19(4):227–245

Challinor AJ, Koehler AK, Ramirez-Villegas J et al (2016) Current warming will reduce yields unless maize breeding and seed systems adapt immediately. Nat Clim Chang 6:954–958. https://doi.org/10.1038/nclimate3061

CIAT (2005) Fast tracking of nutritionally-rich bean varieties. Highlights: CIAT in Africa no. 24

Context Network (2016) Kenya early generation seed study: country report. USAID Bureau of Food Security, Washington, DC

De Groote H, Oloo F, Tongruksawattana S et al (2016) Community-survey based assessment of the geographic distribution and impact of maize lethal necrosis (MLN) disease in Kenya. Crop Prot 82:30–35. https://doi.org/10.1016/j.cropro.2015.12.003

De Ron AM, Rodiño AP, Santalla M et al (2016) Seedling emergence and phenotypic response of common bean germplasm to different temperatures under controlled conditions and in open field. Front Plant Sci 7:1087. https://doi.org/10.3389/fpls.2016.01087

EGS Africa Communiqué (2016) Communiqué on commercial and sustainable supply of early generation seed of food crops in sub-Saharan Africa. EGS Africa Communiqué, Addis Ababa Available at: http://www.issdseed.org/resource/communiqué-promoting-commercial-and-sustainable-supply-early-generation-seed-food-crops-sub

FSNWG (2013) Maize lethal necrosis disease (MLND): a snapshot. Food Security and Nutrition Working Group, FAO Sub-Regional Emergency Office for Eastern & Central Africa

Louwaars NP, Simon de Boef W (2012) Integrated Seed Sector Development in Africa: a conceptual framework for creating coherence between practices, programs, and policies. J Crop Improv 26:39–59. https://doi.org/10.1080/15427528.2011.611277

McGuire S, Sperling L (2016) Seed systems smallholder farmers use. Food Secur 8:179–195. https://doi.org/10.1007/s12571-015-0528-8

MoALF (2016) Climate risk profile for Nakuru, Kenya County Climate Risk Profile Series. The Kenya Ministry of Agriculture, Livestock and Fisheries, Nairobi

Muthoni RA, Andrade R (2015) The performance of bean improvement programmes in sub-Saharan Africa from the perspectives of varietal output and adoption. In: Walker TS, Alwang J (eds) Crop improvement, adoption and impact of improved varieties in food crops in sub-Saharan Africa. CGIAR and CABI, Oxfordshire

Myers SS, Smith MR, Guth S et al (2017) Climate change and global food systems: potential impacts on food security and undernutrition. Annu Rev Public Health 38:259–277. https://doi.org/10.1146/annurev-publhealth-031816-044356

PABRA (2018) Where we work. Pan-Africa Bean Research Alliance, Nairobi, Available at: www.pabra-africa.org/where-we-work/

Ramirez-Villegas J, Thornton PK (2015) Climate change impacts on African crop production, CCAFS Working Paper no. 119. CGIAR Research Program on Climate Change, Agriculture and Food Security (CCAFS), Copenhagen Available at: www.ccafs.cgiar.org

Rubyogo JC, Sperling L, Muthoni R et al (2010) Bean seed delivery for small farmers in sub-Saharan Africa: the power of partnerships. Soc Nat Resour 23:285–302. https://doi.org/10.1080/08941920802395297

Scoones I, Thompson J (2011) The politics of seed in Africa's green revolution: alternative narratives and competing pathways. IDS Bull 42:1–23. https://doi.org/10.1111/j.1759-5436.2011.00232.x

Sperling L, McGuire SJ (2010) Persistent myths about emergency seed aid. Food Policy 35:195–201. https://doi.org/10.1016/j.foodpol.2009.12.004

Spielman DJ, Smale M (2017) Policy options to accelerate variety change among smallholder farmers in South Asia and Africa south of the Sahara. International Food Policy Research Institute, Washington, DC

Walker TS, Alwang J (2015) Crop improvement, adoption and impact of improved varieties in food crops in sub-Saharan Africa. CGIAR and CAB International, Montpellier

Wangara IN, Kimani PM (2007) Resistance of nutrient-rich bean varieties to major biotic constraints in Kenya. Afr Crop Sci Conf Proc 8:2087–2090

9

RTB Crop Seed Systems and Climate Change

Monica L. Parker, Jan W. Low, Maria Andrade, Elmar Schulte-Geldermann, and Jorge Andrade-Piedra

9.1 The Significance of RTB Crops for Food and Income Security Under Climate Change

Throughout the humid African tropics, root, tuber and banana (RTB) crops are the most important food staple. Approximately 300 million people in developing countries depend on RTB value chains (namely cassava, potato, sweetpotato, bananas and yams) for food security and income (Thiele et al. 2017). Indeed, foods derived from RTB crops contribute significantly to caloric needs, from nearly 25% in Nigeria to close to 60% in the Democratic Republic of Congo (RTB 2016). Being bulky and perishable, RTB crops are commonly grown for local consumption (Bentley et al. 2016).

The potential of RTB production to contribute to food security in sub-Saharan Africa (SSA) has not yet been realized due to low productivity. Underdeveloped seed systems have been unable to disseminate clean seed of climate-smart varieties of RTB crops. Potato yields in most of SSA have stagnated at 8–15 t/ha, largely as a consequence of limited access to quality seed (Demo et al. 2015). In Kenya, Uganda and Ethiopia, nearly 75% of the potato fields are contaminated with

M. L. Parker (✉) · J. W. Low · E. Schulte-Geldermann
CGIAR Systems Organization Research Program on Roots, Tubers and Bananas (RTB) and International Potato Center (CIP), Nairobi, Kenya
e-mail: m.parker@cgiar.org; j.low@cgiar.org; e.schulte-geldermann@cgiar.org

M. Andrade
CGIAR Systems Organization Research Program on Roots, Tubers and Bananas (RTB), International Potato Center (CIP), Maputo, Mozambique
e-mail: m.andrade@cgiar.org

J. Andrade-Piedra
CGIAR Systems Organization Research Program on Roots, Tubers and Bananas (RTB), International Potato Center (CIP), Lima, Peru
e-mail: j.andrade@cgiar.org

Ralstonia solanacearum (a long surviving, soil-borne bacterial pathogen), and less than 5% of farmers have access to quality seed (Gildemacher et al. 2009). However, rates of food production can double, and possibly triple, without expanding the area under production, by developing seed systems that deliver abiotic and biotic stress-tolerant varieties.

We present two case studies that describe the introduction of climate-smart varieties of potato in Kenya and orange-fleshed sweetpotato (OFSP) in Mozambique, and the associated challenges in their delivery through seed systems.

9.2 Challenges to RTB Seed Systems

Unlike true seed crops, RTB crops are vegetatively propagated crops (VPCs) and their seed systems have received limited investment. Since VPCs tend to remain true to varietal type for generations, farmers tend to save seed over several years. However, there is a problem with this approach; multiplying the VPC seed without acquiring fresh seed to flush through diseased stock can risk degeneration—the process when pests and diseases accumulate over successive cycles of propagation (Bentley et al. 2016). More efficient seed systems that deliver climate-smart varieties and reduce the spread of disease are required to reduce the yield gap in RTB crops.

As shown in Table 9.1, there are challenges to encouraging investment along RTB seed systems, such as the bulky and perishable nature of the planting material. Investment must therefore be focussed near the seed users who are often in isolated, rural areas. Furthermore, the low multiplication ratios mean seed production is more expensive and requires more time than for grain crops.

The benefits of climate-smart varieties can only be realized by addressing weaknesses in the delivery chain through functioning seed systems, directly linking seed systems as a key tool to address climate change. The complexity of the production and logistics systems must also be expertly addressed in order to speed up the delivery of well adapted varieties to markets.

9.3 Case Studies

9.3.1 Improving Access to Quality Seed of Climate-Smart Potato Varieties in Kenya

Potato (*Solanum tuberosum* L.) is a key staple and fast expanding commercial crop in SSA with more than 1.6 million hectares under production and five million potato farmers (FAOSTAT 2017). In SSA and other tropical regions, potato production is limited to the cooler highlands that lie between 1600 and 3000 m above sea level (masl), and where night temperatures drop below the 16–18 °C required for

Table 9.1 Key characteristics of propagation material of potato and sweetpotato, as compared to maize

Characteristics	Maize	Potato	Sweetpotato[g]
Consumed plant part	Seeds	Tubers	Roots
Most common propagation material	Seeds	Tubers	Vine cuttings
Multiplication ratio	1:300	1:7.5–10	1:3 (a vine may yield 2 or 3 cuttings 30 cm long)
Bulkiness	20 kg/ha	2000 kg/ha[b]	Approx. 666 kg/ha depending on variety and stage of wilting (33,300 cuttings of 25–30 cm)
Storability of harvested seed	Up to 1 year	Up to 6 months	2–3 days
Seed cost (USD/ha)	USD16 to USD27[a]	Up to 50–70% of the total production cost[c]: USD2,527/ha (Chile[d]); USD818/ha (Idaho[e]); USD1090/ha (Peru[f])	Highly variable. For Tanzania: USD2 bundle of 300 vines (900 cuttings), circa USD76/ha
Main pest and diseases causing seed degeneration	Seed degeneration is due to contamination by pollen from other varieties	Potato virus X, potato virus Y, potato leafroll virus, *Ralstonia*, *Rhizoctonia*, *Pectobacterium*, *Spongospora*, *Globodera*, *Meloidogyne*, *Tecia*, *Symmetrischema*, *Phthorimaea*, etc.[c]	Viruses: a complex sweetpotato chlorotic stunt virus and sweetpotato feathery mottle virus transmitted by whitefly and aphids. Weevils also damage and are transmitted through seed, namely *Cylas brunneus* and *C. puncticollis*

Adapted from Bentley et al. (2016)
[a]USD0.80–USD1.00/kg for open-pollinated subsidized maize in Nigeria, USD1.33/kg for private-sector hybrid (Mele and Guéi 2011). Certified maize seed is sold for roughly the same price in Peru, according to the INIA website www.inia.gob.pe/prod-servicios/semillas
[b]Struik and Wiersema (1999)
[c]Thomas-Sharma et al. (2016)
[d]Ministerio de Agricultura de Chile (2013) 1 USD = 554 Chilean pesos
[e]Patterson (2014)
[f]Victor Suárez, personal communication. Varieties Canchán and Yungay in Julcán province, La Libertad department in 2013. 1 USD = 2.75 Peruvian Sol
[g]Kwame Ogero, personal communication

tuberisation (Haverkort and Harris 1987). However, highland farmers are at risk of unpredictable rainfall and increasing temperatures caused by climate change and variability that affect farm productivity under rain-fed conditions. Potato growing is highly susceptible to precipitation variation and 575 mm is the minimum rainfall required per cropping season to obtain reasonable yields of 20 t/ha. Erratic rainfall in Kenya during the 2016–2017 drought reduced yields obtained by seed potato multipliers by 56%, from 15 to 7 t/ha. This was after a reduction in rainfall from a seasonal mean of 737 to 126 mm (International Potato Center 2017a). Potential future impacts of climate change will exacerbate this trend (Zemba et al. 2013).

In Kenya, certified seed production meets approximately 5% of demand, which has slowly increased from 0.6% in 2009 (International Potato Center 2016). The majority of farmers obtain seed from informal sources or save a portion of their harvest as seed for several generations. This is the case in most potato-producing countries in SSA, where certification protocols are not put into practice. The low yields plaguing this region (8–15 t/ha compared to realistic yields of 20–30 t/ha obtainable under smallholder farmer conditions) are largely a consequence of farmers' limited access to quality seed of biotic and abiotic stress tolerant climate-smart varieties (Demo et al. 2015).

9.3.1.1 Climate-Smart Varieties

Climate change can be a major threat to potato production systems in Africa. In many of the drier potato growing regions, climate change causes yields to decline as a result of water and heat stress. Yields will decrease even further where there is no possibility of irrigation, to the extent that growing potatoes will become impossible. Traditional potato growing areas are also at risk of increasing temperatures; hence varieties need to be heat tolerant. To adapt the potato to overcome these challenges, breeding efforts by CIP prioritize resilience to the most likely future abiotic and biotic stresses: heat tolerance, water use efficiency, earliness and disease tolerance. In a series of adaptive participatory trials in several SSA countries, some climate-smart potato clones have shown great adaptability to erratic weather conditions. With 15–20% less precipitation and a temperature increase of 2–3 °C under the scenarios of climate change, these clones have shown greater tolerance to drought and heat without yield losses (International Potato Center 2017b). This reduces the risk of yield losses due to climate change, and offers farmers in mid-altitude regions the possibility to integrate potato into their agrifood system.

From 2013 to 2015, 15 clones were evaluated for water-stress tolerance over three seasons (2013–2015) at three locations ranging from 1300 to 1700 masl, where seasonal precipitation averaged 295 mm (range 210–414 mm) and yielded significantly greater than the existing varieties (Table 9.2, International Potato Center 2017b). In 2016 and 2017, five of these biotic and abiotic-stress tolerant clones with water-stress tolerance and enhanced resistance to late blight and viral diseases were officially released in Kenya, specifically: Unica, Lenana, Wanjiku, Chulu and Nyota.

9.3.1.2 Complexity of the Seed Potato Production System

Seed potato goes through physiologically different forms and rounds of bulking before arriving at the final product. The first generation (G0) is the product of tissue culture (TC) plantlets (the foundation and conservation material) in the laboratory.

Table 9.2 Performance of potato clones in water stressed conditions at average precipitation of 295 mm (range from 210 to 414 mm) across three seasons and three locations between 1300 and 1700 masl in Kenya

Cluster by % age above mean of existing varieties	Yield t/ha	Number of clones
Greater than 40%	22.9	1
Greater than 30%	20.7	5
Greater than 20%	19.4	5
Greater than 10%	18.3	4
Mean of existing varieties	15.5	

The TC plantlets are transferred to a screen-house to produce minitubers (G1) using sand hydroponics or aeroponics. The minitubers are then planted in the field to produce G2 seed in standard seed sizes. The next phase in the seed production process involves bulking tubers. After two to three generations of field multiplication, the seed can be certified. In those countries without operational certification systems, seed multipliers obtain starter material from the National Agricultural Research System (NARS). Informal systems rely on seed multipliers multiplying certified seed for an additional one or two seasons to make quality seed locally available to farmers (Fig. 9.1).

The seed potato planting rate is 2 t/ha and seeds are often sold at farm gates, which means the expansion of improved seed systems is vital to ensuring farmers can access quality seed. Agro-dealers do not distribute seed potato due to its bulk and perishability, and few businesses have invested in certified seed because of high resource and human capacity requirements. To fill the gap in the supply of quality seed at the local level, informal seed multipliers (ISM) are now beginning to diversify.

9.3.1.3 Diversifying Seed Potato Systems

The supply of certified seed in Kenya is limited, therefore many farmers use unmarketable ware potatoes as seed, which they source either on farm or from local markets. This perpetuates the cycle of low yields, as there is no input of quality seed to flush out the diseased (Bentley et al. 2016). To improve localised access to quality seed, ISM in four Kenyan counties (Elgeyo-Marakwet, Nandi, Meru and Uasin Gishu counties) were trained in seed potato multiplication, quality control and record keeping, to support their seed production businesses. The ISMs invested in certified seed potato as starter material which would then multiply. Their transport costs were covered initially, and reduced as the ISM's businesses developed.

In the first 18 months, 220 ISM sold 322 tonnes of quality seed, enough for 1700 farmers to plant 160 ha. The ISM's mean gross margins over three seasons of seed potato sales ranged from 2000 to 4000 USD/ha (International Potato Center 2017c).

Establishing these ISM seed businesses also greatly benefitted the farmers, who travelled significantly shorter distances (reduced from 110 to 3 km) to access qual-

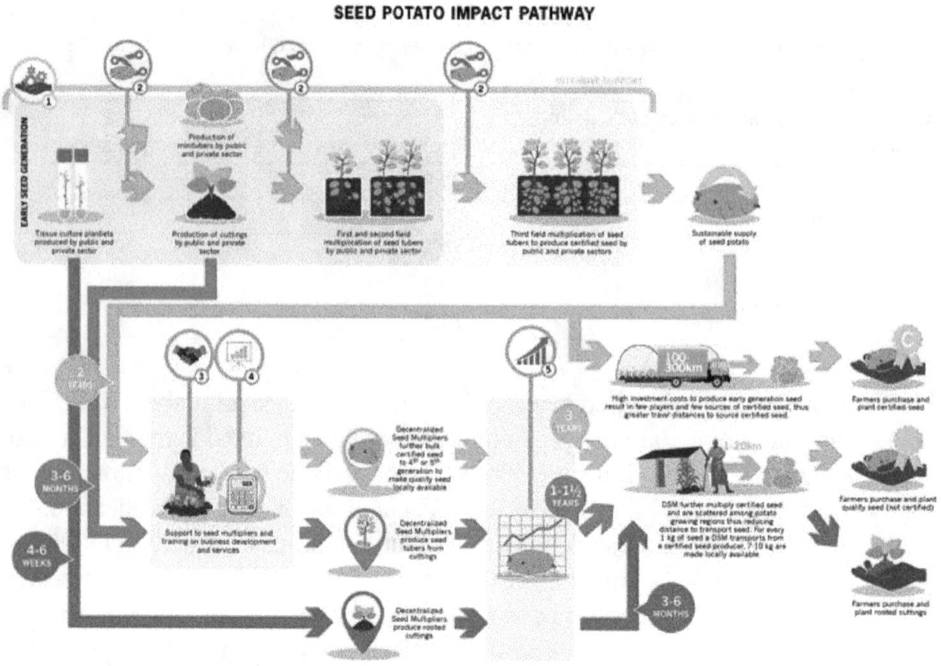

SEED POTATO IMPACT PATHWAY

Fig. 9.1 Seed potato production system showing diverse entry and exit points to engage in seed production that suits various farmer and entrepreneurial profiles

ity, certified seed in Kenya (International Potato Center 2017c). In Meru, the preliminary data also showed that farmers are benefitting from the ISM. Their yields doubled after just one season using the quality seed, averaging 19.2 t/ha compared to 9.4 t/ha using traditional seed (unpublished data).

9.3.1.4 Using Apical Cuttings to Boost Potato Seed Systems

An apical cutting is similar to a nursery-grown seedling, except that it is produced through vegetative means. Rather than allowing TC plantlets to mature and produce minitubers in the screen-house, apical cuttings are produced from the plantlets. Once rooted, the cuttings are planted in the field to produce field seed tubers, followed by one to three successive generations of field multiplication.

In current production systems, apical cuttings can be used in place of minitubers (Bryan 1981). While the latter are more versatile—minitubers can be stored until planting and are easy to transport—apical cuttings are more productive and reduce the time needed to complete the production cycle by one season.

Using apical cuttings in seed systems is a relatively new concept in Kenya, gaining acceptance among stakeholders largely due to productivity gains over seed systems that use minitubers. Within 1 year of the initial trial to test the performance of apical cuttings in the field, two private sector enterprises have invested in producing

Table 9.3 Season 2 on-farm assessment of productivity of apical cuttings to produce seed potato tubers

Variety/spacing	Mean # tubers/cutting	Mean # tubers >20 mm/cutting
Dutch Robyjin[a]	**12.0**	**10.4**
15 × 20	10.7	8.6
20 × 25	9.3	8.0
75 × 30	16.1	14.6
Tigoni	**17.0**	**15.6**
15 × 20	13.4	11.8
20 × 25	15.2	13.9
75 × 30	22.4	21.1
Unica	**9.5**	**8.8**
15 × 20	7.7	7.3
20 × 25	8.2	7.7
75 × 30	12.5	11.6

[a]Highlighted rows are mean tuber yield for the variety across all spacings

them, and the seed potato unit at the National Potato Program has adopted the technology into their seed production system. The body that regulates seed certification has also endorsed apical cuttings and is integrating the technology into seed potato certification protocol.

Progressive farmers and ISM hosted two trials to assess the productivity of apical cuttings over two seasons. This was the first time after one on-station trial to assess productivity, and while the results from the first season were highly variable, they mostly achieved the expected multiplication rate of eight tubers/cutting (data not presented).

Productivity improved from season one to season two, with the mean tuber multiplication rate surpassing the target of eight tubers >20 mm/cutting, averaging 8.8–15.6 tubers >20 mm/cutting (Table 9.3).

9.3.1.5 Productivity Obtained by Informal Seed Multipliers

Additionally, 40 ISM trialed cuttings to produce seed potato. In their first season of production, ISM yields surpassed the expected eight tubers/plant (Table 9.4).

High rates of productivity (between 8 and 10 and up to 15+ tubers per cutting) means seed sales from the cuttings can become profitable after two seasons of multiplication and farmers can access earlier generation seed. Seed tubers produced from cuttings can also be multiplied on farm for a further few seasons without risking significant seed degeneration, provided good agricultural practices are followed.

Table 9.4 Productivity seed potato tubers from apical cuttings obtained by informal seed multipliers (ISM) in their first season of production

Mean number of tubers/cutting				
Variety	Kibiricha network[a]	Kiirua network[a]	Nkuene network[b]	Abothoguchi network[c]
Tigoni	11.9	8.3	11.0 (8.2–13.8)	8.1 (4.2–13.1)
Unica	22.9	18.4	–	7.4 (2.5–13.0)
Konjo	25.5	24.1	14.1 (10.2–19.9)	–
Dutch Robyn	13.0	9.1	–	–

[a]Data are mean of 10 ISM between the two networks
[b]Brackets are minimum and maximum values among 12 ISM
[c]Brackets are minimum and maximum values among 13 ISM

9.3.2 *Adapting Sweetpotato Varieties and Seed Systems Combatting Drought and Food Insecurity in Mozambique*

Mozambique has experienced 13 significant drought years between 1979 and 2016[1] and represents the challenge across much of SSA, where an estimated 2.3 million people needed humanitarian assistance between January and March 2017 (FSIN 2017). Levels of chronic undernutrition are high among children under five in the region, with 71.2% estimated to be vitamin A deficient (VAD) (Aguayo et al. 2005).

High levels of VAD in young children prompted researchers to introduce beta-carotene rich sweetpotato into Mozambique in the late 1990s, because one root (125 g) of an OFSP variety can meet a young child's daily vitamin A needs (Low et al. 2017). Sweetpotato (*Ipomoea batatas* L.) has long been a staple in Mozambique, but the dominant varieties are white-fleshed with no beta-carotene, which the body converts into vitamin A. Initial efforts focused on testing contending varieties from around the world, resulting in the release of nine OFSP varieties in 2000. In 2002 these varieties were widely distributed in southern Mozambique as a post-flood disaster recovery initiative. They performed well in southern and central Mozambique until three seasons of consecutive drought hit in 2005.

Among the most popular of these first-generation varieties was the American-bred *Resisto*, which outyielded local varieties, matured earlier (at 4 months), had a deep orange flesh, moderate dry matter (24%) and the smooth oblong shape favoured by marketers. In the dry season, when farmers plant a second crop in valley bottoms with residual moisture. Resisto produced more roots than the dominant, reputedly drought tolerant local variety *Canasumana*, but it had no vines left at the time of harvest. In contrast, *Canasumana* had abundant foliage left (Low et al. 2001). In tropical areas, sweetpotato is largely propagated from vine cuttings of the previous crop, therefore not maintaining sufficient quantities of vigorous vines resulted in a shortage of *Resisto* planting material the following season.

[1]Significant drought in parts of the country in 1979, 1981, 1987, 1990, 1998, 2001, 2005, 2007, 2008, 2010, 2015, 2016.

In recognition that OFSP was well liked by the population, especially young children, but that better adapted varieties were needed, funds were raised to support breeding in Mozambique. As over 50% of sweetpotato production was lost (both white- and orange-fleshed) in the prolonged 2005 drought, the first step was to collect all landraces throughout the country that had survived. In total, 147 accessions (both landraces and improved materials) were characterized morphologically and molecularly. The best (in terms of high yield performance under water-stressed and non-water stressed condition) were selected as parents to develop drought-tolerant OFSP.

Breeding varieties to survive drought is a complex process. Drought can occur at any point in the development cycle of the crop, and the varieties selected need to perform well under water-stressed and non-water stressed conditions (Andrade et al. 2016a; Makunde et al. 2017). For a variety to be permanently adopted, it needs to have vigorous vines and roots left in the ground at harvest (a traditional source of planting material) must sprout well at the beginning of the rains. With regards to taste, a floury texture is preferred, a characteristic associated with dry matter contents of 28% or above.

The standard protocol historically used for many sweetpotato breeding programs required a variety to develop over a period of 8 years, including: the crossing of the new parents and generation of seed; the growing out of clones from those seeds; a selection process over a number of years, specifically evaluating the variety's agronomic and organoleptic characteristics with active farmer participation. The Accelerated Breeding Scheme (ABS), unlike this traditional approach, exploits the fact that each clone is a potential variety and has more sites earlier in the breeding cycle, including one stress environment for the trait of interest (in this case a drought-prone site). The ABS reduces the breeding cycle from 8 to 4–5 years (Grüneberg et al. 2015).

By applying the ABS, 15 new drought-tolerant OFSP varieties were released in Mozambique in 2011 (Andrade et al. 2016b). An additional four OFSP varieties were released in 2016 (Andrade et al. 2016c). Some of these varieties have widely-adapted and others performed well in specific agroecologies with a range of maturity periods. Many farmers prefer the six improved early-maturing varieties, which are ready in 3–4 months, because they enable them to manage rainy seasons of unpredictable lengths (Alvaro et al. 2017). Some of the later maturing varieties are deeper rooting which can be advantageous, because when the soil dries and cracks, weevils can reach the roots and the deeper the root, the harder it is for the weevil to reach it (Low et al. 2009).

There is strong evidence to suggest that combining OFSP introduction with community-level nutrition education increases the intake of vitamin A in young children and their mothers, and reduces VAD in children under 5 years of age (Low et al. 2007; Hotz et al. 2012; Brauw et al. 2013). However, a major challenge in drought prone areas has been ensuring quality planting material is available when the rains begin (Fig. 9.2). In bimodal areas, vine retention is not a major issue because farmers often use cuttings from an existing crop to start a new one. This explains the larger per capita production of sweetpotato in East and Central Africa

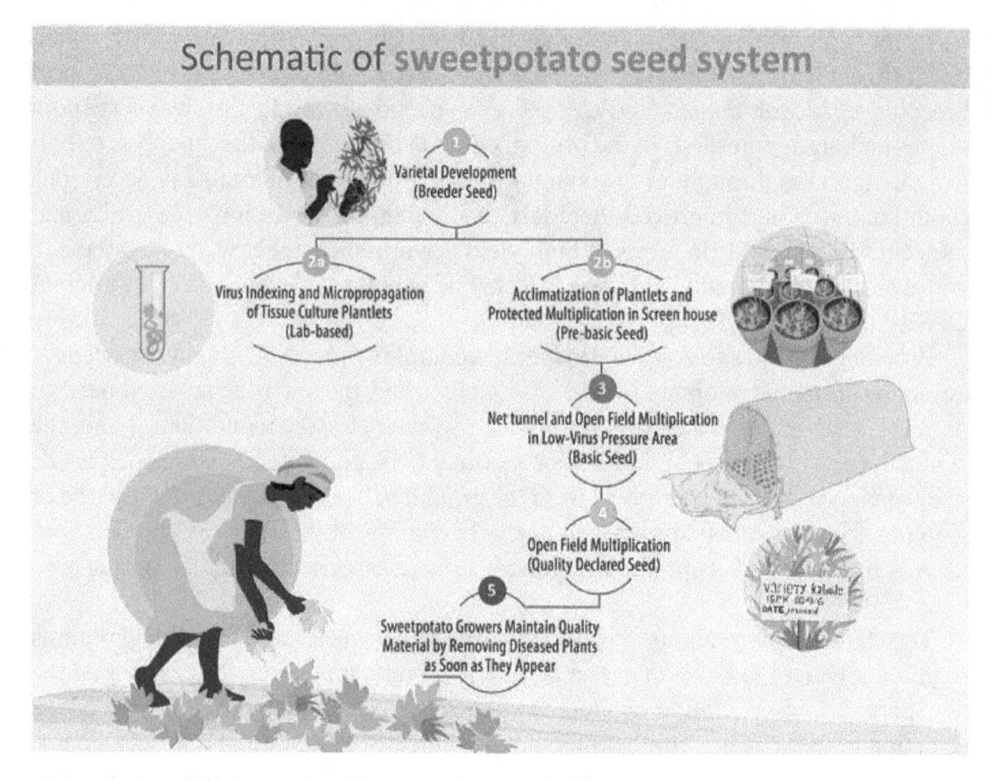

Fig. 9.2 Steps to produce sweetpotato seed

than in unimodal Southern Africa. In drought-prone areas, some farmers with access to valley bottoms with residual moisture use this land for a second crop. Other farmers water small plots near their homes, and some invest in irrigation. The most common method, however, is to leave roots in the ground, ready to sprout when the rains start. The drawback is that the roots are often attacked by weevils or other diseases, which reduces both the quantity and quality of subsequent root output (Gibson et al. 2011).

A method known as Triple S (Storage in Sand and Sprouting) improves upon this traditional practice. Developed in drought-prone areas of Uganda, the method selects pest-free roots at harvest, layers them in a container of sand and stores them in the home for up to 7 months. Some 6–8 weeks before the rains are expected to start, the sprouted roots are planted in a nursery and watered twice a week, producing approximately 40 cuttings per root. Being ready to plant when the rains begin, they enable yield gains ranging from 25% to 300% (Namanda et al. 2013). This technology is now being adapted to local conditions and promoted in six other countries, including Mozambique. It is a low-cost, knowledge-based technology that enables farmers to adapt to drought, and the technique can also be used to store larger roots for consumption for an additional 2–4 months.

9.4 Implications for Development

As described in the potato and sweetpotato case studies, adapting smallholder farming to climate change can be achieved by growing varieties that can cope with high temperatures, erratic rainfall patterns, and even drought. However, functional seed systems are essential for delivering such varieties and providing healthy seed. Research is revolutionizing this adaption to climate change, from new breeding approaches (e.g. ABS), multiplication techniques (e.g. apical cuttings) and on-farm seed management techniques (e.g. Triple S), to new approaches for engaging with specialized seed producers, seed users, markets and regulatory agencies. The clear links between climate change, improved varieties and seed systems illustrate the importance of interdisciplinary collaborations to ensure that scientific, technical, socio-economic and gender aspects are considered in such interventions.

Given the need for strict quality control to manage the high risk of seed degeneration in VPCs, developing seed systems to deliver climate-smart varieties requires a multi-stakeholder approach, especially if support for a project is limited. Sustaining seed systems beyond project life is a key challenge that can be addressed through well-targeted partnerships that drive the process while supporting those who use the system with the technologies to deliver them.

References

Aguayo VM, Kahn S, Ismael C, Meershoek S (2005) Vitamin A deficiency and child mortality in Mozambique. Public Health Nutr 8(1):29–31. https://doi.org/10.1079/PHN2004664

Alvaro A, Andrade MI, Makunde GS, Dango F, Idowu O, Grüneberg W (2017) Yield, nutritional quality and stability of orangefleshed sweetpotato cultivars successively later harvesting periods in Mozambique. Open Agric 2(1):464–468. De Gruyter Open. https://doi.org/10.1515/opag-2017-0050

Andrade MI, Naico A, Ricardo J, Eyzaguirre R, Makunde GS, Ortiz R, Grüneberg WJ (2016a) Genotype × environment interaction and selection for drought adaptation in sweetpotato (*Ipomoea Batatas* [L.] Lam.) in Mozambique. Euphytica 209(1):261–280. https://doi.org/10.1007/s10681-016-1684-4

Andrade MI, Ricardo J, Naico A, Alvaro A, Makunde GS, Low JW, Gruneberg WJ (2016b) Release of orange-fleshed sweetpotato (*Ipomoea batatas* [l.] Lam.) cultivars in Mozambique through an accelerated breeding scheme. J Agric Sci:1–11. https://doi.org/10.1017/S002185961600099X

Andrade MI, Alvaro A, Menomussanga J, Makunde GS, Ricardo J, Grüneberg WJ, Eyzaguirre R, Low J, Ortiz R (2016c) 'Alisha','Anamaria','Bie','Bita','Caelan','Ivone','Lawrence','Margarete', and 'Victoria'Sweetpotato. Hortscience 51(5):597–600

Bentley J et al (2016) *Case Studies of Roots, Tubers and Banana Seed Systems.* https://doi.org/10.4160/23096586RTBWP20163

de Brauw A, Eozenou P, Gilligan D O, Hotz C, Kumar N, Meenakshi JV (2013) Biofortification, crop adoption and health information: impact pathways in Mozambique and Uganda. In: Agricultural and Applied Economics Association, 2013 annual meeting, August 4–6, 2013, Washington, D.C. https://doi.org/10.13140/2.1.1751.4243

Bryan J et al (1981) 'Single-Node Cuttings: A Rapid Multiplication Technique for Potatoes.', in *Rapid Multiplication Techniques.* International Potato Center (CIP), Lima, Peru (CIP slide training series, I), p. 10pp

Chile MA (2013) Ficha Técnico-Económica de Papa de Guarda

Demo P, Lemaga B, Kakuhenzire R, Schulz S, Borus D, Barker I, Woldegiorgis G, Parker ML, Schulte-Geldermann E (2015) Strategies to improve seed potato quality and supply in Sub-Saharan Africa: experience from interventions in five countries. In: Potato and sweetpotato in Africa: transforming the value chains for food and nutrition security, DABI, Wallingford, pp 155–67

FAOSTAT (2017) 2017. http://www.fao.org/faostat/en/#home

FSIN (2017) Food Security Information Network (FSIN) -Home. 2017. http://www.fsincop.net/

Gibson RW, Namanda S, Sindi K (2011) Sweetpotato seed systems in East Africa. Afr Crop Sci Conf Proc 10:449–451

Gildemacher PR, Demo P, Barker I, Kaguongo W, Woldegiorgis G, Wagoire WW, Wakahiu M, Leeuwis C, Struik PC (2009) A description of seed potato systems in Kenya, Uganda and Ethiopia. Am J Potato Res 86(5):373–382. https://doi.org/10.1007/s12230-009-9092-0

Grüneberg WJ, Ma D, Mwanga ROM, Carey EE, Huamani K, Diaz F, Eyzaguirre R et al (2015) Advances in sweetpotato breeding from 1992 to 2012. In: Potato and sweetpotato in Africa: transforming the value chains for food and nutrition security. CABI, Wallingford, pp 3–68. https://doi.org/10.1079/9781780644202.0003

Haverkort AJ, Harris PM (1987) A model for potato growth and yield under tropical highland conditions. Agric For Meteorol 39(4):271–282. Elsevier. https://doi.org/10.1016/0168-1923(87)90020-7

Hotz C, Loechl C, De Brauw A, Eozenou P, Gilligan D, Moursi M, Munhaua B, Van Jaarsveld P, Carriquiry A, Meenakshi JV (2012) A large-scale intervention to introduce orange sweet potato in rural Mozambique increases vitamin A intakes among children and women. Br J Nutr 108(1):163–176. https://doi.org/10.1017/S0007114511005174

International Potato Center (2016) The status of potato value chains in Kenya. International Potato Center, Lima, Peru

International Potato Center (2017a) Accelerated value chain development program. Root crops quarter 3 of year 2 report, International Potato Center, Lima, Peru

International Potato Center (2017b) Evaluation and selection of heat and drought tolerance of CIP potato germplasm, International Potato Center, Lima, Peru

International Potato Center (2017c) Feed the future Kenya, accelerated value chain development program root crops year 2 annual report. International Potato Center, Lima, Peru

Low JW, Arimond M, Osman N, Osei AK, Zano F, Cunguara B, Selemane M L, Abdullah D, Tschirley D (2001) Towards sustainable nutrition improvement in rural Mozambique: addressing macro- and micro-nutrient malnutrition through new cultivars and new behaviors. Pdfs.semanticscholar.org, Michigan State University, pp 1–40

Low JW, Arimond M, Osman N, Cunguara B, Zano F, Tschirley D (2007) A food-based approach introducing orange-fleshed sweetpotatoes increased vitamin A intake and serum retinol concentrations in young children in rural Mozambique. J Nutr 137(5):1320–1327. Drop American Society for Nutrition

Low J, Lynam J, Lemaga B, Crissman C, Barker I, Thiele G, Namanda S, Wheatley C, Andrade M (2009) Sweetpotato in Sub-Saharan Africa. In: The sweetpotato. Springer Netherlands, Dordrecht, pp 359–390. https://doi.org/10.1007/978-1-4020-9475-0_16

Low JW, Mwanga ROM, Andrade M, Carey E, Ball A-m (2017) Tackling vitamin A deficiency with biofortified sweet potato in Sub-Saharan Africa. Glob Food Sec 14:23–30

Makunde GS, Andrade MI, Ricardo J, Alvaro A, Menomussanga J, Gruneberg W (2017) Adaptation to mid-season drought in a sweetpotato (*Ipomoea Batatas* [L.] Lam) germplasm collection grown in Mozambique. Open Agric 2(1). https://doi.org/10.1515/opag-2017-0012

Namanda S, Amour R, Gibson RW (2013) The triple S method of producing sweet potato planting material for areas in Africa with long dry seasons. J Crop Improv 27(1). Taylor & Francis Group):67–84. https://doi.org/10.1080/15427528.2012.727376

Patterson PE (2014) *2014 Cost of Potato Production for Idaho With Comparisons to 2013.* University of Idaho

RTB (2016) RTB CGIAR research program on root, tubers and bananas. Proposal 2017–2022. Lima: CIP III (July 2016)

Struik PC, Wiersema SG (1999) Seed potato technology. Wageningen University Press, Wageningen

Thiele G, Khan A, Heider B, Kroschel J, Harahagazwe D, Andrade M, Bonierbale M et al (2017) Roots, tubers and bananas: planning and research for climate resilience. Open Agric 2(1). https://doi.org/10.1515/opag-2017-0039

Thomas-Sharma S, Abdurahman A, Ali S, Andrade-Piedra JL, Bao S, Charkowski AO, Crook D et al (2016) Seed degeneration in potato: the need for an integrated seed health strategy to mitigate the problem in developing countries. Plant Pathol 65(1):3–16. https://doi.org/10.1111/ppa.12439

Van Mele P, Bentley JW, Guei RG (2011) *African Seed Enterprises: Sowing the Seeds of Food Security*. Wallingford, UK: CAB International

Zemba AA, Wuyep SZ, Adebayo AA, Jahknwa CJ (2013) Growth and yield response of Irish potato (*Solanum Tuberosum*) to climate in Jos-South, Plateau State, Nigeria. Global J Hum Soc Sci 13(5):13–18

10

Perennial NOC Species and Hidden Hunger

Ian K. Dawson, Stepha McMullin, Roeland Kindt, Alice Muchugi, Prasad Hendre, Jens-Peter B. Lillesø, and Ramni Jamnadass

10.1 Introduction

Sub-Saharan Africa (SSA) has areas of high 'hidden hunger' so improving food nutritional quality is crucial (von Grebmer et al. 2014). One method, that is supported by governments in the region (Covic and Hendricks 2016), is the diversification of food systems. In the context of climate-change-related challenges, this approach may have significant benefits compared to alternative methods such as biofortification, as diversity can promote resilience to more variable environmental conditions that negatively affect individual crops (Ray et al. 2012). Diversity-based resilience is, for example, possible through mechanisms including risk spreading and positive stabilising interactions in production (Altieri et al. 2015; for a wider discussion on diversity–stability relationships see Thibaut and Connolly 2013). One crop diversification approach recommended for the region is promoting 'new and orphan crops' (NOC) that include many perennial foods. These are novel or traditional crops that—although important to consumers and farmers—have largely been neglected by researchers and businesses (Dawson et al. 2018b). They are, however, often nutrient-rich and frequently have properties that support their integration into existing food systems, potentially countering increasing reliance on a narrow range of calorie-rich but nutritionally limited foods (Khoury et al. 2014).

Despite their apparent potential, analysis of the contributions of perennial NOC to the resilience of African food systems is limited. In this paper, we help to fill this

I. K. Dawson (✉)
World Agroforestry Centre (ICRAF), UN Avenue, Nairobi, Kenya

Scotland's Rural College (SRUC), Edinburgh, UK

S. McMullin · R. Kindt · A. Muchugi · P. Hendre · R. Jamnadass
World Agroforestry Centre (ICRAF), UN Avenue, Nairobi, Kenya

J.-P. B. Lillesø
Forest & Landscape Denmark, University of Copenhagen, Copenhagen, Denmark

knowledge gap through two approaches applied to eastern and southern African nations. First, we process community-level production data sets to see if perennial NOC species can help fill seasonal gaps in diets subnationally, as part of 'crop portfolios'. Second, using country-level data sets on perennial and annual crop production that are freely available through FAOSTAT (FAOSTAT 2017; see Garibaldi et al. 2011; Khoury et al. 2014 as quoted in the current paper for other examples of the use of these or derived data sets), we explore year-on-year variability in yields for a subset of nations in the region. Beyond this, we explore constraints to perennial NOC integration into the region's food systems, again making use of FAOSTAT data; we also outline some current approaches to overcoming these barriers. We reference the work of the African Orphan Crops Consortium (AOCC 2018) which focuses on production-based NOC interventions, including the training of African plant breeders to use advanced crop improvement methods. Although crop improvement is essential, we outline coinvestments that are also necessary, including in delivery systems to supply improved and adapted planting material of perennial NOC to farmers.

10.2 Can Perennial NOC Contribute to the Resilience of Eastern and Southern African Food Systems?

10.2.1 Perennial NOC Foods and Crop Portfolios

To explore if perennial NOC foods can support seasonal gaps in diets that may be vulnerable to climate-change-related alterations in weather patterns, we have surveyed food systems in rural locations in a range of African countries. The methods for this research have been described elsewhere (McMullin et al. 2017), but involve characterising socio-ecological site-specific food production and consumption information, including the seasonal availability of foods. Food insecurity among farming households is also estimated.

This research shows that perennial NOC foods can be highly important for supporting local consumers' diets in food insecure months at a subnational level in African countries, as illustrated by the case of Siaya County in southwestern Kenya (Fig. 10.1). In addition to filling harvest gaps, the portfolio is adjusted to address certain nutrient gaps, such as in pro-vitamins A and C that are often lacking in diets in SSA and whose absence has significant detrimental health consequences. Selected crop portfolios are then recommended to farmers for specific locations.

This analysis is based on consumption being tightly linked with what foods are produced locally. While such a connection is often found in subsistence farming (Powell et al. 2015), incomes and market access affect the relationship (Sibhatu et al. 2015). These aspects require further exploration, while across-season measurements are required to account for the effects of changing weather patterns on crop phenologies, to better model climate change effects.

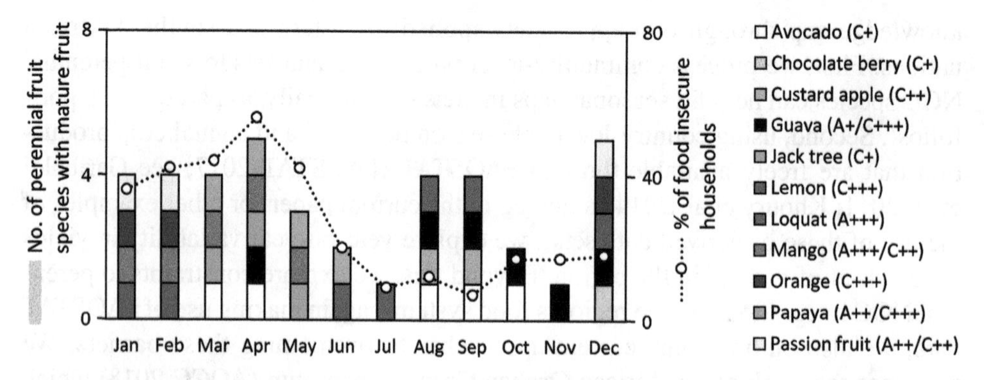

Fig. 10.1 Monthly mapping of harvest periods of prioritised perennial fruits overlaid with information on household food insecurity for Siaya County in Western Kenya. Analysis was based on a survey of 275 farm households. Pro-vitamin A and C content of different fruits is indicated (+++ = high, ++ = medium, + = low). Perennial crops that fruited in the most food insecure month (April) and that had high or medium levels of both pro-vitamins A and C were mango and papaya, both of which are classified as orphan crops in the SSA region (AOCC 2018). Promoting these perennial orphan species in Siaya County may be an effective means of ensuring access to vitamins A and C in food insecure periods. Source: ICRAF Fruiting Africa Project

10.3 Perennial Foods and FAOSTAT Yield Stabilities

To explore the usefulness of FAOSTAT data sets for assessing the ability of perennial foods to support the resilience of food systems in eastern and southern Africa, we extracted annual yield time series for the years 1961–2014 for a range of 12 annual and perennial crops (or groups of crop species) from ten countries in the region (Table 10.1). The crops chosen are representative of food production and, in the case of the chosen perennials, are either NOC (cashew, coconut and the pooled FAOSTAT crops mango, mangosteens and guava [these last crops are considered a single crop in FAOSTAT reporting] are all entries on the AOCC's crop list; AOCC 2018) or are proxies for NOC for which production data are not yet available. The non-orphan perennials chosen as proxies were coffee, date and orange. This is because, like the majority of perennial NOC considered by the AOCC (2018), the crop product is based on fruit production. These crops can support diets directly or provide incomes to buy food.

We converted yield data for these crops into fractional year-on-year yield changes and applied a logarithmic transformation for scaling purposes, based on the approaches outlined in Dawson et al. (2018a) (see also this reference for some of the caveats in the use of FAOSTAT data sets that we do not describe fully here, but that include the different levels of accuracy in reporting for different types of crop; a measure to take this into account in our analysis involved the exclusion of country–crop combinations if there were many identical yield values given over the time series). This analysis provides a measure of yield stability over the last half-century for individual country–crop combinations that can be summed across nations to provide overall estimates of the stability of annual and perennial crop production.

Table 10.1 Ten eastern and southern African countries for extraction of FAOSTAT (2017) production (yield and value) data for the 12 indicated crops for the time period 1961–2014. Half of the crops chosen were annual and half perennial. Perennial crops were either orphan crops (AOCC 2018) or proxies. Where country reporting boundaries have changed within the assessed time period, information was compiled within common notional boundaries. Crops with the least stable production (yield) characteristics are listed (see text for further explanation)

Country	Annual (A) crops	Perennial (P) crops	Least stable production (yield) (1st, 2nd)
Ethiopia PDR (or Ethiopia + Eritrea)	Chick peas, maize, millet, sorghum	Coffee (green)	Sorghum, chick peas
Kenya	Maize, millet, potatoes, sorghum	Cashew nuts (with shell), coffee (green), [Mangoes + mangosteens + guavas][a], oranges	Millet, sorghum
Malawi	Chick peas, groundnuts (with shell), maize, potatoes, sorghum	Coffee (green), [Mangoes + mangosteens + guavas]	Sorghum, maize
Mozambique	Groundnuts (with shell), maize, potatoes, sorghum	Cashew nuts (with shell), coconuts	Maize, sorghum
Somalia	Groundnuts (with shell), maize, sorghum	Coconuts, dates, oranges	Sorghum, maize
Sudan (former) (or Sudan + South Sudan)	Chick peas, groundnuts (with shell), millet, potatoes, sorghum	Dates, oranges	Chick peas, millet
Uganda	Chick peas, groundnuts (with shell), maize, millet, potatoes, sorghum	Coffee (green)	Coffee (green), chick peas
United Republic of Tanzania	Chick peas, groundnuts (with shell), maize, millet, potatoes, sorghum	Cashew nuts (with shell), coconuts, coffee (green), [Mangoes + mangosteens + guavas]	Cashew nuts (with shell), maize
Zambia	Groundnuts (with shell), maize, millet, potatoes, sorghum	Oranges	Maize, sorghum
Zimbabwe	Groundnuts (with shell), maize, millet, potatoes, sorghum	Oranges	Maize, millet

[a][] Considered as a single crop in FAOSTAT

In theory, perennial species could offer greater yield stability than annuals over time series because crop establishment—which is based on an individual season's weather at the time of planting—is a 'one-off' event for perennials, after which production can continue over a number of years. This is unlike the situation for annual crops for which production depends on the right conditions for establishment in each new season (e.g. the right amount of rainfall, without drought or flood). Conversely, many perennial crops have some degree of dependence on animal pollinators, while a large number of staples rely on wind- or self-pollination. For this reason, perennials' yields may be more susceptible to the vagaries of weather that influence the behaviour of animal pollinators (Garibaldi et al. 2011).

Our analysis (Fig. 10.2a) indicates that for individual countries in eastern and southern Africa the tested crops with the least stable yields vary, depending on the country. Of the 68 country–crop combinations that we analysed, however, while 21 combinations involved perennial crops (31%) and the remainder were for annuals, only 2 of the 20 least stable country–crop combinations involved perennial species (10%), suggesting that perennial crops overall display more stable production characteristics than annuals (see also Table 10.1). This is supported by an analysis of absolute deviations in transformed year-on-year yield changes that averages results across nations and crops, where overall deviations are lower for perennials than annuals (Fig. 10.2b). Our current analysis does not further explore the reasons for this stability; but, in some cases, it may reflect greater investments in production for what are sometimes valuable perennial commodities rather than intrinsic differences in their stability compared to annuals.

As expected, based on different production ecologies, individual country profiles of crops (Fig. 10.2a) indicate that directions in yield change for any particular year-to-year interval vary depending on the crop. This raises the prospect of actively designing compensatory crop combinations, where crops with different responses are deliberately combined to support resilience to variable seasonal conditions. To explore this issue further, we took the two countries with the highest number of crops, Kenya (N = 8) and Tanzania (N = 10) and, for each crop–crop combination in each nation, regressed transformed fractional year-on-year yield changes against each other. The results demonstrated that most comparisons had positive associations (40 of all comparisons, summing for both countries), indicating that yields for a pair of crops increase or decrease in the same direction over tested yearly intervals. However, in 33 cases the association was negative, indicating that yield for one member of a pair of crops increased and yield for the other decreased over yearly intervals.

The majority of positive associations indicates that most crops respond similarly to climatic conditions for a particular season; but the negative associations also indicate the possibilities for deliberate planning of compensatory crop combinations on a country-specific basis. Applying an initial probability test to regressions of paired comparisons only revealed a few to be of statistical significance ($P \leq 0.05$) (Fig. 10.3); but, in the case of Tanzania, the one significant negative correlation observed was for an annual-perennial crop pair (potato–coffee, Fig. 10.3b). This raises the prospect that perennial crops could have a particularly important role in defining compensatory crop combinations. A more complete analysis would, however, compare a wider range of countries and crops. In addition, it would explore weather data over the time period to try and identify the causal factors behind yield changes for specific crops, to establish the mechanisms involved and possible stabilising responses on an individual crop basis.

Because of the caveats associated with the use of FAOSTAT data sets for such analyses (Dawson et al. 2018a), alternative across-species crop production data sets

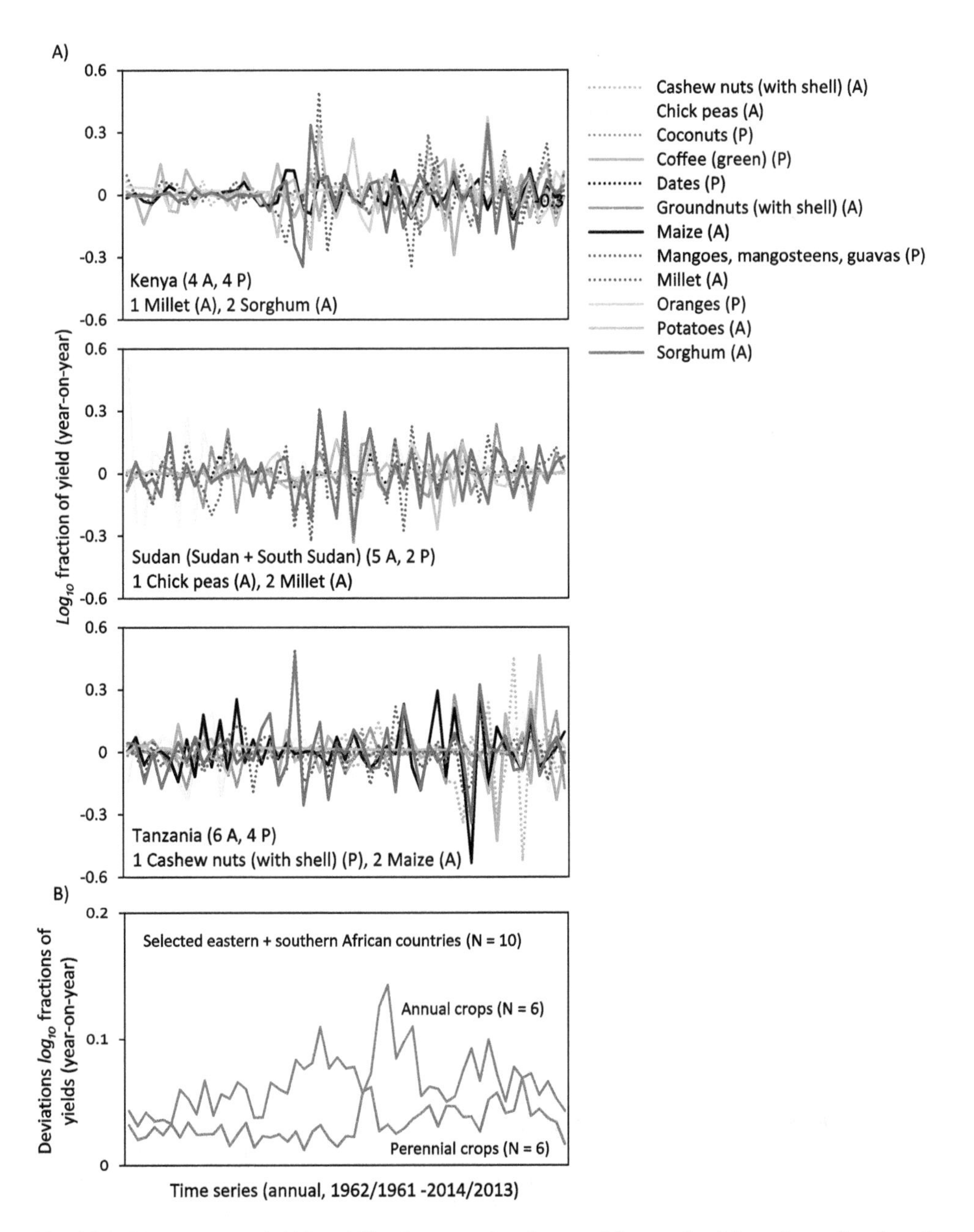

Fig. 10.2 Year-on-year yield instability for annual and perennial crops for (**a**) a subset of analysed countries from eastern and southern Africa and (**b**) averaged for each crop across nations and then across crops, based on FAOSTAT (2017) data. In (**a**), the *y*-axis represents a logarithmic transformation of fractional year-on-year yield changes that is an indication of the instability of production. The two crops with the least stable production characteristics for each country, as measured by the greatest amplitude in consecutive values along the *y*-axis, are given (see Table 10.1 for this information for all 10 tested countries). In (**b**), the same values for instability are used as in (**a**), but with the sign of the year-on-year change in yield removed to allow summing across countries and crops (see Table 10.1 for crops and countries included in (**b**) calculations)

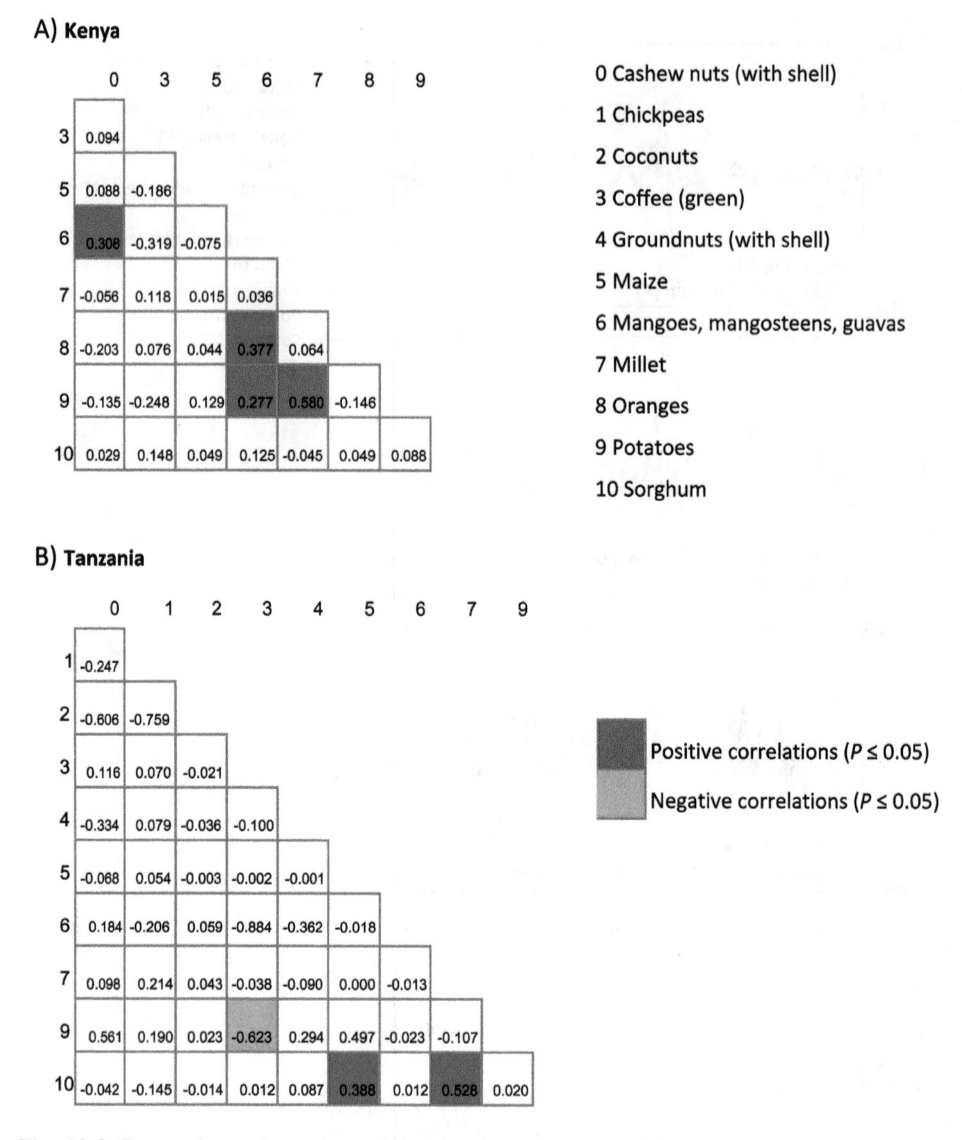

Fig. 10.3 Regressions of transformed fractional year-on-year yield changes (see Fig. 10.2a) for pairs of crops in (**a**) Kenya and (**b**) Tanzania, based on FAOSTAT (2017) data. Values in the matrix indicate the strength of the correlation, with matrix cells in red and blue indicating statistically significant positive and negative correlations ($P \leq 0.05$), respectively, in initial tests (without correcting for the total number of tests)

should also be explored and, if necessary, de novo collections of crop yields made. Such data collection should ideally be on a subnational basis, since changing patterns of production can be site-specific. Our analysis of FAOSTAT data should, therefore, only be considered as an exploratory starting point.

10.4 What Measures Are Needed to Drive Perennial NOC Integration into Eastern and Southern African Food Systems?

As illustrated above, analyses of the timings of food production in crop portfolios, and of FAOSTAT data sets, support the role that perennial crops, including perennial NOC, can play in supporting resilient, nutritious food systems in eastern and southern Africa at subnational and country levels. This is provided proper consideration is also given to crop features such as reproductive mechanisms and nutritional compositions. In common with production systems globally, however, there is an increasing reliance on less diverse and less nutritious foods in the region, with most research efforts focused on a few major annual crops (Khoury et al. 2014). This is demonstrated by trends in gross production value for annual and perennial crops in eastern and southern Africa, which show greater relative value accruing to annuals over the last half-century (Fig. 10.4a, based on the same sets of crops analysed in Table 10.1 and Fig. 10.2b). On the other hand, while increases in yields of annual crops have, relatively speaking, fallen behind in eastern and southern Africa, compared to the world as a whole (Fig. 10.4b), this is not as evident for perennial crops (where yield increases generally have been lower, Fig. 10.4c), which could suggest that in the latter case there are opportunities for the region to take a lead in engaging with consumer markets if production improvements can be introduced there.

As already noted, however, production research on NOC has generally been neglected (Dawson et al. 2018b). To support production improvements in SSA, therefore, the AOCC (AOCC 2018) was set up to develop advanced breeding methods and related resources. The 101 NOC considered by AOCC were prioritised to be of nutritional importance to local consumers in the region, and half of them are perennial species. The advanced methods being developed are based on next generation breeding technologies that include genomic selection and marker–trait associations to expe-

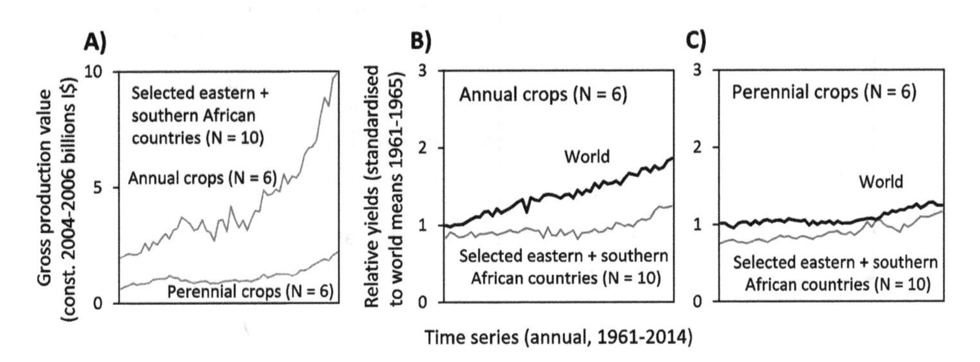

Fig. 10.4 Trends in (**a**) gross production value and (**b, c**) yields in annual and perennial crops. In (**a**), values are given in summed billions of constant international US Dollars for the annual and perennial crops and eastern and southern Africa countries listed in Table 10.1. In (**b, c**), yields from the same crops and African countries are standardised to world means for these crops for the period 1961–1965 (Source: FAOSTAT 2017)

dite the breeding cycle (Hickey et al. 2017). The 101 chosen NOC typically have not passed through intensive domestication to improve traits related to yield, quality and labour intensity, that are particularly important to SSA producers and consumers. This is especially so for the perennial species on the AOCC list, many of which are essentially indigenous wild species. This means that there are large pools of genetic variation that can be exploited in improvement programmes that could, in addition to the above traits, also specifically consider environmental adaptation to conditions such as elevated temperatures, increased carbon dioxide levels, increased salinity, drought and flooding that are associated with climate change (Dawson et al. 2011).

A key part of the AOCC initiative is building capacity to improve NOC via the training of African plant breeders in modern genetic improvement approaches. To date, more than 80 breeders have been trained through the AOCC's UC Davis African Plant Breeding Academy, with more than half coming from eastern and southern Africa (in particular, Ethiopia, with 13 trained breeders). Related initiatives that train plant breeders also exist elsewhere in SSA (e.g. the West African Centre for Crop Improvement; WACCI 2018) which, together with the Academy, provide a significant increase in the region's capacity to drive production improvements in African crops that include NOC—although most of the breeders currently work on annual rather than perennial crops. This training needs to be combined with funded crop improvement programmes that include field trial evaluation. In the case of perennial NOC improvement programmes in eastern and southern Africa, current field trials include those of ICRAF and partners on a range of species (Fig. 10.5).

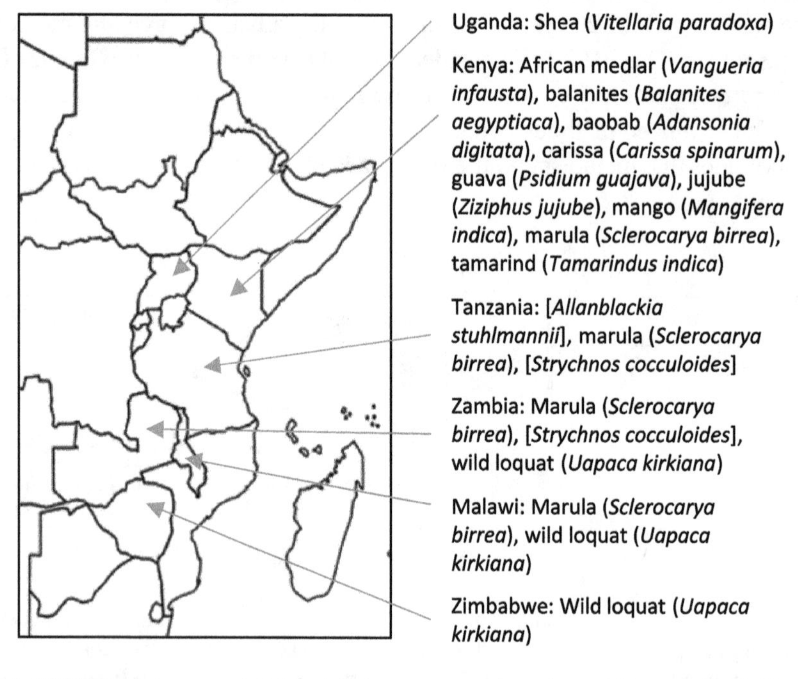

Fig. 10.5 ICRAF's and partners' field trials in eastern and southern African nations of perennial NOC that are among the 101 crops considered by the African Orphan Crops Consortium (AOCC 2018). These trials also function as living gene banks. [] indicates a related species to an AOCC crop

These trials also function as living gene banks that maintain resources for future genetic improvement and climate adaptation. Because some of these trials duplicate species and germplasm sources across locations, they can provide information on responses to environmental variation that is crucial for devising climate adaptation strategies. However, further field experiments are required to properly understand variation in AOCC species.

As well as the need for greater funding for perennial NOC improvement programmes, another key issue is the need to develop delivery systems that can provide smallholder farmers in SSA with improved perennial NOC genotypes. Delivery to smallholders of improved annual crops is also weak in the subcontinent and similarly requires improvement (see other papers presented in this book), but there are specific additional issues faced by perennial trees that require adjusted approaches to annual crops. These include: the wide variety of tree species involved; the range of different possible germplasm sources available; the form in which material is planted by farmers (generally not as seed but as seedlings); the time that trees take to mature; the large amount of offspring that can be produced by any one tree; and the generally low planting densities that are applied during cultivation, due to the large size of mature individuals (Lillesø et al. 2017).

Developing improved germplasm and supporting the participation of small-scale commercial providers, operating at local levels accessible to farmers, have been identified as key for improving current tree crop delivery systems (Lillesø et al. 2011). Attention to climate-change trends and how these affect planting is also crucial, especially as the longevity of trees mean that measurable changes in climate at specific sites are possible within the life cycle of single generations (Alfaro et al. 2014). Climate planning requires that existing local providers are linked to suppliers that operate over greater geographic distances, such as national tree seed centres. The latter must coordinate the long-distance transfers, often working across countries, which are required to cope with the scale of climate-change trends. These tree seed centres must then also interact effectively with local-level networks that have lower transaction costs to reach farmers efficiently with climate-adapted tree planting material.

An effective delivery system for perennial crops requires a reorientation of the current roles of different actors in germplasm supply (Lillesø et al. 2011, 2017). For example, non-governmental organizations need to move away from using donor support to supply tree planting material 'for free' to farmers—which inadvertently out-competes local commercial suppliers and is an unstable short-term approach, because of the vagaries of donor funding—to providing business and technical training to support local supplier enterprises. Such reorientation is supported in SSA by ICRAF and partners through training and decision-support tools (Kindt et al. 2006). In addition, modelling 'seed zones' in current and future climates (Kindt et al. 2016) helps to direct the larger scale translocations that are required for adaptation, as illustrated by the case of the perennial NOC marula in East Africa in Fig. 10.6. In this case, particular geographic sources of germplasm are likely to be better adapted under future climates, meaning that these should be the focus of current collection and multiplication (compare Fig. 10.6a, b).

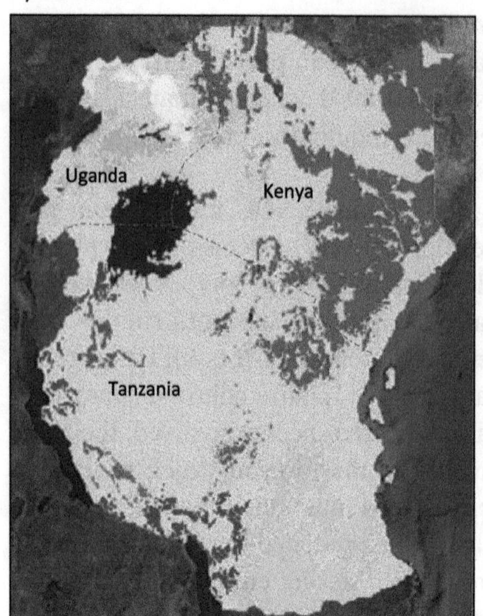

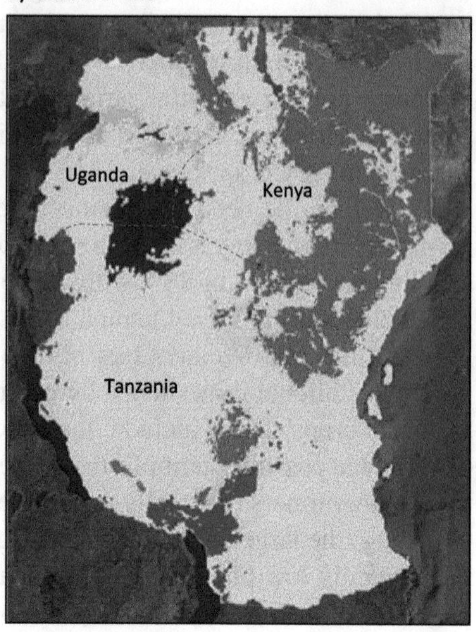

Fig. 10.6 Modelling 'seed zones' in (**a**) current and (**b**) potential future climates for marula (*Sclerocarya birrea*) in East Africa. Details on mapping, which was based on potential natural vegetation and bioclimatic distances from germplasm sources, are given by Kindt et al. (2016). Future climatic conditions centre on 2055 (2041–2070) and represent the mean of ten general circulation models for the Representative Concentration Pathway of 4.5 Wm^{-2}. The figure shows the conditions associated with current seed zones on a future climate map and predicts that in the future germplasm from current yellow and light blue zones will be of limited utility in the region, while the area adapted to germplasm from the current red zone will in future expand

10.5 Implications for Development

As African nations' economies transform, as people move from the countryside to the cities, and as new production, processing and communication technologies develop, there are new opportunities to integrate perennial NOC into food systems. These arise through new ways of processing and producing food and for informing consumer choices, as well as through the greater incomes of at least some consumers (Dawson et al. 2018b). This integration of perennial NOC aligns closely with the United Nations Sustainable Development Goals to reduce poverty, promote the accessibility and use of nutritious foods, and contribute to the food security of growing cities.

In this paper, we have indicated for eastern and southern Africa how greater use of perennial NOC foods could help support food system resilience at subnational and national levels. We have also given examples of production-based interventions to promote the integration of these crops into food systems that take account of climate change. As well as supporting production improvements by better selection and breeding through initiatives such as the AOCC, a focus on delivering planting material to farmers is essential. New approaches to more effectively reach farmers with planting material are required, such as a greater emphasis on business and technical training of local suppliers, including small-scale tree nurseries.

Of course, farmers simply producing more NOC will not in itself lead to the enhanced nutrition of consumers in SSA, as there are many additional interconnecting factors that determine individuals' diets, including culture, economics, policies and agro-environments. This means that a systems-oriented approach is crucial in future research, in which the many additional current barriers limiting NOC integration are properly considered, including market constraints and consumers' behaviour. The creation of interdisciplinary research and development teams to address multiple system-level constraints, across geographic scales, and targeted to different future challenges of which climate change is only one, is thus a priority.

References

AOCC (2018) African Orphan Crops Consortium. https://www.africanorphancrops.org/. Accessed 12 Jan 2018

Alfaro RI, Fady B, Vendramin GG et al (2014) The role of forest genetic resources in responding to biotic and abiotic factors in the context of anthropogenic climate change. For Ecol Manag 333:76–87

Altieri MA, Nicholls CI, Henao A et al (2015) Agroecology and the design of climate change-resilient farming systems. Agron Sustain Dev 35:869–890

Covic N, Hendricks S (2016) Introduction. In: Covic N, Hendriks S (eds) Achieving a nutrition revolution for Africa: the road to healthier diets and optimal nutrition. ReSAKSS Annual Trends and Outlook Report 2015. International Food Policy Research Institute, Washington, DC

Dawson IK, Vinceti B, Weber JC et al (2011) Climate change and tree genetic resource management: maintaining and enhancing the productivity and value of smallholder tropical agroforestry landscapes. A review. Agrofor Syst 81:67–78

Dawson IK, Attwood SJ, Park SE et al (2018a) Contributions of biodiversity to the sustainable intensification of food production. Thematic study to support the State of the World's Biodiversity for Food and Agriculture. Food and Agriculture Organization of the United Nations, Rome in press

Dawson IK, Hendre P, Powell W et al (2018b) Supporting human nutrition in Africa through the integration of new and orphan crops into food systems: placing the work of the African Orphan Crops Consortium in context. ICRAF Working Paper. International Centre for Research in Agroforestry, Nairobi

FAOSTAT (2017) Food and Agriculture Organization Corporate Statistical Database. www.fao.org/faostat/. Accessed 20 Nov 2017

Garibaldi LA, Aizen MA, Klein AM et al (2011) Global growth and stability of agricultural yield decrease with pollinator dependence. Proc Natl Acad Sci U S A 108:5909–5914

von Grebmer K, Saltzman A, Birol E et al (2014) 2014 Global hunger index: the challenge of hidden hunger. Deutsche Welthungerhilfe, Bonn with the International Food Policy Research Institute, Washington, DC

Hickey JM, Chiurugwi T, Mackay I et al (2017) Genomic prediction unifies animal and plant breeding programs to form platforms for biological discovery. Nat Genet 49:297–1303

Khoury CK, Bjorkman AD, Dempewolf H et al (2014) Increasing homogeneity in global food supplies and the implications for food security. Proc Natl Acad Sci U S A 111:4001–4006

Kindt R, Lillesø J-PB, Mbora A et al (2006) Tree seeds for farmers: a toolkit and reference source. International Centre for Research in Agroforestry, Nairobi

Kindt R, van Breugel P, Lillesø J-PB et al (2016) Future tree seed zonation in East Africa determined by potential natural vegetation and bioclimatic distance. International Centre for Research in Agroforestry, Nairobi

Lillesø J-PB, Graudal L, Moestrup S et al (2011) Innovation in input supply systems in smallholder agroforestry: seed sources, supply chains and support systems. Agrofor Syst 83:347–359

Lillesø J-PB, Harwood C, Derero A et al (2017) Why institutional environments for agroforestry seed systems matters. Dev Policy Rev 2017:1–24. https://doi.org/10.1111/dpr.12233

McMullin S, Njogu K, Wekesa B et al (2017) Developing fruit tree portfolios for filling food and nutrition gaps: guidelines and data collection tools. International Centre for Research in Agroforestry, Nairobi

Powell B, Thilsted SH, Ickowitz A et al (2015) Improving diets with wild and cultivated biodiversity from across the landscape. Food Secur 7:535–554

Ray DK, Ramankutty N, Mueller ND et al (2012) Recent patterns of crop yield growth and stagnation. Nat Commun 3:1293

Sibhatu KT, Krishna VV, Qaim M (2015) Production diversity and dietary diversity in smallholder farm households. Proc Nat Acad Sci USA 112:10657–10662

Thibaut LM, Connolly SR (2013) Understanding diversity–stability relationships: towards a unified model of portfolio effects. Ecol Lett 16:140–150

WACCI (2018) West African Centre for Crop Improvement. https://www.wacci.edu.gh/. Accessed 12 Jan 2018

11

Issues of On-Farm Trials: Tricot Approach

Carlo Fadda and Jacob van Etten

11.1 Introduction

Choice of crop varieties plays an important role in climate adaptation (Ceccarelli et al. 2010). One of the most important options farmers have to adapt arable farming to future climates is adjusting the crop varieties they use to new climates as they emerge (IPCC 2014). Also, a portfolio of two or more varieties can substantially buffer the impact of climate variation between seasons (Nalley and Barkley 2010; Di Falco et al. 2007).

Several barriers, however, stand in the way of a more effective use of intra-specific crop diversity for climate adaptation. Variety recommendations are often based on station trial data—hardly reflecting variety performance in low-input agriculture—and are seldom based on climate analysis (Abay and Bjørnstad 2009). This means farmers often reject the new varieties they try because of poor performance (Ceccarelli and Grando 2007). In addition, many varieties are released based on their potential for broad adaptation. This approach offers a good average potential yield over many localities but will not maximise yield at any given place (Ceccarelli and Grando 2007). In addition, breeding rarely relies on genebank material, focusing instead on elite varieties with limited allelic diversity. Genebanks hold thousands of varieties of major crops that have (co-)evolved under natural and human selection for thousands of years and have the potential to host alleles for adaptation to various biotic and abiotic stresses (Vavilov and Dorofeev 1992). Yet, these promising, diverse materials are rarely used. Introducing sets of diverse materials into areas where modern varieties have not yet made an impact is a possible first step in

C. Fadda (✉)
Bioversity International, Addis Ababa, Ethiopia
e-mail: c.fadda@cgiar.org

J. van Etten
Bioversity International, Turrialba, Costa Rica
e-mail: j.vanetten@cgiar.org

support of climate adaptation. This does not only serve to identify initial populations for breeding programmes, but can also identify farmers varieties (or varieties bred for other areas) that may prove superior and can therefore be disseminated directly. For example, in Ethiopia we found in a durum wheat trial that the best farmer variety outperformed the best modern variety with a yield difference of 20% (Mengistu et al. 2018).

The use of improved or modern varieties (MVs) has limitations in Africa (Salami et al. 2010). MVs often require a high quantity of external inputs to fulfil their potential. On African low-input farms in high-risk areas, landraces may be chosen over MVs by local farmers because of their better adaptation, higher market value and better end-product quality (Ceccarelli et al. 2010). In addition, the cultivation of a small set of MVs over large areas lowers the genetic diversity at a landscape scale, with detrimental effects on the resilience of agro-ecosystems (Cabell and Oelofse 2012).

At present, both public and private efforts fail to insert varietal diversity for climate adaptation into local farming systems in a rational way (Ceccarelli 2015). On-farm testing is crucial to determine farmer knowledge and preferences (Mancini et al. 2017). Such tests can also identify suitable germplasm for breeding and can be linked to improved dissemination of the genetic material to local communities through more efficient seed systems (Thomas et al. 2012). Current on-farm testing is usually done with a limited set of elite materials, which are compared to the current market-leader variety. These trials require constant attention from technical personnel. As a result, the testing is relatively costly, especially in marginal areas where technical personnel must travel long distances. These trials are therefore kept relatively small and thus have limited statistical power. In some cases no formal statistical inference is done, and decisions are made based on tallies of farmer votes and simple averages of yield data. These trials allow the release of a small number of varieties backed by limited evidence of their value under farm conditions (Abay and Bjørnstad 2009).

For climate adaptation of African smallholder agriculture, a different approach is needed. The best approach would be a hybrid system in which the quantitative aspects of conventional trials are combined with the benefits of participatory on-farm methods. This would ensure that a diverse range of useful genetic material reaches farmers. A system in which farmers play a more active role would accelerate genetic gain and access to variety diversity, thus contributing to system resilience (Badstue et al. 2012).

In this chapter, we present a possible solution to a number of the problems of on-farm trials: the triadic comparisons of technologies, or "tricot". Following a citizen science philosophy, this approach increases farmer ownership of trials and uses smart, simple data collection formats to help scale on-farm testing (van Etten et al. 2016). The tricot approach involves cost-effective, large-scale, repeated participatory evaluation of varieties under farm conditions using novel material from national gene banks or other sources (advanced lines from breeding programmes, varieties bred for other areas). Van Etten et al. (2016) provide a detailed discussion on how the tricot approach simultaneously builds on and differs from previous participatory approaches in crop improvement. The approach was designed to overcome a number of specific challenges in participatory crop improvement, including the need for scaling, cost reduction, data standardisation, and taking into account heterogeneity in environments and farmer preferences.

The tricot approach is especially suited for climate adaptation. Combining the resulting geo-referenced variety evaluation data with environmental data and climatic data, the approach distinguishes different responses of crop varieties to seasonal climatic conditions. The data can then be translated into concrete variety recommendations that reflect current farm conditions, stabilise yields, and track climate change over time. We illustrate the approach with an example, using simulated (yet realistic) data.

11.2 Analyzing Data from On-Farm Trials Using the Tricot Approach

We performed a series of simple simulations to illustrate the methods and results of the tricot approach. In our simulation we use realistic data that mimic the data collected in a number of countries, including India, Nicaragua, Honduras and Ethiopia. Farmers provide feedback based on different traits. These traits are selected together with farmers and technical personnel in focus groups. The traits can include yield, pest and disease resistance, phenological characteristics, plant vigour and more.

In the trials, each farmer receives three different varieties and is trained on how to set up the experiment in terms of plot layout and management. They plant the seeds, and, as the crop grows, they rank the varieties for each of the traits. Farmers do not know the names of the varieties in their set, which are randomly allocated to them from a larger portfolio, generally at least 10–20 per trial, at times previously selected from a much larger set—up to 400 in a recent application in Ethiopia (Mancini et al. 2017). Farmers are asked to fill out the forms during the season based on the traits they are evaluating. At the end of the season, farmers complete the forms by providing their assessments of productivity and the quality of the final product, as well as an overall performance judgment (Steinke and Van Etten 2016; Van Etten et al. 2016). The entire process is supported by the digital platform ClimMob (http://climmob.net/), which takes the user through a structured process of trial design, electronic data collection, analysis and automatic reporting.

The overall performance of crop varieties was analysed using Plackett-Luce trees, using R software (Hothorn and Zeileis 2015; Zeileis et al. 2008; Turner et al. 2018). Publicly available soil and climate data can be linked to the trial dataset by using location data (latitude and longitude) and the planting date of each farm. The Plackett-Luce model can use these data to distinguish between groups of environments with different patterns of variety performance.

We simulated two examples. In each, 500 farmers ranked a set of 3 varieties taken from a set of 20 varieties. Varieties are assigned in a randomised and balanced

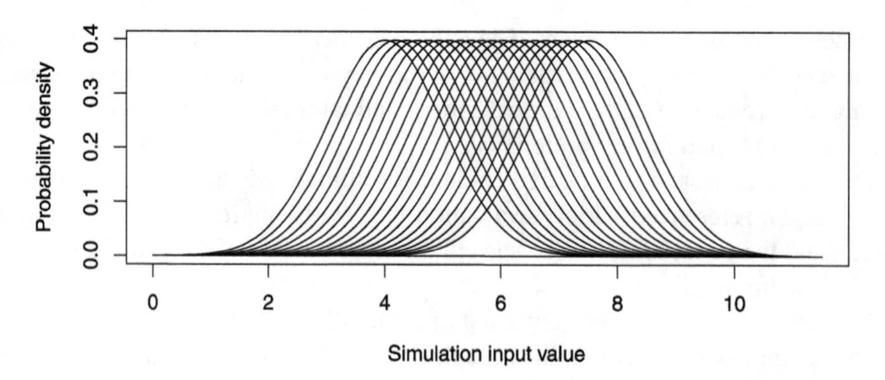

Fig. 11.1 Probability distributions used for the simulation for 20 varieties. Normal distributions with standard deviation of 1 and means separated by an interval of 0.185 (value from Steinke et al. 2017 for "challenging trait")

way. In Example 1, farmers draw random values from a normal distribution separated by a small interval and rank the varieties based on these values (see Fig. 11.1). This simulates the error that farmers make, following our findings on the accuracy of farmer observations in these trials for a relatively difficult trait (Steinke et al. 2017). An interesting feature of this ranking approach is that it also works for more elusive traits that depend on farmers' preferences, such as the taste of the product or farmers' overall evaluation of each variety, which is eventually what determines variety adoption.

In Fig. 11.2, the results of the first simulation can be seen. The original input values of the simulation are on the x axis and the Plackett-Luce model estimates are on the y axis. As the graph shows, the PL model is able to reconstruct the values very closely, with a correlation of 0.994 with the original values. In a few cases, the model does not retrieve the right order. Variety 13 is ranked lower than Variety 12 and Variety 18 is ranked lower than Variety 17. In other words, only 10% of the varieties are shifted by one position. There is very little information loss. However, there are important features that reveal the limitations of the PL model. The y scale represents the log-odds of winning from Variety 1, the variety arbitrarily chosen as our reference. The scale of the model parameters does not have an absolute zero. Variety 1 has a parameter value of zero, but any other variety can be chosen as the reference variety. In reality the underlying mean value for Variety 1 is 4. The original value cannot be retrieved from the model. In other words, the index given by the PL model has a meaning only relative to the other varieties, even though there is a strong linear relationship with the underlying latent variable.

In Example 2, we added a complication in that 250 of the 500 farmers experienced a drought condition, which made two varieties increase their mean. However, there is also an error in the measurement of the drought condition. We then applied a Plackett-Luce tree model to this artificial dataset, to visualise how it distinguishes between the two groups of farmers and their variety rankings. Also, we show how to derive variety recommendations from the model outputs to respond to climate risk.

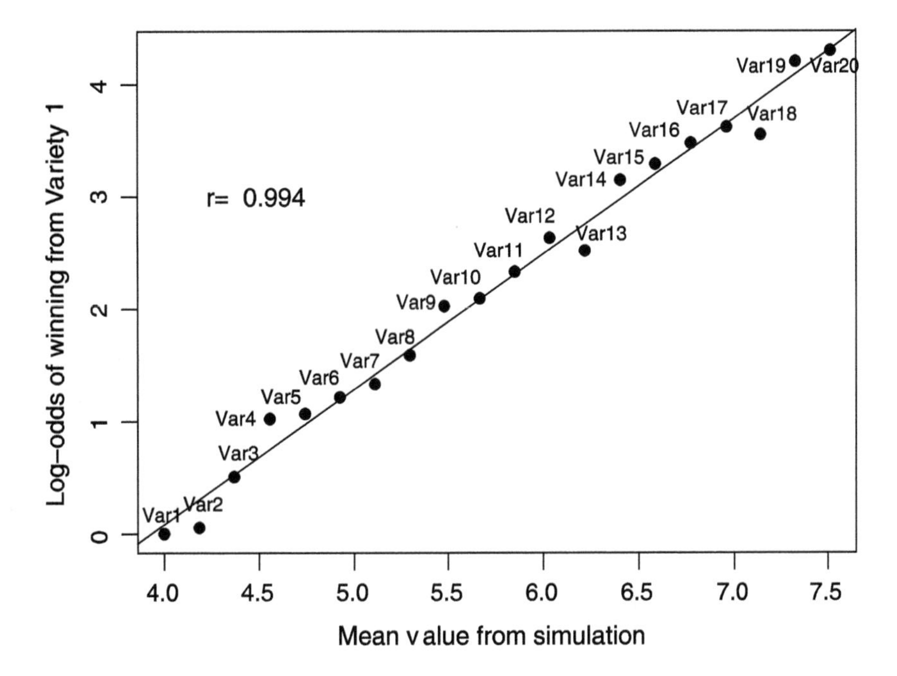

Fig. 11.2 Results of Example 1. Relation between simulation input values and Plackett-Luce output values. Based on a simulation of 500 farmers, each ranking 3 varieties out of a set of 20

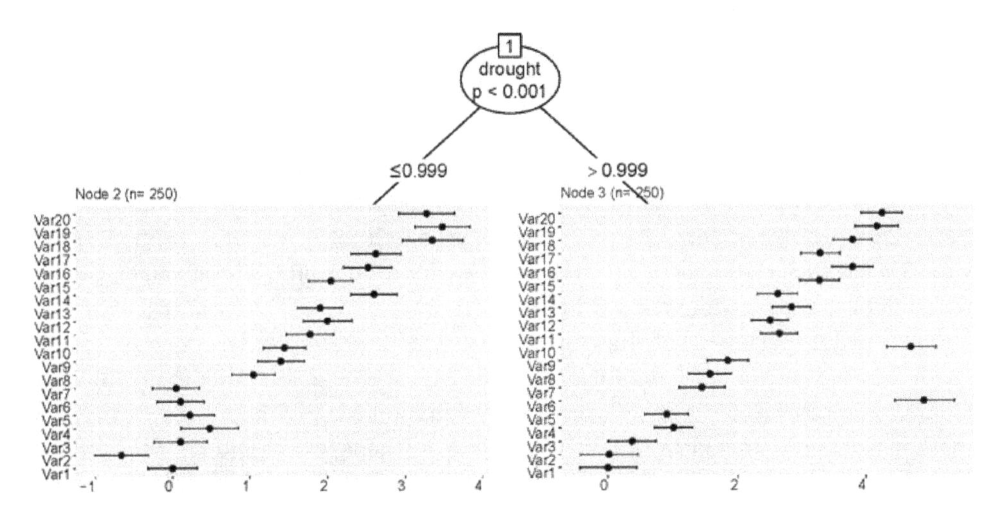

Fig. 11.3 Results of Example 2. Half of the farmers were selecting under a drought condition, in which Varieties 6 and 10 excelled. The Plackett-Luce tree distinguishes correctly between the two groups

In Fig. 11.3, we show the results of a Plackett-Luce tree, applied on data with a covariate representing seasonal rainfall. The Plackett-Luce tree determines how to use the drought variable to split the data.

In real analyses, we derive covariates from a geospatial weather dataset, using the GPS point of each farm and the planting date to retrieve data from the right place and time. We then generate multiple variables from these data (number of consecutive dry days, number of days with temperature above a certain threshold, total rainfall accumulated, etc.). We also use variables related to the terrain and soil. The Plackett-Luce model picks the best predictor and determines the best point where to split the data (for example, making a group with less than four consecutive dry days and another group with more than 4 days).

In our second example, the model splits the set of farmers in two equal groups. In the simulation, 250 farmers were assigned to each condition. It correctly identifies the drought resistant varieties—Varieties 6 and 10—which jump out in the right part of the graph. The graph also shows 95% confidence intervals around the parameter estimations, which give an idea of the certainty we have that the varieties are really different.

The results of our simulation illustrate how the tricot approach can distinguish between different varieties and has the power to evaluate the variety by climatic conditions. In a simulation of different environmental conditions, it is clear that the performance of the different varieties varies based on those basic climatic conditions. This influences the evaluation of the farmers, who provide different feedback.

11.3 Deriving Variety Recommendations from On-Farm Trials

The outputs from the Plackett-Luce model and the Plackett-Luce tree are shown on a log-scale in reference to winning from a particular variety. These values are a bit abstract, but as shown in Fig. 11.2, the values are linearly related to the underlying trait values. The PL model can also produce probabilities of winning from all other varieties for each of the varieties, which are easier to interpret. These values can be used to construct portfolios of varieties.

We illustrate variety portfolio construction with an example. To construct robust portfolios, we use theory from financial asset management, adapting the method of Dembo and King (1992) to relative losses (probabilities of being the best). The method is closely related to Conditional Value at Risk (Testuri and Uryasev 2004). This is a state-of-the-art metric now widely used in banks, which was previously applied by Sukcharoen and Leatham (2016) to variety portfolio construction.

For simplicity, we focus on a smaller example, with four varieties in two seasonal climate scenarios. In Table 11.1 we show a possible output from a Plackett-Luce tree, which can be interpreted as a payoff matrix for the construction of robust portfolios.

We generated another table from this, Table 11.2, showing the relative opportunity loss. We obtained these values by dividing the values by the highest value in each column, to first get the so-called competitive ratio. We subtract the competitive ratio from 1 to get relative opportunity loss values. Different types of seasonal

Table 11.1 Imaginary example of probability of winning from all other varieties for two different seasonal climate scenarios

Variety	Dry (P = 0.6)	Wet (P = 0.4)
Var1	0.30	0.25
Var2	0.27	0.27
Var3	0.25	0.25
Var4	0.18	0.23

Table 11.2 Relative opportunity loss of each variety in each seasonal climate

Variety	Dry (P = 0.6)	Wet (P = 0.4)
Var1	1–0.30/0.30 = 0.00	1–0.25/0.27 = 0.07
Var2	1–0.27/0.30 = 0.10	1–0.27/0.27 = 0.00
Var3	1–0.25/0.30 = 0.17	1–0.25/0.27 = 0.07
Var4	1–0.18/0.30 = 0.40	1–0.23/0.27 = 0.15

Table 11.3 Regret calculation for a portfolio of 50% Variety 2 and 50% Variety 3

Variety	Dry (P = 0.6)	Wet (P = 0.4)	Expected regret
Var2 (0.5 share)	0.6 * (0.10 * 0.5)2 = 0.0015	0.4 * (0.00 * 0.5)2 = 0.0000	0.0015
Var3 (0.5 share)	0.6 * (0.17 * 0.5)2 = 0.0042	0.4 * (0.07 * 0.5)2 = 0.0006	0.0047
Expected regret	0.0057	0.00055	0.0062

climate happen with different probabilities. In our example, we have determined with long-term weather data or seasonal climate forecasting that the probability of dry conditions during the growing season is 0.6 and of a wet condition 0.4.

From this table, we can calculate the expected regret for a particular portfolio of varieties. We square the regret per variety per scenario to give more emphasis to higher regret values, following Dembo and King (1992). For example, a portfolio with 50% Variety 2 and 50% Variety 3, would give a regret of 0.0062 (Table 11.3).

Expected regret will never become zero, because we can never beat a perfect forecast by choosing a good portfolio. But we can get very close. We can pick an optimally robust portfolio by minimising the expected regret. We can calculate the proportions of each of these varieties in an optimally robust portfolio through a simple optimisation, which can be done in Microsoft Excel. In this case, the optimal portfolio has 67% of Variety 1 and 24% of Variety 2 (and small contributions from Varieties 3 and 4), achieving a regret value of 0.0014, more than four times less than the portfolio we looked at above. More study is needed to determine the best portfolio design method on the basis of this type of data. There are various ways to parameterise the model further. However, our main point here was to demonstrate that it is possible to construct rational variety portfolios from this type of data. This portfolio construction approach can also be used to construct crop portfolios for climate resilience.

11.4 Contribution of the Tricot Approach

Our simulation exercises show that the tricot approach is statistically robust and allows us to identify the varieties or portfolios of varieties that are preferred by farmers in different environments. Each farm constitutes a mini-experiment in which most of the conditions are not constant. The tricot approach does not try to eliminate the variability between farmers' management practices, soil types, seasons and preferences, but rather makes statistical use of such information to provide recommendations that work in each place and are robust to climate risk.

The approach can also determine if varieties perform differently under different environmental conditions. This has the potential to significantly contribute to the improvement of seed systems by allowing the delivery of varieties based on seasonal climate forecasts or on prevailing conditions in different environments. When working in a complex topography such as those found in Ethiopia, one can expect important differences in conditions among villages, depending on altitude, rainfall and other factors. The tricot approach can help to deliver the best seeds based on the actual climatic conditions of a particular village.

The tricot approach also can cover a higher number of varieties than usual on-farm testing approaches. It can engage with a larger community of farmers than a conventional participatory variety selection (PVS), and the larger number of farmers provides considerable statistical power, resulting in more data points. In addition, the tricot approach could be combined with genomic data to increase the predictive power of the model (Jean-Luc Jannink, personal communication).

Lastly, even without determining absolute levels of yield or other variables, the tricot approach can deliver variety recommendations for risk-reducing portfolios, which adds another tool for climate adaptation. In the literature, limited applications of crop variety portfolio design can be found, mainly for well-endowed production environments in the US and Mexico (e.g., Nalley and Barkley 2010, among others). Our simulation shows that it is possible in principle to generate crop variety portfolio recommendations for marginal environments through participatory trials at scale.

11.5 Implications for Development

Under the current agricultural model, climate change will cause a reduction of yields for many crops in many parts of Africa. Farmers in these environments need accelerated seed-based innovation to cope with climate change. It seems logical, therefore, to diversify in ways that will enhance productivity at any given locality by quickly delivering varieties that are tested by the farmers. Such an approach will significantly increase the adoption rate. As Ceccarelli (2015) has argued, success of plant breeding should be measured based on the technologies that are adopted by the farmers and not by the number of released varieties.

Modern plant breeding has not yet reached many marginal environments and is unlikely to reach them in the next couple of decades through conventional approaches. Research has shown that all over the world, farmers often prefer to grow traditional varieties despite the availability of more productive technologies (Jarvis et al. 2011). The reasons for this include the traditional varieties' better adaptation to prevailing climatic and soil conditions, taste preferences, market preferences, nutritional value, resistance to pests and diseases, and reduced risks. Many times new technologies are not adopted because of costs or lack of accessibility—or simply because they do not match farmers' needs.

Engaging farmers directly in the development of new technologies has many benefits (Beza et al. 2017). It decentralises crop improvement efforts, reduces costs, enhances the efficiency of plant breeding, and shortens the time frame for new varieties to be released. It also increases adoption rates, and allows the adoption of a portfolio of varieties that will enhance resilience in the face of climate unpredictability. Perhaps most importantly, it will maximise yields at any given location rather than promoting a good average variety.

Another significant advantage is the use of material conserved in national or international gene banks. This injects into the production systems novel alleles for adaptation to a number of stresses, biotic and a-biotic, that are lost when only elite lines are used in the process of breeding. Many of the farmers' varieties conserved in the gene banks have been exposed to different pests and pathogens and different climatic conditions, and as a result have developed alleles that allow them to adapt to a multitude of conditions. The gene banks, designed mainly as recipient of material to be eventually distributed by breeders, need to rethink their role and become a source of new traits for farmers and production systems. A shift in the functioning of the gene banks is already underway in Ethiopia and Uganda, where gene banks are delivering material to the farmers and are helping to manage community seed banks.

Once such genetic material is out in the production system, the new alleles in the varieties selected by farmers may prove very important for breeders. If the varieties are also investigated using genotyping approaches, this would allow the identification of quantitative trait loci (QTLs) for relevant traits and promote marker-assisted participatory breeding.

The tricot approach has the potential to contribute to making seed systems more dynamic when demand and supply are put in contact—e.g., using the ClimMob platform—and more diversified because more varieties per crop will be delivered in a location-specific way. A more integrated seed system will allow both informal and formal contributions to the sustainability and resilience of farming system. It also creates the space for an intermediate seed system in which local seed cooperatives or community seed banks are involved in the production of preferred seeds that are identified through the tricot approach. Legislation is being developed in Africa (including in Tanzania, Uganda and Ethiopia) in which seed production rules are designed to allow those local actors to multiply and sell seeds of certain varieties of a sufficiently good quality (Quality Declared Seeds or QDS). This legislation favors the promotion of varieties developed using the tricot approach. Also, variety release procedures could benefit from scaled on-farm testing using the tricot approach.

How does this work contribute to Climate-Smart Agriculture? Climate-Smart Agriculture means different things to different people (Chandra et al. 2018). The smartness in our approach does not come through technical prioritisation exercises that guide investments towards certain "climate-smart" agricultural practices that are guaranteed to confer climate-related benefits. We have serious doubts about this approach. The Green Revolution settled on seeds largely because more knowledge-intensive approaches were more difficult to realise in the absence of well-developed extension systems (Fitzgerald 1986; Harwood 2009). As a result, "smartness" had be put into scientifically-bred seeds as the vehicles that would reach farms. Farmers would not need to learn, they simply had to start using the new seeds. This worked, but it worked best where the ground was already prepared, in production areas that most resembled modern temperate-climate agriculture, where agriculture was commercial in outlook, used high levels of inputs or irrigation water, and worked in relatively homogeneous environments (Fitzgerald 1986). Mechanisation and increased use of bulky fossil inputs characterised these farming systems, rather than knowledge intensification.

In our approach, which focuses specifically on marginal areas, we do not pretend that agricultural science can inject smartness into farming using seeds or other "climate-smart" technologies as the vehicle. "Climate-smart technologies" do not exist literally, if at all. It is subject to the fallacy of misplaced concreteness. In the end, smartness is about how people do things, how farmers are involved in constantly assessing the local appropriateness of technologies, how farmers, extension agents and researchers create new linkages that enhance information generation and exchange, and how these different ways of doing are then leading to new types of knowledge, seeds, and technologies. These end products may symbolise people's collective smartness, but do not replace it. The desired smartness (or better, wisdom) emerges as a systemic property of reconfigured seed and knowledge systems in which knowledge and technology is generated and exchanged in ways that are in pace with accelerated climate and socio-economic change, more equitable, and more attentive to environmental and social diversity and needs.

References

Abay F, Bjørnstad A (2009) Specific adaptation of barley varieties in different locations in Ethiopia. Euphytica 167(2):181–195. https://doi.org/10.1007/s10681-008-9858-3

Badstue ALB, Hellin J, Berthaud J (2012) Re-orienting participatory plant breeding for wider impact. Afr J Agric Res 7(4):523–533

Beza E, Steinke J, van Etten J et al (2017) What are the prospects for large-N citizen science in agriculture? Evidence from three continents on motivation and mobile telephone use of resource-poor farmers participating in "tricot" crop research trials. PLoS One 12(5):e0175700

Cabell JF, Oelofse M (2012) An indicator framework for assessing agroecosystem resilience. Ecol Soc 17(1):18. https://doi.org/10.5751/ES-04666-170118

Ceccarelli S (2015) Efficiency of plant breeding. Crop Sci. https://doi.org/10.2135/cropsci2014.02.0158

Ceccarelli S, Grando S (2007) Decentralized-participatory plant breeding: an example of demand driven research. Euphytica 155(3):349–360. https://doi.org/10.1007/s10681-006-9336-8

Ceccarelli S, Grando M, Maatougui M et al (2010) Plant breeding and climate changes. J Agric Sci 148(6):627–637. https://doi.org/10.1017/S0021859610000651

Chandra A, McNamara KE, Dargusch P (2018) Climate-smart agriculture: perspectives and framings. Clim Pol 18(4):526–541

Dembo RS, King AJ (1992) Tracking models and the optimal regret distribution in asset allocation. Appl Stoch Model Bus Ind 8(3):151–157

Di Falco S, Chavas JP, Smale M (2007) Farmer management of production risk on degraded lands: the role of wheat variety diversity in the Tigray region. Ethiop Agric Econ 36(2):147–156

Fitzgerald D (1986) Exporting American agriculture: the rockefeller foundation in Mexico, 1943–53. Soc Stud Sci 16(3):457–483

Harwood J (2009) Peasant friendly plant breeding and the early years of the green revolution in Mexico. Agric Hist 83(3):384–410

Hothorn T, Zeileis A (2015) partykit: a modular toolkit for recursive partytioning in R. J Mach Learn Res 16:3905–3909 Available from: http://jmlr.org/papers/v16/hothorn15a.html

IPCC (2014) In: Core Writing Team, Pachauri RK, Meyer LA (eds) Climate change 2014: synthesis report. Contribution of working groups I, II and III to the fifth assessment report of the intergovernmental panel on climate change. IPCC, Geneva 151 pp

Jarvis DI, Hodgkin T, Sthapit BR et al (2011) An heuristic framework for identifying multiple ways of supporting the conservation and use of traditional crop varieties within the agricultural production system. Crit Rev Plant Sci 30(1–2):125–176. https://doi.org/10.1080/07352689.2011.554358.

Mancini C, Kidane YG, Mengistu DK et al (2017) Joining smallholder farmers' traditional knowledge with metric traits to select better varieties of Ethiopian wheat. Sci Rep 7(1). https://doi.org/10.1038/s41598-017-07628-4

Mengistu DK, Kidane YG, Fadda C et al (2018) Genetic diversity in ethiopian durum wheat (*Triticum turgidum* var durum) inferred from phenotypic variations. Plant Genet Resour 16(1):39–49

Nalley LL, Barkley AP (2010) Using portfolio theory to enhance wheat yield stability in low-income nations: an application in the Yaqui Valley of Northwestern Mexico. J Agric Resour Econ 35:334–347

Salami A, Kamara AB, Brixiova Z (2010) Smallholder agriculture in East Africa: trends, constraints and opportunities. African Development Bank, Tunis Available from: http://core.ac.uk/download/pdf/6590805.pdf

Steinke J, van Etten J (2016) Farmer experimentation for climate adaptation with triadic comparisons of technologies (tricot). A methodological guide. Bioversity International, Rome

Steinke J, van Etten J, Mejía Zelan P (2017) The accuracy of farmer-generated data in an agricultural citizen science methodology. Agron Sustain Dev 37:32

Sukcharoen K, Leatham D (2016) Mean-variance versus mean–expected shortfall models: An application to wheat variety selection. J Agric Appl Econ 48(2):148–172

Testuri CE, Uryasev S (2004) On relation between expected regret and conditional value-at-risk. In: Rachev ST, Anastassiou GA (eds) Handbook of Computational and Numerical Methods in Finance. Birkhäuser, Boston, pp 361–372

Thomas M, Demeulenaere E, Dawson J et al (2012) On-farm dynamic management of genetic diversity: the impact of seed diffusions and seed saving practices on a population-variety of bread wheat. Evol Appl 5(8):779–795. https://doi.org/10.1111/j.1752-4571.2012.00257

Turner H, Kosmidis I, Firth D, van Etten J (2018) PlackettLuce: plackett-luce models for rankings (R package). https://cran.r-project.org/package=PlackettLuce

Van Etten J, Beza E, Calderer L et al (2016) First experiences with a novel farmer citizen science approach: crowdsourcing participatory variety selection through on-farm triadic comparisons of technologies (tricot). Experimental Agriculture, Online. https://doi.org/10.1017/S0014479716000739

Vavilov N, Dorofeev VF (1992) Origin and geography of cultivated plants. Cambridge University Press, Cambridge, UK

Zeileis A, Hothorn T, Hornik K (2008) Model-based recursive partitioning. J Comput Graph Stat 17(2):492–514

Part III
Technologies and Climate Smart Agriculture

12

CSA in Africa and its Evidence Base

Todd S. Rosenstock, Christine Lamanna, Nictor Namoi, Aslihan Arslan, and Meryl Richards

12.1 Investments in CSA

More than 500 million USD will soon be invested in climate-smart agriculture (CSA) programmes across sub-Saharan Africa, a non-trivial fraction of which is targeted for East and Southern Africa. CSA is increasingly endorsed and promoted by national, regional, continental and global institutions (e.g., governments, the Regional Economic Communities of the African Union, the New Partnership for Africa's Development, international non-governmental organizations and the Green

T. S. Rosenstock (✉)
World Agroforestry Centre (ICRAF), Kinshasa, Democratic Republic of Congo

CGIAR Research Program on Climate Change, Agriculture and Food Security (CCAFS), Wageningen, Netherlands
e-mail: T.Rosenstock@cgiar.org

C. Lamanna · N. Namoi
World Agroforestry Centre (ICRAF), Nairobi, Kenya
e-mail: C.Lamanna@cgiar.org; N.Namoi@cgiar.org

A. Arslan
International Fund for Agricultural Development (IFAD), Rome, Italy
e-mail: a.arslan@ifad.org

M. Richards
CGIAR Research Program on Climate Change, Agriculture and Food Security (CCAFS), Wageningen, Netherlands

Rubenstein School of Environment and Natural Resources, University of Vermont, Burlington, VT, USA
e-mail: meryl.richards@uvm.edu

Climate Fund). The aim is to help smallholder farmers (1) sustainably increase productivity and incomes, (2) adapt to climate variability and change and (3) mitigate climate change where possible (FAO 2013). With planned investments, political will and implementation capacity, CSA is emerging as a mechanism for coherent and coordinated action on climate change adaptation and mitigation for agriculture.

Farm- and field-level management technologies are a core component of most planned CSA investments (Thierfelder et al. 2017; Kimaro et al. 2015). Farm-level technologies represent a broad category of direct activities that farmers can undertake on their fields, in livestock husbandry, or through management of communal lands. Climate-smart actions may include both the adoption of new/improved inputs and new/improved application methods, such as adopting drought resistant crop varieties, reducing stocking rates of animals, changing harvesting and postharvest storage techniques (Lipper et al. 2014). The vast number of farm-level options that might meet CSA objectives coupled with the large number of possible outcomes that fit under the three pillars of CSA, has led many development practitioners, scientists and governments to the question: "What is CSA and what is not CSA?" (Rosenstock et al. 2015a).

This question, however, presents a false dichotomy. By definition, CSA is context specific and subject to the priorities of farmers, communities and governments where it is being implemented. Until now, little empirical evidence has been provided to systematically evaluate which CSA practices work where (see Branca et al. 2011 for a first attempt). Instead, CSA is often supported with case studies, anecdotes, or aggregate data, which paint an incomplete picture of both the potential and challenges of CSA (e.g., FAO 2014; Neate 2013). The lack of comprehensive information on CSA is not surprising, given the fact that it includes a wide diversity of solutions at the farm production and rural livelihood levels. Consequently many interventions that increase productivity are labelled as "CSA" without evidence on the other two objectives of CSA, at least one of which would need to be also documented to qualify any intervention as CSA. Although "triple win" interventions at the field level may be the exception rather than the rule, evidence has to be provided on all objectives to support policies and programmes that may wish to promote CSA (Arslan et al. 2017).

There is an urgent need to provide decision-makers—including investors—with information to help them design programmes and policies, as well as to increase the effectiveness of development programming. In response and in this paper, we have conducted a quantitative and systematic review to map the evidence published in peer-reviewed literature on the effectiveness of technologies and management practices to achieve the objectives of increased productivity, resilience and mitigation for the five countries in East and Southern Africa: Tanzania, Malawi, Mozambique, Zimbabwe and Zambia. Our systematic map sets the benchmark on what data and evidence are available on how farm and field management practices affect indicators of CSA outcomes.

12.2 A Systematic Approach

This systematic map relies on a data set compiled as part of the CSA Compendium "The Compendium". The Compendium created search terms relevant to one of 102 technologies including new inputs and farm management practices (58 agronomic, 15 agroforestry, 19 livestock, 5 energy, or 5 postharvest management practices) on more than 57 outcomes in productivity, resilience or mitigation, such as, yields, gender differentiated labour use, or soil organic carbon, respectively. Studies were included based on four inclusion criteria: (1) conducted in a tropical developing country, (2) included conventional control practice and a practice being suggested as CSA, (3) contained primary data on the impacts on at least one of the indicators of interest and (4) conducted in the field (i.e., no modelling studies). Lists of the search terms for practices and outcomes and additional details on the inclusion criteria can be found in the systematic review protocol (Rosenstock et al. 2015b).

Studies were identified by searching the Web of Science and Scopus databases using search terms indicative of practices and outcomes. Our search found 150,367 candidate studies, 7497 of which were included in the final Compendium library based initially on abstract/title reviews and then full text reviews. Out of these, 313 studies were conducted in one of the five countries. Data were compiled into an Excel database manually from each study. Data retrieved from the selected studies include information on location, climate, soils, crops, livestock species and outcome values for both conventional (non-CSA control) and treatment practices. Frequency and distribution of components in the data set (i.e., practices, outcomes and products) are analysed by summary statistics.

12.3 The Evidence

More than 150 studies met our inclusion criteria for this paper and were included in the data set analysed here. The data set contains 12,509 data points that compare a conventional practice with a potential CSA practice in a specific time and place. For example, the comparison of conservation agriculture versus conventional agriculture at Chitedze Agricultural Research Station, Malawi in 2007 (see Thierfelder et al. 2013). Studies were unevenly distributed across the five countries with a tenfold difference in the number of studies conducted in the most studied country (Tanzania) versus the least studied country (Mozambique) (Fig. 12.1). The studies were primarily conducted on research stations where 58% of data was generated compared with 42% on farmers' fields or in household surveys. This is significant because research on station under scientist-controlled conditions often outperforms the same practice in farmers' fields due to the higher quality of implementing the practices and historical management of the site (Cook et al. 2013). Thus, the evidence will generally reflect the upper bound of what can be achieved by farmers.

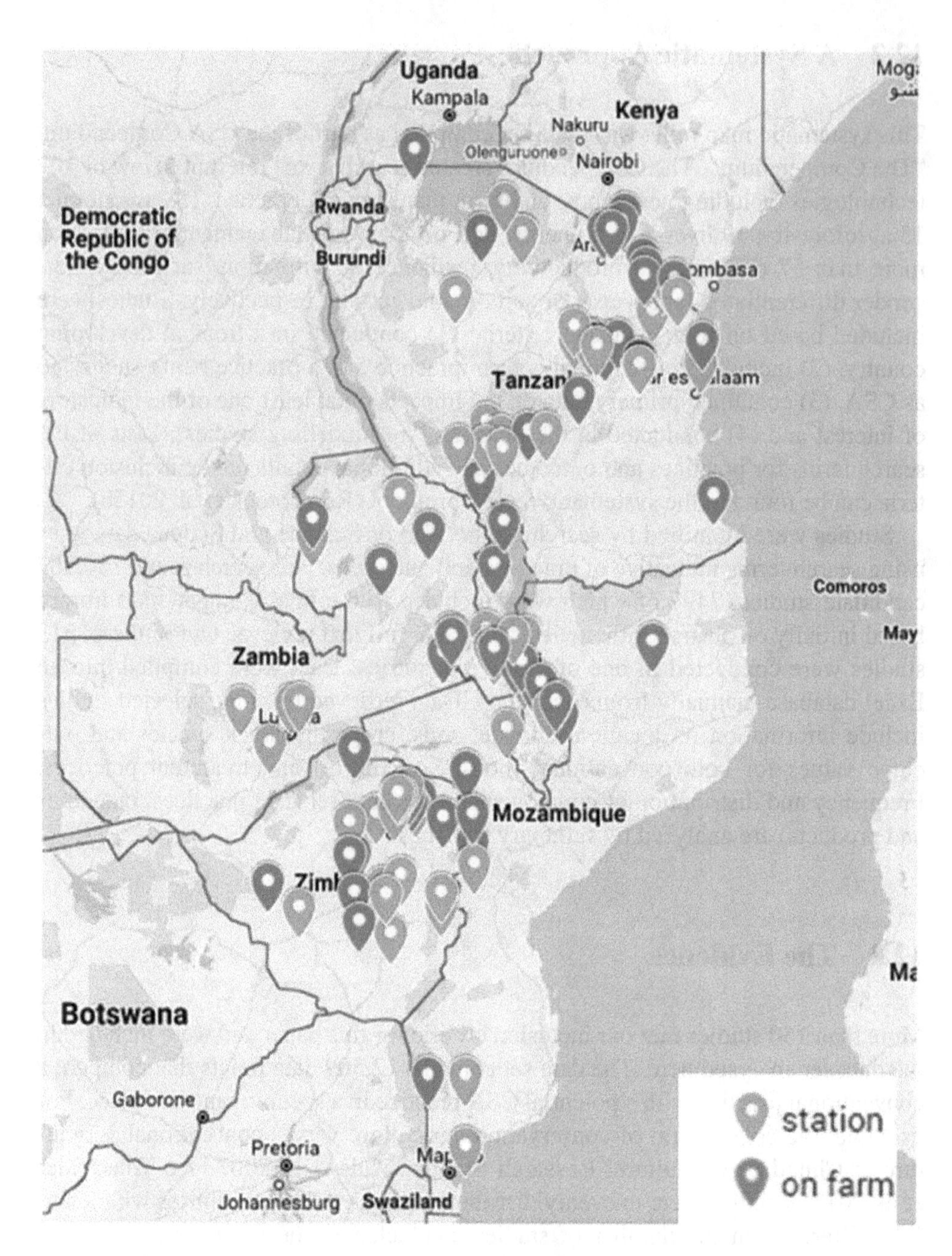

Fig. 12.1 Location of the studies included in this systematic map

Studies were clustered in a few locations and agroecologies within each country. This is unsurprising given the investments and infrastructure necessary to conduct field research. However, geographical clustering further indicates the potential for skew in the available evidence. With clustering, it is unlikely that the full range of CSA options are analysed, which limits the utility of the work to help decision-makers to choose among various options. Key gaps in agroecologies include coastal

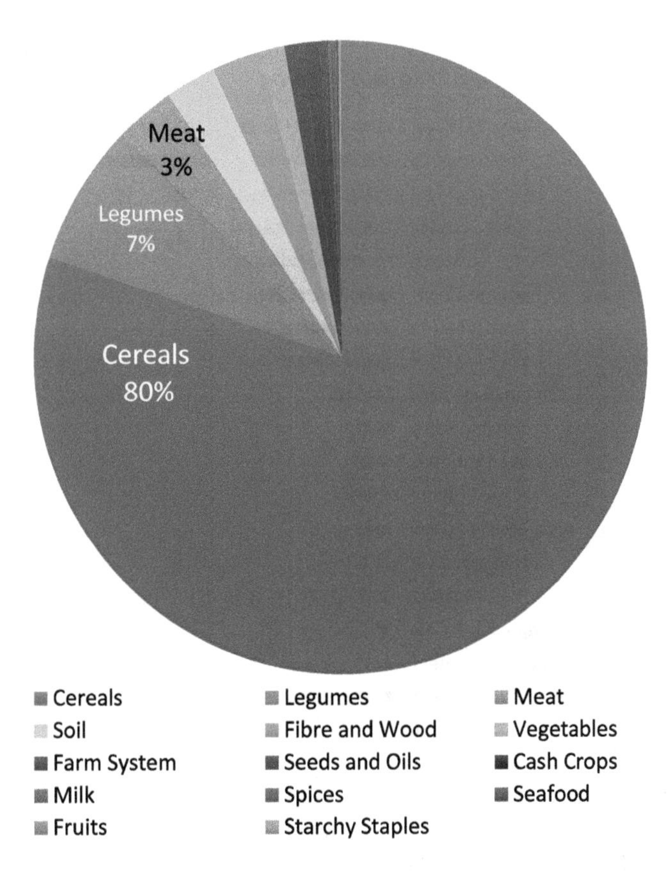

Fig. 12.2 Products included in the data set. The majority of data is on cereal crops, specifically maize. This creates gaps in our knowledge of lesser studied products. For example, only 2% of the data is on other cereals

and semiarid zones. Future analyses of these data should examine if the distribution of practices and agroecologies reflects key criteria such as percentage of the population that relies on the production of the agricultural output studied for food security, etc.

While the data set contains information on 39 agricultural products such as milk, pulses, spices, cotton, etc., the vast majority of data comprise only a handful of products. For example, data on maize accounts for 78% of the data set (Fig. 12.2). Pulses were second but made up only 7% of the data set. In contrast, many products (21) make up less than 2% of the data set. Therefore, we know a lot about maize production in the region but much less about other products. This presents a challenge for investments in CSA, because many of the proposed actions intend to diversify smallholder fields and farms, but this data set suggests a lack of information on crops other than maize. It also indicates that there is little evidence on switching to crops that may be more resilient or better suited to future climates, such as sorghum (0.8% of data set) and millets (no data available in these countries, despite its importance in the drylands of the region). However, it should be noted that crop switching is often studied through modeling efforts and therefore would not have been selected as part of this assessment. Regardless, there is a need for more empirical studies on maize alternatives, particularly given that maize yields

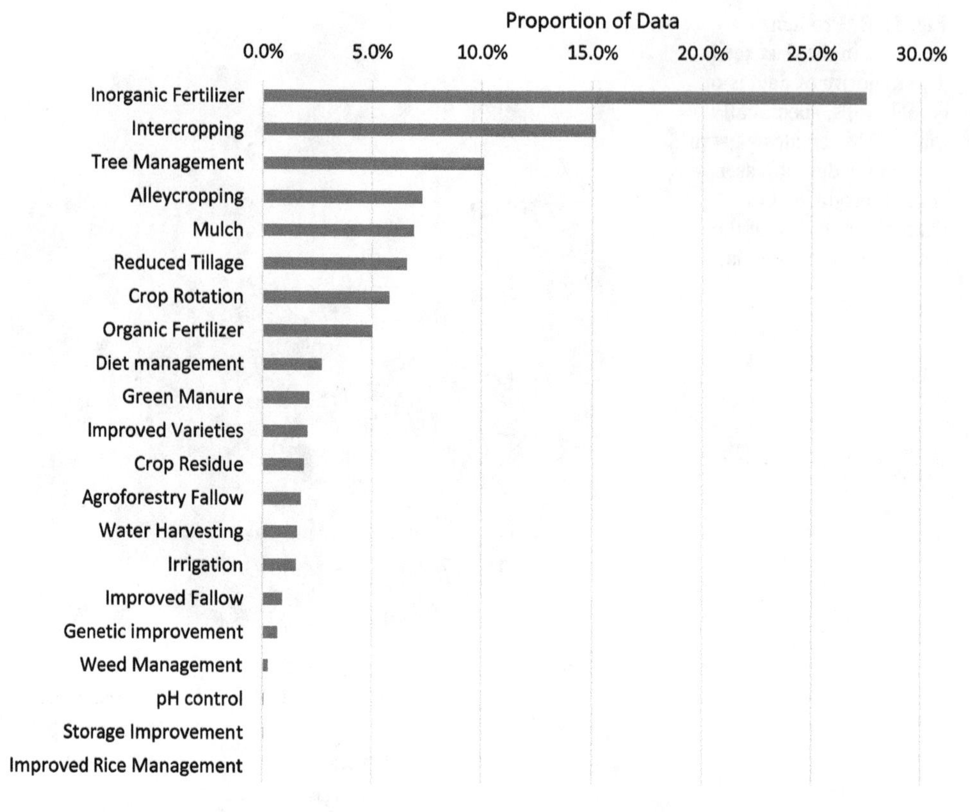

Fig. 12.3 Extent of data available for 21 practices (and 63 subpractices – not shown) in the data set

are projected to decline with climate change in the region, especially in Malawi (Ramirez-Villegas and Thornton 2015).

Existing evidence is also limited on integrated crop and livestock systems, because 93% of the data were on crops while only 3.5% on livestock. Almost all of the data on livestock were on improved diets, with a little on improved breeds. Some of the most commonly mentioned regional livestock adaptation strategies, such as pasture management technologies and animal housing, are absent from the data set. This is an important gap to be filled as these technologies are also relevant for the mitigation pillar of CSA.

Data on practices are similarly skewed with a few practices accounting for a significant percentage of the data set on 63 CSA practices. For example, studies of inorganic fertilizers are the most common (27.5% of data) and almost 3500 individual data points involved the addition of nitrogen alone (Fig. 12.3). However, this is due in part to the difference in how research is performed in different fields. Agronomic field trials on fertilizers typically use multiple types of fertilizers at many rates (e.g., 0, 20, 40, 80 kg/ha) over at least 3 years and sometimes decades (e.g., Akinnifesi et al. 2006, 2007; Matthews et al. 1992). On the other hand, studies on livestock feeding practices typically analyse a few alternative diets over just one or two short periods (e.g., Gusha et al. 2014; Mataka et al. 2007; Sarwatt et al. 2002). Despite most data being on a relatively small number of practices, significant data are available for practices of high interest to the development community. For example, 28% of the data is on practices that diversify production systems such as rotations, intercropping and agroforestry (e.g., Myaka et al. 2006; Munisse et al. 2012; Thierfelder et al. 2013; Nyamadzawo et al. 2008; Chamshama et al. 1998). Therefore, some information exists to reduce the uncertainty about implementing such interventions. Other commonly studied practices include mulching, organic fertilizers and reduced tillage.

Common recommendations for CSA interventions include packages of technologies, such as conservation agriculture or systems to intensify rice production. When multiple practices are adopted together, they can have synergistic or antagonistic effects on CSA outcomes. A significant majority (72%) of our data is from practices done in combination with at least one other CSA practice (e.g., agroforestry + mulching, intercropping + manure). This provides insights into how practices operate alone or in combination, which helps in making decisions and recommendations on best practices under specific conditions.

Lastly, we analysed the distributions of outcomes. The first striking pattern is that 82% of data are related to the productivity pillar – yields, incomes, etc. (Fig. 12.4b). Contrastingly, resilience outcomes make up only 17.5% of the data, which is primarily related to soil quality (11.4%) and input-use efficiencies (4.5%). This means that there is scant evidence on many other indicators, especially those that are believed to impart some level of resilience. It is also indicative of the difficulty in defining resilience indicators in the literature. Finally, only 0.5% of the data set is related directly to mitigation outcomes, such as greenhouse gas emissions or total carbon stocks. Thus, there are major gaps in our understanding of how potential CSA practices affect resilience and mitigation outcomes across various contexts in East and Southern Africa. There is almost a complete lack of data on mitigation, which requires urgent action to calibrate low emission trajectories.

One of the fundamental goals of CSA is to produce win-win or win-win-win outcomes across productivity, resilience and mitigation. However, our data set suggests that it is only possible to analyse win-win outcomes, given the dearth of information on mitigation. That is because most studies only examine a single pillar, about 32% study two pillars and less than 1% study all three (Fig. 12.4a). This is a critical insight into the evidence base of CSA because it shows the lack of co-located (in the same study) research across pillars. It is often not possible to extrapolate results on the same practice between sites because outcomes can be significantly influenced by local context (e.g., Pittelkow et al. 2015a, b; Bayala et al. 2012). Given the general lack of co-located research across CSA outcomes, aggregation techniques such as the Compendium and meta-analyses, can be used to gain insights into multiple outcomes from practices, including looking into potential trade-offs between different objectives.

It was not a surprise that most studies on potential CSA practices examine yields and soil health, as they are the basis of agronomic research. Perhaps the biggest surprise in the data set is that there is a significant amount of economic information

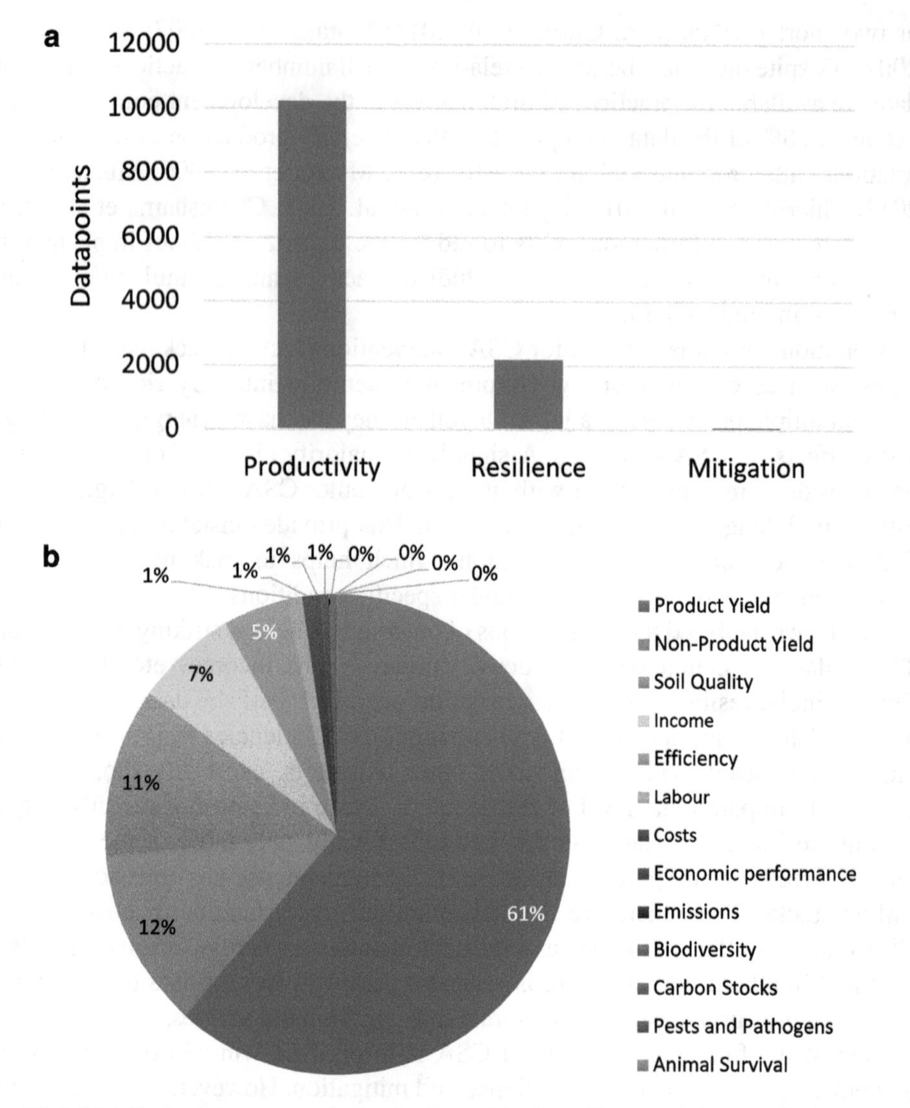

Fig. 12.4 Distribution of outcomes in the database: (**a**) what pillars are being studied? (**b**) which outcomes are being studied?

available. Nearly 20% of the papers presented economic information, derived from farm enterprise budgets, including indicators such as net returns, variable costs, net present value, etc. This subset of the data provides key information on the costs and benefits for the farmer in adopting CSA, information often missing in the discussion around programming and policy for interventions. These data will be used in future studies in combination with agronomic information to address this gap to the extent possible.

12.4 Implications for Practitioners

Our systematic map provides a first appraisal of the evidence base to assess the contributions of a wide set of field level technologies to CSA objectives in East and Southern Africa. Despite more than 50 years of agricultural research, this database shines a light on potential skew in our knowledge base. It also identifies key areas for future investments in research. Although the database may not be as comprehensive as desired due to shortcomings on the number of agroecologies, products or outcomes included, it does provide a wide range of information on many products, practices and outcomes, and therefore reduces the uncertainty of making decisions in the countries reflected in the analysis presented here. Over the next 6 months, the authors will conduct a quantitative meta-analysis—a statistical approach to combine information across studies—to help identify best interventions (and combinations thereof) during the design phase of programmes and policies.

Acknowledgements The CSA Compendium—of which this analysis is a part—has been principally funded by the CGIAR Research Program on Climate Change, Agriculture and Food Security's (CCAFS) Flagship 2 – Climate-smart practices. Supplementary financial support has been provided by The United Nations Food and Agriculture Organization of the United Nations (FAO), CCAFS Flagship 3 – Low-emissions development, the International Fund for Agricultural Development (IFAD) and the Center for International Forestry Research (CIFOR)'s Evidence-Based Forestry programme. The Compendium has benefitted from the support of more than 20 PhD students, MSc students and consultants during the effort.

References

Akinnifesi FK, Makumba W, Kwesiga FR (2006) Sustainable maize production using gliricidia/maize intercropping in southern Malawi. Exp Agric 42(4):441–457

Akinnifesi FK, Makumba W, Sileshi G et al (2007) Synergistic effect of inorganic N and P fertilizers and organic inputs from *Gliricidia sepium* on productivity of intercropped maize in Southern Malawi. Plant Soil 294(1–2):203–217

Arslan A, Cavatassi R, Grewer U et al (2017) The theory of change for CSA: a guide to evidence-based national implementation. Module C11 of the CSA Sourcebook Version 2. Food and Agriculture Organization of the United Nations. Available via. http://www.fao.org/climate-smart-agriculture-sourcebook/enabling-frameworks/module-c11-evidence-based-implementation/c11-acknowledegements/en/

Bayala J, Sileshi GW, Coe R et al (2012) Cereal yield response to conservation agriculture practices in drylands of West Africa: a quantitative synthesis. J Arid Environ 78:13–25

Branca G, McCarthy N, Lipper L et al (2011) Climate-smart agriculture: a synthesis of empirical evidence of food security and mitigation benefits from improved cropland management. Mitig Clim Chang Agric Ser 3:1–42

Chamshama SAO, Mugasha AG, Kløvstad A et al (1998) Growth and yield of maize alley cropped with *Leucaena leucocephala* and Faidherbia albidain Morogoro. Tanzania Agrofor Syst 40(3):215–225

Cook S, Cock J, Oberthür T et al (2013) On-farm experimentation. Better Crop Plant Food 97(4):17–20

FAO (2013) Climate-smart agriculture sourcebook. Food and Agriculture Organization of the United Nations. Available from: http://www.fao.org/docrep/018/i3325e/i3325e.pdf

FAO (2014) FAO success stories on climate-smart agriculture. Food and Agriculture Organization of the United Nations. Available from: http://www.fao.org/3/a-i3817e.pdf

Gusha J, Manyuchi CR, Imbayarwo-Chikosi VE et al (2014) Production and economic performance of F1-crossbred dairy cattle fed non-conventional protein supplements in Zimbabwe. Trop Anim Health Prod 46(1):229–234

Kimaro AA, Mpanda M, Rioux J et al (2015) Is conservation agriculture 'climate-smart' for maize farmers in the highlands of Tanzania? Nutr Cycl Agroecosyst 105(3):217–228

Lipper L, Thornton P, Campbell BM et al (2014) Climate-smart agriculture for food security. Nat Clim Chang 4(12):1068–1072

Matthews RB, Lungu S, Volk J et al (1992) The potential of alley cropping in improvement of cultivation systems in the high rainfall areas of Zambia II. Maize production. Agrofor Syst 17(3):241–261

Munisse P, Jensen BD, Quilambo OA et al (2012) Watermelon intercropped with cereals under semi-arid conditions: an on-farm study. Exp Agric 48(3):388–398

Myaka FM, Sakala WD, Adu-Gyamfi JJ et al (2006) Yields and accumulations of N and P in farmer-managed intercrops of maize–pigeonpea in semi-arid Africa. Plant Soil 285(1):207–220

Neate PJ (2013) Climate-smart agriculture success stories from farming communities around the world. CGIAR Research Program on Climate Change, Agriculture and Food Security (CCAFS) and the Technical Centre for Agricultural and Rural Cooperation (CTA), Wageningen

Nyamadzawo G, Chikowo R, Nyamugafata P et al (2008) Soil organic carbon dynamics of improved fallow-maize rotation systems under conventional and no-tillage in Central Zimbabwe. Nutr Cycl Agroecosyst 81(1):85–93

Pittelkow CM, Liang X, Linquist BA et al (2015a) Productivity limits and potentials of the principles of conservation agriculture. Nature 517(7534):365–368

Pittelkow CM, Linquist BA, Lundy ME et al (2015b) When does no-till yield more? A global meta-analysis. Field Crop Res 183:156–168

Ramirez-Villegas J, Thornton PK (2015) Climate change impacts on African crop production. CCAFS Working Paper no. 119. CGIAR Research Program on Climate Change, Agriculture and Food Security (CCAFS), Copenhagen

Rosenstock TS, Lamanna C, Arslan A et al (2015a) What is the scientific basis for climate-smart agriculture? CGIAR Research Program on Climate Change, Agriculture and Food Security, Copenhagen

Rosenstock TS, Lamanna C, Chesterman S et al (2015b) The scientific basis of climate-smart agriculture: a systematic review protocol. CCAFS Working Paper no. 138. CGIAR Research Program on Climate Change, Agriculture and Food Security (CCAFS), Copenhagen. Available from: http://www.worldagroforestry.org/downloads/Publications/PDFS/WP16086.pdf

Sarwatt SV, Kapange SS, Kakengi AMV (2002) Substituting sunflower seed-cake with *Moringa oleifera* leaves as a supplemental goat feed in Tanzania. Agrofor Syst 56(3):241–247

Thierfelder C, Cheesman S, Rusinamhodzi L (2013) Benefits and challenges of crop rotations in maize-based conservation agriculture (CA) cropping systems of southern Africa. Int J Agric Sustain 11(2):108–124

Thierfelder C, Chivenge P, Mupangwa W et al (2017) How climate-smart is conservation agriculture (CA)?– its potential to deliver on adaptation, mitigation and productivity on smallholder farms in southern Africa. Food Secur 9:1–24

13

CSA and its Multidimensionality

Anthony A. Kimaro, Ogossy G. Sererya, Peter Matata, Götz Uckert, Johannes Hafner, Frieder Graef, Stefan Sieber, and Todd S. Rosenstock

13.1 Introduction

Persistent and resilient food insecurity afflicts smallholder farmers throughout much of East and Southern Africa, including Tanzania, where more than 80% of people in rural areas are involved in agriculture and charcoal production (Rioux et al. 2017; Mwampamba 2007). With such a large proportion of the population involved, agriculture in Tanzania acts as an economic driver of the national economy and presents a way out of poverty (Hansen et al. 2018). However, rural livelihoods in Tanzania are at risk. Farmland is near universally rain-fed, and already susceptible to droughts and weather variability. While current predictions indicate

A. A. Kimaro (✉)
World Agroforestry Centre (ICRAF), Tanzania Programme, Dar es Salaam, Tanzania

Leibniz Centre for Agricultural Landscape Research (ZALF), P. Müncheberg, Germany
e-mail: a.kimaro@cgiar.org

O. G. Sererya
Ministry of Natural Resource and Tourism, Morogoro, Tanzania

P. Matata
Tumbi Agricultural Research Institute and Training Institute, Tabora, Tanzania

G. Uckert · J. Hafner · F. Graef · S. Sieber
Leibniz Centre for Agricultural Landscape Research (ZALF), P. Müncheberg, Germany
e-mail: uckert@zalf.de; johannes.hafner@zalf.de; fgraef@zalf.de; stefan.sieber@zalf.de

T. S. Rosenstock
World Agroforestry Centre (ICRAF), Tanzania Programme, Dar es Salaam, Tanzania

ICRAF/CGIAR Research Program on Climate Change, Agriculture and Food Security (CCAFS), Kinshasa, Democratic Republic of the Congo
e-mail: T.Rosenstock@cgiar.org

that, on average, Tanzania will have future rainfall totals approximately equivalent to today, seasons may shift and the predictability of precipitation will decline (International Center for Tropical Agriculture and World Bank 2017). There is already a need to help farmers to prepare for today's variable precipitation. And coping with today's conditions will help farmers adjust better to changes in the future.

Increasing the rate of adoption of improved agricultural technologies can help build resilience to weather-related risks. For example, Kimaro et al. (2015) examined the resilience of productivity across four seasons within conventional and conservation agriculture in the highlands of central Tanzania and found higher yields and lower interannual variation across all permutations of conservation agriculture in comparison to the control. Furthermore, rainwater use efficiency (RUE) and soil moisture retention were found to be higher in conservation farming and intercropping practices in Tanzania versus traditional practices (Kizito et al. 2016). These results suggest that improved technologies can increase resilience, especially in areas with lower than average rainfall and persistent drought.

However, the adoption of improved technology may affect more than just the resilience of the farming system. It may also affect the system's productivity, including both yields of edible and non-edible crop products and incomes (Charles et al. 2013). Furthermore, it may change the environmental sustainability of the production system; for instance, the climate change mitigation potential, by either sequestering carbon in biomass and/or soils and/or reducing greenhouse gas emissions (Kaonga and Bayliss-Smith 2009). The multidimensionalities of impacts with agricultural change are fundamental to climate-smart agriculture (CSA), which aims to achieve three goals simultaneously: sustainably increase production, improve resilience and mitigate climate change.

Despite multiple goals, rarely are CSA practices evaluated in ways that cross more than one of these three objectives (Rosenstock et al. this volume). This is important for development practitioners because it limits the evidence with which to evaluate potential trade-offs and increases the likelihood of unintended consequences with development programming (Lamanna et al. 2016). Comprehensive information that addresses multiple objectives is needed to evaluate changes in agricultural systems. That, however, is easier said than done, because research is typically undertaken for specific purposes without these three factors in mind, and the costs of multi-indicator measurements may be prohibitive.

We present data from three previously unpublished experiments in two regions of Tanzania: two near Dodoma and one near Tabora. Dodoma has a semiarid climate with a unimodal rainfall regime (7 to 8-month dry period) and mean annual precipitation of 560 millimetres (mm) (Kimaro et al. 2009). Tabora is subhumid with mean annual precipitation of 928 mm (Nyadzi et al. 2003). The experiments in both sites use pigeonpea -based intercropping systems. Here, we present examples of how scientists can investigate CSA in multidimensional assessments.

13.2 Production and Mitigation Benefits of Agroforestry and Intercropping Practices in Dodoma

In arid and semiarid areas of Tanzania, food crops and fuelwood are both the product of agricultural landscapes. Thus, issues of food, fuel and climate are inherently linked and may be best addressed together. Agroforestry—specifically, shelterbelt, *G.sepium* intercropping, and border plantings of fuelwood and food crops—has been promoted to address these concerns simultaneously. In theory, this technology may be climate-smart. Growing trees and crops together has been shown to have positive, negative and no effect on crop productivity (Coe et al. 2016). For instance, intercropping maize with 'fertiliser trees' such as *G. sepium* and/or pigeonpea (*Cajanus cajan*) improves land productivity, soil fertility and enhances the ability of the land to capture and store rainfall, creating resilient cropping systems (Sileshi et al. 2011; Kimaro et al. 2016). Lastly, production of fuelwood reduces collection from natural areas as well as deforestation and degradation (Ramadhani et al. 2002). The mitigation benefits may be further enhanced when coupled with improved cook stove (ICS) technologies that increase the efficiency of fuelwood use. Thus, assessing the synergies of on-farm wood production using agroforestry along with ICS technology increases our understanding of the multidimensional impacts of CSA; yet the impacts of these technologies have often been evaluated separately. We conducted studies to evaluate the CSA benefits of on-farm wood supply and its efficient use by ICS as well as crop yields under agroforestry and maize–pigeonpea intercropping in Kongwa and Chamwino districts, Dodoma, Tanzania.

The first study assessed wood supply from agroforestry technologies (shelterbelts, boundary tree planting, contours planting, and *Gliricidia sepium* intercropping), established on nearly 110 farmers' fields, to evaluate the climate-smartness of these technologies in Chamwino (Ilolo village) and Kongwa (Molet, Mlali Laikala and Chitego villages) districts. Fuelwood yield was determined using species-specific biomass equations (Sererya et al. 2017) and household wood consumption was assessed using the kitchen performance test (Uckert et al. 2017). While it has been found that greenhouse gas (GHG) emissions reduced through the use of the ICS (Sererya 2016), the offset of carbon dioxide emissions by using fuelwood produced on-farm was used to assess the mitigation impacts of ICS and agroforestry technologies. Crops production in alleys between shelterbelts was determined through the systematic sampling of small plots.

We found evidence that agroforestry met some components of CSA. Maize grain yield in the alleyways between shelterbelt strips ranged from 2.3 to 3.2 tons per hectare (t ha^{-1}). Crop yields declined slightly in shelterbelt areas under the influence of trees, but were similar in yield to that obtained in maize monoculture in Dodoma (Kimaro et al. 2009). Wood biomass production in shelterbelt, farm boundaries, intercropping and on contour bounds ranged from 0.5 to 8 t ha^{-1}, depending on the species and spacing adopted (Table 13.1). This amount of wood can sustain a five-member family for 4–6 years when using the traditional three-stone firewood (TSF) stove and ICS, respectively (Table 13.1). Relative to the TSF, households using ICS

Table 13.1 Wood yields and consumption time (months) for different agroforestry technologies in Chamwino and Kongwa Districts, Tanzania

Technology	Tree species	Spacing (m)	Wood (t ha^{-1})	ICS[a]	TFS[a]
Boundary	*Acacia polyacantha*	2 × 2	4.41	3.5	2.4
	Eucalyptus camadulensis	2 × 2	7.70	6.1	4.2
Woodlots	*Grevillea robusta*	2 × 2	2.64	2.1	1.4
	Senna siamea	3 × 3	1.01	0.8	0.6
	Melia azadirachta	4 × 4	0.84	0.7	0.5
Shelterbelt	*Grevillea robusta*	3 × 3	0.46	0.4	0.3
	Gliricidia sepium	1 × 2	2.08	1.6	1.1
Intercropping	*Gliricidia sepium*	3 × 3	1.34	1.1	0.7

[a]Duration of time (years) it will take for a household of five members to complete the amount of wood produced on-farm. The estimate is based on a household consumption rate of 5 kg per day when using the traditional TSF (Uckert et al. 2017)

consumed 23% less firewood, which resulted in a reduction in fuelwood collection time (32%) as well as cooking time (20%) (Uckert et al. 2017). However, firewood and time consumption vary between different foods cooked (Hafner et al. 2018). The reduction of GHG emissions (carbon monoxide, carbon dioxide and particulate organic matter) by the ICS technology, relative to TSF, ranged from 60% to 62% (Sererya 2016). The costs of fuelwood used in ICS and TSF in Dodoma is estimated at Tanzanian Shilling (TZS) 15,984 (United States Dollar (USD) 7.2) and TZS 32,940 (USD 14.8), respectively (Sererya 2016). Based on these estimates, the economic benefits (in terms of cost savings) of on-farm wood supply ranged from USD 90 to 750 ha^{-1}, depending on the tree species and planting spacing adopted. These results suggest that diversification of production (crops and wood) options and income sources through agroforestry contribute in building community resilience (adaptive capacity) as noted by Charles et al. (2013).

We did not measure directly the resilience benefits of the agroforestry systems. Neither the interannual variability of production nor explicit indicators of proxies for resilient agroecosystems (soil carbon, biodiversity, resource efficiency etc.) were available. The former because of the short timeframe of the research and the latter because the research was designed for other purposes. Increasing the duration of research would have helped provide more robust evidence, as would collecting a wider range of indicators. This agrees with early assessments of the literature available in Tanzania (Lamanna et al. 2016) and, therefore, we suggest research protocols for CSA need to be more inclusive to capture specific measures of resilience.

'Mother' and 'baby' research designs were used in Malawi to examine effects across heterogeneous conditions (Snapp 2002). In this study, the mother trial (N = 15)—or replicated on-farm experiments—were laid out in a randomised complete block design and were managed by researchers. Baby trials (N = 275)—or farmer-managed demonstrations of maize-pigenopea intercropping—took place in farmers' fields to allow for participatory evaluation of the technology. Mother trial had five treatments including the control, and the baby trial had maize–pigeonpea intercropping and maize monoculture as a control.

Table 13.2 Maize grain yields (t ha^{-1}) in different intercropping combinations with pigeonpea (PP) at Mlali and Chitego villages, Kongwa district, Dodoma, Tanzania

	2015^{2}		2016	
Maize–PP ratio[a]	Mlali	Chitego	Mlali	Chitego
MM	2.04a	3.25a	2.92a	3.53a
1M:1PP	1.21a	2.26ba	2.53ba	2.99a
1M:2PP	1.46a	1.24b	1.77b	2.35a
2M:1PP	1.39a	3.19a	2.14ba	2.70a
Mean	1.52	2.49	2.34	2.89

[a]Planting ratios tested were: alternate rows of maize and pigeonpea (1M:1M), one maize row and two pigeonpea rows (1M:2PP), two maize rows and one pigeonpea row (2M:1PP) and monocultures of maize (MM) and pigeonpea as controls

Table 13.3 LER for maize (M) and pigeonpea (PP) intercropping at Mlali and Chitego villages, Kongwa district, Dodoma, Tanzania

	2015		2016	
Maize–PP ratio[a]	Mlali	Chitego	Mlali	Chitego
2M:1PP	1.13	1.56	1.21	1.17
1M:1PP	1.12	1.47	1.46	1.53
1M:2PP	1.32	1.15	1.54	1.28

[a]Planting ratios tested were: alternate rows of maize and pigeonpea(1M:1M), one maize row and two pigeonpea rows (1M:2PP), two maize rows and one pigeonpea row (2M:1PP) and monocultures of maize and pigeonpea as controls

Results of this intercropping experiment showed key challenges in understanding what is and isn't CSA. Productivity in farmer-managed baby trials in three villages (Laikala, Mlali and Chitego) ranged from 1.2 to 3.2 t ha^{-1} (>150%), suggesting variations in site and weather conditions. Laikala and Mlali are lower potential sites due to greater degradation while Chitego is a higher potential site for crop production (Kimaro et al. 2015). Overall, maize yield in baby trials across sites was 50% higher than the farmer practice yield of 1.5 t ha^{-1} in the same areas (Kimaro et al. 2012). However, productivity benefits were by no means universal across all planting arrangements and agroecologies. Apart from an intercropping combination—on a one-to-one (1:1) ratio—maize grain yield was reduced by pigeonpea intercropping (Table 13.2). This yield suppression of one component in the mixture was offset when considering farm-level productivity, as reflected by the land equivalent ratio (LER) of greater than one (Table 13.3). Moreover, the intercropping arrangement with higher legume proportions of pigeonpea than maize (1:2 ratio of maize to pigeonpea) was more beneficial to farmers at Mlali village, a lower potential site (LER = 1.46) than in Chitego village, a high potential site (LER = 1.24); but only in the year of poor precipitation and yields (Table 13.3). These findings demonstrate the importance of adopting research protocols that have sufficient temporal and spatial representation to get less spurious results. In this trial, pigeonpea—a drought-resistant crop relative to maize—determines farm-level productivity benefit within the mixture under harsh conditions; reflecting improved resilience due to diversification with drought-tolerant crops. Accordingly, we found the 1:1 arrangement (maize/pigeonpea)—the common farmer practice—to be less sensitive to site and year heterogeneity, suggesting greater resilience. Lastly, the use of LERs to quantify value for the farmer allows the combination of multiple farm outputs and, thus, provides a way to compare monoculture to polycultures. Our results show mixed results when describing yield, but clear benefits of intercropped agroforestry in terms of increased productivity and decreased variance when quantified with LERs (a more comprehensive measure), highlighting the importance of selecting appropriate indicators when studying CSA.

13.3 Production and Resilience Benefits of Cassava-Based Intercropping Practices in Tabora

Cassava is the third most important food crop, after maize and rice, in Urambo and Uyui districts, Tabora region. It is also a more drought-resistant crop than maize and rice (de Oliveira et al. 2017). Most farmers use this crop as a safety net for food shortages, especially in years with prolonged drought. At the same time, intercropping has potential to mitigate soil fertility issues. The added biomass to soils under the cassava-based intercropping system often improves fertility, acidity and soil structure, especially when leguminous species such as pigeonpea are used (Makumba et al. 2009). Thus, cassava–legume intercropping was tested as a strategy for diversifying production and/or income sources as well as building biological quality in these villages.

We evaluated the resilience and productivity aspects of cassava farming under monoculture, intercropping and rotations with pigeonpea in Mbola, Itebulanda and Utenge villages. The research followed a mother–baby plot approach (Snapp 2002), with a researcher-managed plot in each village. Then, ninety farmer-managed plots (baby plots) were set up across the three villages. There was also an on-station experiment at the research farm of the Agricultural Research Institute (ARI) in Tumbi. At the research and mother plots, each treatment was replicated three times in a randomised complete block design; while treatments at the baby plots were unreplicated. Measures of yield, soil moisture and RUE were used as indicators of productivity and resilience. Mitigation was not estimated.

Yields of intercropping treatments (*Canavalia*, Cowpea and Pigeonpea), by comparison to the control, were reduced by 78.5%, 58% and 43% respectively in the research site at the ARI (Table 13.4). The greater reduction in yield in mother plots provides some indication of the differences between research and farmer-managed implementations of these trials. Similar results were also noted for cassava yields intercropped with pigeonpea (50%) and cowpea (60%) by farmers in their baby trial. Such a difference has broad implications for our understanding of the ability of management practices to generate resilience and livelihood benefits, as the vast majority of the data available to evaluate the climate-smartness of technologies was generated on the research stations (Rosenstock et al. this volume). Intercropping

Table 13.4 Growth and yield of cassava at the on-farm ('mother') trials in Urambo and Uyui districts, Tabora, Tanzania

Treatments	Survival (%)	Yield (t ha^{-1})	RUE (kg^{-1} ha^{-1} mm^{-1})
Cassava + Cannavalia	84.3a[2]	2.0c	4.0c
Cassava + Cowpea	88.2ab	3.9b	8.1cb
Cassava + Pigeonpea	86.9ab[a]	5.3b	10.9b
Cassava monoculture	89.2b	9.3a	19.4a

[1]RUE = Rainwater Use Efficiency.[2]Means within a column bearing similar letter(s) are not statistically different at 5% level of probability based on the Duncan's multiple range test (n = 3)

effects on RUE were similar to those on yield (Table 13.4). Comparatively low soil moisture content in intercropping treatments compared to monoculture (data not shown) suggests competition on soil moisture, which resulted in reduced yield and RUE. Apparently, monocultures of drought-tolerant crops, like cassava, provide a promising strategy to enhance farm production and to build resilience, while minimising the negative effects of intercropping. The most promising crop combinations need to be identified after more seasons (crop rotations).

This study has only been conducted for one season so far. However, it already illustrates the importance of offering CSA options from a farmer-centric perspective. Preliminary results suggest that cassava is sensitive to competition, and yields may be adversely affected by intercropping, especially in seasons with low and sporadic precipitation, like in 2017. Thus, despite the best intentions, cassava intercropping may not be climate-smart in this area and, perhaps, farmers are better off by diversifying into cassava monocultures, cassava–legume rotation if they want to diversify out of maize.

13.4 Implications for Development

This chapter analyses the benefits and trade-offs of three agroforestry and intercropping practices in two agroecologies to build evidence for CSA scaling in Tanzania. The analysis involved on-farm wood supply using shelterbelts, intercropping and contours technologies as well as crops production and the resilience effects of pigeonpea -based intercropping systems in semiarid Dodoma and subhumid Tabora. Integrating on-farm wood production and ICS contributed to meeting the multi-objectives of CSA through improved wood supply to meet household annual demand and reducing GHG emissions (less than 60% relative to TSF) as well as productive time lost in cooking and searching for firewood. Moreover, crop diversification at the appropriate intercropping combinations enhanced crop yield (maize and pigeonpea) and agroecosystem resilience as noted by higher LER in the 1:1 ratio across sites. Plant combinations with higher proportions of pigeonpea conferred greater resilience, especially in seasons with less precipitation, which demonstrates the significance of selecting for drought-resistant crops and appropriate farm management practices (i.e., planting combinations/density) in building resilient farming systems. The suppression of yields of intercropped cassava (43–79%) during the first season suggests that the benefits of intercropping may take time to be realised and/or may be comprised by the poor selection of companion crops and farm management practices, such as intercropping, crop rotation, and plant spacing. Trends of CSA benefits in research and farmer-managed experiments were similar, although absolute values of yields (maize, cassava or fuelwood) were higher in research plots. Thus, participatory evaluation of technology is critical for validating and downscaling research results under farmer management conditions and for farmers to appreciate the benefits of CSA prior to wide scaling. Overall results of our analysis of CSA benefits illustrate key principles when considering the multidimensionality of CSA, including the need to: select appropriate indicators, ensure designs are robust for heterogeneity, examine trade-offs, and conduct participatory evaluation of CSA on farmers' field sites. Together, these factors provide more robust evidence for CSA programming and help practitioners and policymakers to be on the lookout for such issues and support evidence-based scaling initiatives. Unfortunately, so many practices and technologies have been labeled CSA in the past few years that some would say it is just rebranding. Accounting for the principles highlighted here, and explicitly considering the multidimensionality of CSA objectives in decision-making, will go a long way to improving implementation and achieving outcomes for farmers.

Acknowledgements We thank Nictor Namoi and Mary Ng'endo for conducting focus group discussions in Tabora as well as farmers and extension officers in Chamwino and Kongwa Districts for supporting this study. The participatory CSA trials in Tabora were funded by the United States Department of Agriculture Foreign Agricultural Service and is part of the CGIAR Research Program on Climate Change, Agriculture and Food Security's Partnerships for Scaling Out Climate-Smart Agriculture project. Research in Kongwa district, Dodoma was funded by the United States Agency for International Development's Feed the Future initiative through Africa RISING; while in Chamwino District, Dodoma it was funded by Deutsche Gesellschaft für Internationale Zusammenarbeit (GIZ) as ICRAF support to Trans-SEC and was co-financed by Bundesministerium für Bildung und Forschung (BMBF) and Bundesministerium für wirtschaftliche Zusammenarbeit und Entwicklung (BMZ). The views expressed in this chapter are those of the authors and may not under any circumstances be regarded as stating official positions of the funders.

References

Charles RL, Munishi PKT, Nzunda EF (2013) Agroforestry as adaptation strategy under climate change in Mwanga District, Kilimanjaro, Tanzania. Int J Environ Protect 3(11):29–38

Coe R, Njoloma J, Sinclair F (2016) Loading the dice in favour of the farmer: reducing the risk of adopting agronomic innovations. Exp Agric:1–17. https://doi.org/10.1017/S0014479716000181

Hafner J, Uckert G, Graef F et al (2018) A quantitative performance assessment of improved cooking stoves and traditional three-stone-fire stoves using a two-pot test design in Chamwino, Dodoma, Tanzania. Environ Res Lett 13(2):025002

Hansen J, Hellin J, Rosenstock T , Fisher E, Cairns J, Stirling C, Lamanna C, van Etton J, Rose A, Campbell B (2018) Climate risk management and rural poverty reduction. Agric Syst. https://doi.org/10.1016/j.agsy.2018.01.019.

International Center for Tropical Agriculture, World Bank (2017) Climate-smart agriculture in Tanzania. CSA country profiles for Africa series. CIAT, World Bank, Washington, DC

Kaonga M, Bayliss-Smith TP (2009) Carbon pools in tree biomass and the soil in improved fallows in eastern Zambia. Agrofor Syst 76:37–51

Kimaro AA, Timmer VR, Chamshama SOA (2009) Competition between maize and pigeonpeain semi-arid Tanzania: effect on yields and nutrition of crops. Agric Ecosyst Environ 134:115–125

Kimaro AA, Sileshi GW, Mpanda M et al (2012) Evidence-based scaling-up of evergreen agriculture for increasing crop productivity, fodder supply and resilience of the maize-mixed and agro-pastoral farming systems in Tanzania and Malawi. https://cgspace.cgiar.org/handle/10568/16883?show=full Accessed

Kimaro AA, Mpanda M, Meliyo JL et al (2015) Soil related constraints for sustainable intensification of cereal-based systems in semi-arid central Tanzania. Proceedings of the Tropentag conference, Berlin, 16–18 September 2015. http://www.tropentag.de/2015/abstracts/full/1005.pdf

Kimaro AA, Mpanda M, Rioux J et al (2016) Is conservation agriculture 'climate-smart' for maize farmers in the highlands of Tanzania? Nutr Cycl Agroecosyst 105:217–228

Kizito F, Lukuyu B, Sikumba G et al (2016) The role of forages in sustainable intensification of crop–livestock agro-ecosystems in the face of climate change: the case for landscapes in Babati, Northern Tanzania. In: Lal R et al (eds) Climate change and multi-dimensional sustainability in African agriculture. https://doi.org/10.1007/978-3-319-41238-2_22

Lamanna C, Namoi N, Kimaro A et al (2016) Evidence-based opportunities for out-scaling climate-smart agriculture in east Africa. CCAFS Working Paper no. 172. CGIAR Research Program on Climate Change, Agriculture and Food Security (CCAFS), Copenhagen. https://ccafs.cgiar.org/publications/evidence-based-opportunities-out-scaling-climate-smart-agriculture-east-africa#.Wp61VUx2vIU. Accessed

Makumba W, Akinnifesi FK, Janssen BH (2009) Spatial rooting patterns of gliricidia, pigeonpeaand maize intercrops and effect on profile soil N and P distribution in southern Malawi. Afr J Agric Res 4:278–288

Mongi H, Majule AE, Lyimo JG (2010) Vulnerability and adaptation of rain fed agriculture to climate change and variability in semi-arid Tanzania. Afr J Environ Sci Technol. https://doi.org/10.5897/AJEST09.207

Mwampamba TH (2007) Has the woodfuel crisis returned? Urban charcoal consumption in Tanzania and its implication for present and future forest availability. Energy Policy 35:4221–4234

Nyadzi GI, Janssen BH, Otsyina RM (2003) Water and nitrogen dynamics in rotational woodlots of five tree species in western Tanzania. Agrofor Syst 59(3):215–229

de Oliveira EJ, Morgante CV, de Tarso Aidar S et al (2017). Evaluation of cassava germplasm for drought tolerance under field conditions. Euphytica 213:188. https://doi.org/10.1007/s10681-017-1972-7

Ramadhani T, Otsyina R, Franzel S (2002) Improving household income and reducing deforestation using rotational woodlots in Tabora district, Tanzania. Agric Ecosyst Environ 89(3):229–239

Rioux J, Lava E, Karttunen K (2017) Climate-smart agriculture guideline for the United Republic of Tanzania: a country-driven response to climate change, food and nutrition insecurity, Food and Agriculture Organization Policy Brief. FAO, Rome

Rosenstock TS, Mpanda M, Aynekulu E (2014) Targeting conservation agriculture in the context of livelihoods and landscapes. Agric Ecosyst Environ 187:47–51

Rowhania P, Lobellb DB, Lindermanc M (2011) Climate variability and crop production in Tanzania. Agric For Meteorol 151:449–460

Sererya OG (2016) Economic analysis of improved firewood cooking stove, its implication on rural livelihoods and environmental sustainability in Chamwino and Kilosa districts, Tanzania. MSc dissertation in Environmental and Natural Resource Economics, Sokoine University of Agriculture, Morogoro

Sererya OG, Kimaro A, Lusambo L (2017) Resilience and livelihood benefits of climate smart agroforestry practices in semi-arid Tanzania. www.tropentag.de/2017/abstracts/posters/960.pdf

Sileshi GW, Akinnifesia FK, Ajayia OC (2011) Integration of legume trees in maize-based cropping systems improves rain use efficiency and yield stability under rainfed agriculture. Agric Water Manag 98:1364–1372

Snapp S (2002) Quantifying farmer evaluation of technologies: the mother and baby trial design. https://www.researchgate.net/publication/237370359_Quantifying_Farmer_Evaluation_of_Technologies_The_Mother_and_Baby_Trial_Design. Accessed

Uckert G, Hafner J, Graef F et al (2017) Farmer innovation driven by needs and understanding: building the capacities of farmer groups for improved cooking stove construction and continued adaptation. Environ Res Lett. https://doi.org/10.1088/1748-9326/aa88d5 (in press)

F. Kizito, B. Lukuyu, G. Sikumba, J. Kihara, M. Bekunda, D. Bossio, K.W. Nganga, A. Kimaro, H. Sseguya, B. Jumbo and P. Okori. 2016. The Role of Forages in Sustainable Intensification of Crop-Livestock Agro-ecosystems in the Face of Climate Change: The Case for Landscapes in Babati, Northern Tanzania. In R. Lal et al. (eds.), Climate Change and Multi- Dimensional Sustainability in African Agriculture, https://doi.org/10.1007/978-3-319-41238-2_22

14

Analyzing the Climate-Smartness of Innovations: Developing a Participatory Protocol

Lucas T. Manda, An M. O. Notenbaert, and Jeroen C. J. Groot

14.1 Introduction

In 2010, Tanzania's agricultural sector accounted for approximately 28% of gross domestic product and 24% of exports (Msambichaka et al. 2009). The sector employed around 75% of the population and is regarded as important for the economic growth of the country (Mnenwa and Maliti 2010). Agriculture in Tanzania is characterised by small-scale farms, whose average land area for cultivation is less than 3 ha (Sarris et al. 2006). Smallholder farmers produce both crops and livestock that are used mainly for subsistence (Amani 2005). Tanzanian agriculture depends on rain as the main source of water, while women contribute a large proportion of the labour force in the sector. In Tanzania, maize is the most widely produced crop followed by rice, sorghum, millet and wheat (Rowhani et al. 2011).

Climate change has affected the living standards of people as well as the performance of important sectors of the Tanzanian economy (Tumbo et al. 2011). It has been estimated that there will be an increase in average daily temperature of 3–5 °C and average annual temperature of 2–5 °C in most parts of the country by the year 2050 (Tumbo et al. 2011). Rainfall is expected to decrease in most parts of the southeastern highlands and central parts of the country, whereas an increase of rainfall is expected in most parts of the northeastern highlands as well as the Lake Victoria Basin (Mwandosya et al. 1998). This variation in temperature and precipitation poses a major threat to cereal crops; with a temperature rise of 2 °C, by the

L. T. Manda (✉)
World Agroforestry Centre (ICRAF), Nairobi, Kenya

A. M. O. Notenbaert
International Center for Tropical Agriculture (CIAT), Nairobi, Kenya
e-mail: a.notenbaert@cgiar.org

J. C. J. Groot
Wageningen University and Research, Wageningen, Netherlands

year 2050, causing the following estimated yield reductions: maize 13%, sorghum 8.8% and rice 7.6% (Rowhani et al. 2011). Already, as a result of warming, a decrease in crop yield has been observed in recent years (Lobell et al. 2011). Droughts have been experienced in many parts of the country, and the disappearance of pasture and water in Sukumaland of the Lake Zone region is well documented. This has resulted in pastoralists travelling long distances in the search for grasses and water to nourish their animals (Kangalawe et al. 2007).

In response to the challenges climate change will present, the concept of climate-smart agriculture (CSA) was brought forward by the Food and Agriculture Organization (FAO) of the United Nations (2013). CSA aims to: (a) sustainably increase food production and income; (b) adapt and build resilience to climate variability; and (c) mitigate/reduce and/or remove greenhouse gas emissions from agricultural practices (FAO 2013). Under the CGIAR Research Program on Climate Change, Agriculture and Food Security (CCAFS), agricultural practices that are climate-smart have been promoted in seven villages in Lushoto District, Tanzania. As part of this programme, 14 farms are implementing improved forages; 21 farms are introducing improved drought-tolerant varieties; 6 are employing terracing; 5 are using composting; 15 others are testing tree planting; and 11 more are benefitting from indigenous knowledge of weather forecasting.

There are no interventions that are climate-smart per se. An intervention's climate-smartness depends on whether it leads to food security, adaptation and mitigation benefits in the specific local climatic, biophysical, socio-economic and developmental context (Williams et al. 2015). In the absence of any assessment of the impact of CCAFS's work in Lushoto, this study aimed to assess the climate-smartness of these interventions.

We developed a participatory protocol for assessing the climate-smartness of innovations at farm level. This evaluates the contribution of newly introduced practices to the productivity, resilience and mitigation of agriculture. Our protocol assesses the food security and adaptation pillars only, for two reasons. Firstly, these pillars are deemed the most important by farmers, and are recognised by many stakeholders as the priority in developing countries; while mitigation is often seen as a potential co-benefit. What's more, the impacts of interventions, across food security and adaptation indicators, are easily observable/measurable/estimable by farmers. Measurements of greenhouse gas (GHG) emissions, on the other hand, are costly and difficult to implement. We, therefore, don't expect farmers to be able to make assessments of mitigation potential so, if this is deemed important within CSA evaluations, participatory assessments should be complemented by researcher-led measurements or modelling exercises.

The protocol was specifically designed for ease of adaption and implementation across a variety of regions and farming systems. It can be applied in a monitoring, evaluating and learning process and allows for the better prioritisation of interventions. This chapter describes the protocol and the lessons learned from its pilot in Lushoto.

14.2 Materials and Methods

A literature review resulted in an early list of suitable farm-level indicators for each of the three CSA pillars. The list was then discussed with extension officers of Lushoto District and experts from CIAT and the Selian Agricultural Research Institute before a final list of indicators was agreed upon.

For the food security and adaptation pillars, the indicators were weighted, scored and finally combined into aggregated indices using a weighted sum of the indicators. Weights and scores were elicited via a survey carried out among a selection of CCAFS project farmers. The data collection protocol involved pairwise ranking and scoring, according to a Likert scale. The weights for each indicator were established through pairwise comparison following the Analytic Hierarchy Process outlined in Saaty (1980). Comparisons of the importance of the indicators were entered into a matrix with a 1–9-point scale. Following this, a consistency ratio was calculated for each pillar. When the consistency ratio was greater than 0.10, all comparisons were reviewed and the inconsistent ones re-evaluated (Saaty 1980). The weight of each indicator from each pillar was calculated using a normalised comparison matrix in which each value present in the matrix was divided by the sum of its column.

Based on these weights, the aggregated food security and adaptation indices were calculated following a three-step process. Firstly, the intervention was scored for each indicator within the food security and adaptation pillars. Farmers were asked to assess whether there had been an increase in the indicators since the beginning of the intervention. The scoring of the indicators was performed using a Likert scale, with scores ranging from 0 to 5—a score of 1 meaning that the farmer strongly disagrees, and 5 meaning that the farmer strongly agrees that the indicator has increased since s/he began the intervention. Secondly, these scores were translated into values ranging between −1 and +1, where: 0 means 'no contribution'; −1 means 'reduces overall score strongly'; −0.5 means 'reduces overall score'; 0.5 means 'increases overall score'; and 1 means 'increases overall score strongly'. The final step resulted in a weighted sum per CSA pillar.

Through this process we achieved a farmer-centric evaluation of the interventions. The establishment of indicator weights, based on the farmer's perspective, ensures that the assessment takes into account the indicators that are most relevant to the farmer in his/her own context. The scoring of the indicators is based on the changes the farmers observe on their own farms as a result of the improved practice and allows for a comparison with the farmers' previous or 'business-as-usual' practice.

The protocol was tested among 72 farmers in the climate-smart village of Lushoto. The data was analysed using Microsoft Excel and different CSA interventions assessed for their contribution to adaptation and productivity.

14.3 Results and Discussion

14.3.1 Suitable Farm-Level Indicators

In the literature review and subsequent discussions with extension staff and experts, we identified a total of 14 indicators relevant to CSA in the Lushoto farming community as listed in Table 14.1. The food security pillar of CSA focuses on strategies that aim to ensure food productivity, food availability, food accessibility and food utilisation. In the assessments in Lushoto, we included the following indicators: food production, animal production, income, and consumption. The adaptation pillar of CSA points towards risk reduction, technological adjustments, and information support for environmental management sustaining the proper growth and development of crops and/or animals. In the Lushoto assessments, we included the following ten indicators in the adaptation pillar: skills and knowledge, access to information, crop adaptation, crop diversity, animal diversity, soil protection, income from farm productivity, stability of farm productivity, income stability, and animal adaptation.

Table 14.1 Indicators selected by extension staff and agricultural experts for the food security and adaptation pillars

Pillars	Indicators	References
Food security	Food production	Nambiar et al. (2001), Yegbemey et al. (2014), Rasul and Thapa (2004), Kamanga et al. (2010), López-Ridaura et al. (2002), and Mittal and Bajwa (2015)
	Animal production	López-Ridaura et al. (2002), Chigwa et al. (2015), Descheemaeker et al. (2011), Herrero et al. (2010), Mittal and Bajwa (2015), and Altieri (1999)
	Income	Hayati et al. (2010), Altieri (1999), and Mittal and Bajwa (2015)
	Consumption	Yegbemey et al. (2014), Kamanga et al. (2010), and Smith et al. (2015)
Adaptation	Skills and knowledge	Kimaru-Muchai et al. (2013)
	Access to information	Smith et al. (2015), Hoang et al. (2006), and Odini (2014)
	Crop adaptation	Vignola et al. (2015)
	Crop diversity	Horrigan et al. (2002), Rasul and Thapa (2003), Nambiar et al. (2001), Valet and Ozier-Lafontaine (2014), and Zhu et al. (2000)
	Animal diversity	Nambiar et al. (2001)
	Soil protection	Lusigi (1995), and Snapp et al. (2010)
	Farm productivity	Meul et al. (2012), and Van Passel and Meul (2012)
	Stability of farm productivity	Organisation for Economic Cooperation and Development (2001)
	Income stability	Mishra and Sandretto (2002), and Dose (2007)
	Animal adaptation	Vignola et al. (2015)

The identification and selection of an appropriate set of indicators forms the basis of any useful impact assessment. Often-cited weaknesses include incomplete coverage of many different factors, including: issues, key considerations, processes, and the causes and effects of the interlinked trends (Van Cauwenbergh et al. 2007). To avoid these, the scope of our literature review covered not only CSA, but also sustainable intensification and organic agriculture. In addition, a thorough scrutiny of potential indicators, in terms of measurability, relevance and practicability, was conducted on those that made the long-list (Lebacq et al. 2013; Van Cauwenbergh et al. 2007; Nambiar et al. 2001; Brown 2009). Narrowing down the long-list with local stakeholders ensured that the final list of indicators is grounded in the local context, and relevant to the challenges being faced and the vision for development in the region. The recent efforts by, for example, CCAFS (Quinney et al. 2016) and the World Bank (2016) to review and guide the selection of suitable CSA indicators are likely to further facilitate this process.

14.3.2 Importance of Indicators in the Food Security and Adaptation Pillars

Figures 14.1 and 14.2 present the importance of different indicators as assessed by the Lushoto farming community. Overall, food production was deemed most important in the food security pillar with a weight of 0.39. This was followed by income and consumption with 0.27 and 0.22 respectively. Animal production scored lowest in this pillar, with a weight of 0.11.

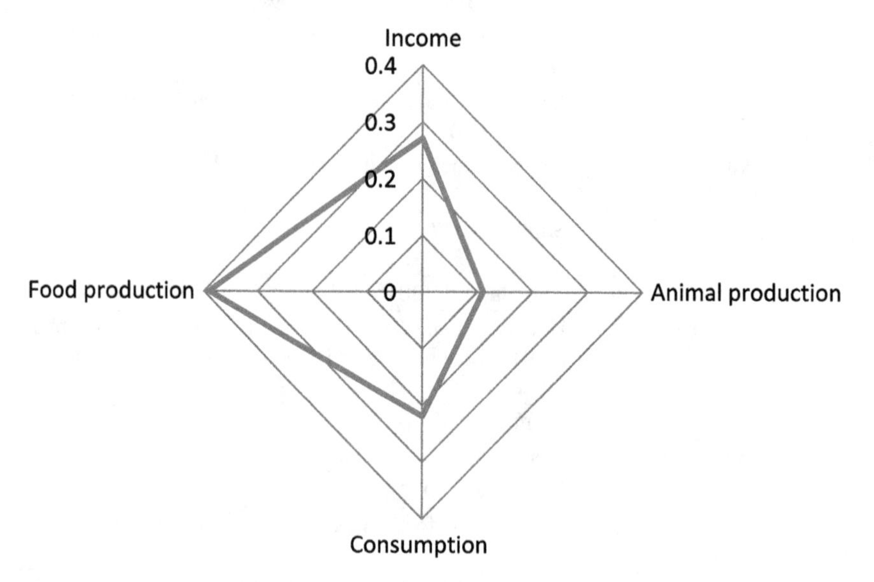

Fig. 14.1 Importance of food security indicators according to small-scale farmers in Lushoto District. Blue line represents indicator weights

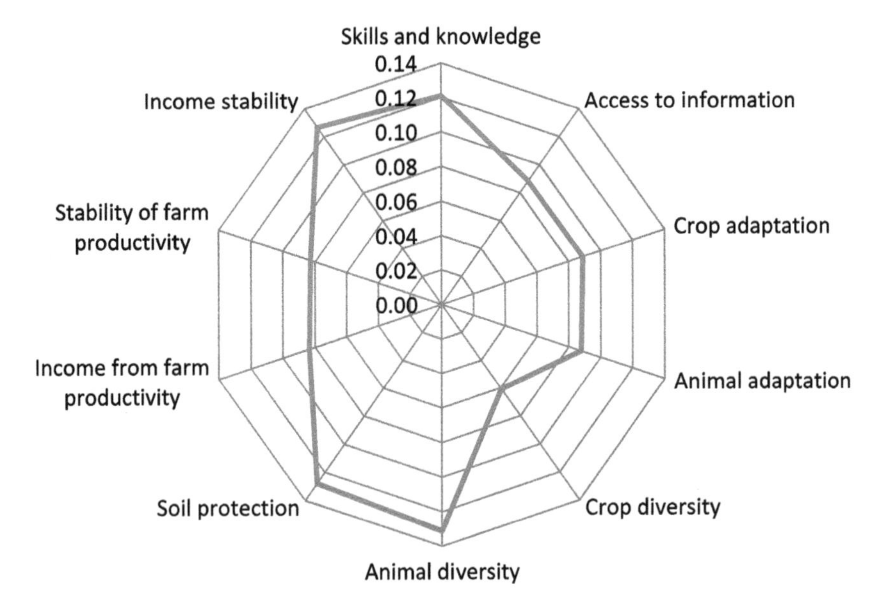

Fig. 14.2 Importance of adaptation indicators as assessed by small-scale farmers in Lushoto District. Blue line represents indicator weights

We associate the first figure with the farmers' priority on allocating resources, where a large proportion of land is allocated for food crops. Likewise, the farmers' priority towards the income indicator was associated with the large proportion of income spent on food items. The importance of the consumption indicator was associated with its direct correlation to food security. The farmers' responses around the animal production indicator points to the higher importance of crop production than livestock for most farmers. A study carried out by Lyamchai et al. (2011) in Lushoto District indeed suggested that 100% of the food crop is produced by smallholder farmers and that crop agriculture is the dominant sector in the area. The findings of this study are also in line with a characterisation survey carried out in western Kenya in which food production was deemed most important by farmers followed by income (Waithaka et al. 2006). Moreover, the study of Shikuku et al. (2016) reported that income and yield were deemed the most important CSA indicators by both male and female farmers in Mbeya, Tanzania. Yet, the study did not specify whether yield is coming from the production of crops and/or animals.

In the adaptation pillar, soil protection, income stability, skills and knowledge were deemed the most important indicators with weights of 0.13 each (Fig 14.2). Our findings concur with those of a study conducted by Shikuku et al. (2016) in the uplands and lowlands of Mbarali and Kilolo Districts, Tanzania in which soil fertility, together with skills and knowledge, were deemed the most important. Surprisingly, regardless of the observed diversified cropping pattern, crop diversity was deemed less important by farmers. Our result in this indicator differed with the study performed in Malawi in which crop diversity was deemed most important by farmers (Cromwell et al. 2001).

The framework incorporates input from farmers through a pairwise comparison of indicator importance and indicator scoring, which involves the careful process of allocating weights (Notenbaert et al. 2010). Here farmers were responsible for allocating the weights of the indicators. In particular, the process represents a challenge when farmers are unable to count and translate their assessments into a 1–9-point scale. This problem necessitated frequent repetition to ensure an acceptable consistency ratio for each pillar. Calculations of the consistency ratio for each pillar were carried out with an expert during the process.

14.3.3 Performance of CSA Interventions Across Two Pillars

Of the six different interventions that were implemented as part of CCAFS's projects and assessed by farmers in terms of their impacts on food security and adaptation, only composting, improved drought-tolerant varieties and improved forages interventions represent true win-win scenarios. This means that they contribute significantly to food security through their ability to increase productivity while ensuring adaptation to climate variability and change. As a result, the Lushoto farmers valued these interventions because they contributed to improving soil fertility and structure, reducing surface runoff, and reclaiming degraded land due to their positive impact on yield and off-season crop agriculture. This result is corroborated by Nyasimi (2017) who mentioned that improved crop varieties and composting were the most commonly implemented CSA interventions by the smallholder farmers.

On the other hand, a clear trade-off is observed between the two pillars when implementing tree planting (Fig. 14.3). According to the farmers, tree planting failed to contribute to food security. This is in contrast with several studies (Murthy et al. 2016; Verchot et al. 2007) which have shown that a combination of beneficial trees on farms tends to increase soil fertility and farm production, while protecting crops from climate risk. In addition, the continued use of these interventions ensures the diversification of farmers' incomes as well as minimising monetary risk. The fact that such evidence is not taken into account by the farmers, points to a weakness in this type of participatory assessment. It is potentially biased as a result of social conditioning and basing results on anecdotes instead of hard evidence (Sen 1999). Participatory assessments, however, elicit the views of the actual beneficiaries and, therefore, ensure the use of locally relevant indicators as well as the assessment of context-specific impacts. It also increases the likelihood of longer term buy-in and farmer-to-farmer promotion of positively assessed interventions. In addition, it can contribute to capacity-building and the empowerment of smallholder farmers in relation to choosing suitable CSA interventions (Williams et al. 2015). In addition, the approach can have broader implications in managing trade-offs in the perceptions of smallholder farmers and policymakers.

The scarceness of win-win interventions, on the other hand, raises a question around whether every activity undertaken by every farmer in every field should generate double or triple wins. According to the FAO (2014), the short answer is no.

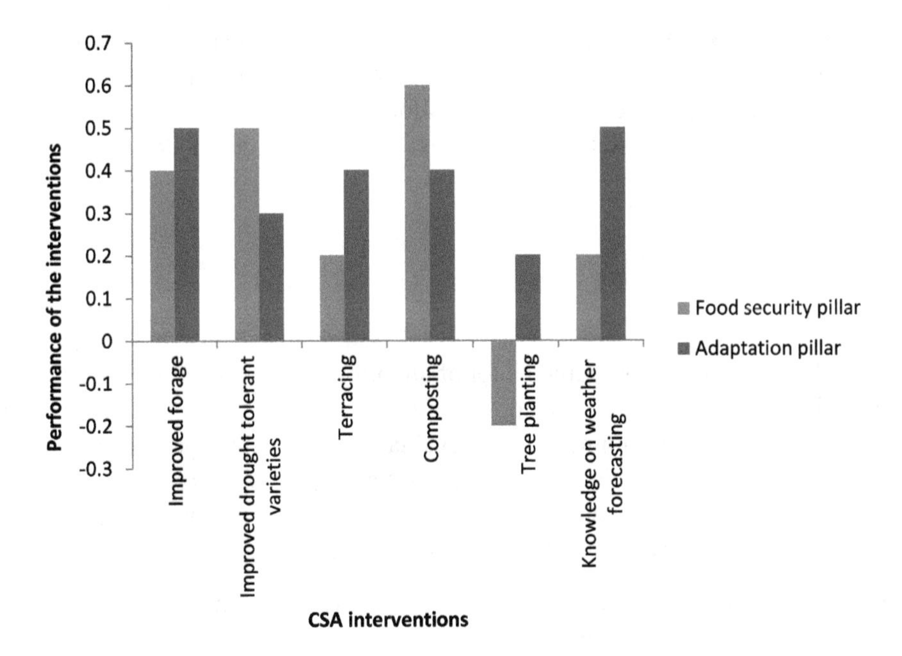

Fig. 14.3 Trade-offs and synergies of the interventions across the two pillars

A CSA policy for agricultural development includes various interventions (on practices, delivery systems/institutions and policies) at various scales (community, landscape, agro-ecological zone, regional and national). The need for adaptation and the potential for mitigation in relation to achieving food security/development vary among these activities and scales and, as a result, so does the ability to capture synergies. Farmers should not only consider CSA as a new set of practices, but also as an integrated approach (Rosenstock et al. 2016; Williams et al. 2015). Likewise, options for an effective combination of interventions would enable smallholder farmers to reap the benefits of both pillars of CSA.

14.4 Implication for Development

Tanzania is experiencing extreme climate change, and the adverse effects have already been reported to affect agriculture and people's livelihoods (United Republic of Tanzania (URT) Ministry of Agriculture 2014, 2015, 2016). As a result, several measures have been taken by the Government to combat the effects of climate change, including the formation of institutions and policies responsible for promoting CSA (Wanzala 2010). These institutions, however, currently lack tools and approaches to assess the performance of the interventions they promote. Our proposed tool could be used as a starting point for assessing the climate smartness of agricultural interventions. Extension officers and other stakeholders can be trained to carry out regular assessments and get insights, based on farmers' opinions on any interventions. These insights then need to be fed back into the planning process and used to inform adjustments to the current interventions or the design of future investments. The application of such a monitoring, evaluation and learning process has the potential to increase the effectiveness of a wide range of CSA initiatives in the country.

14.5 Conclusion

Around the world, CSA has gained a lot of attention; while a variety of agricultural interventions has been hypothesized to contribute to food security, adaptation and mitigation. Assessment, monitoring and evaluation are integral parts of CSA planning and implementation. They are crucial for making decisions on the use of financial, natural and human resources. CSA options should therefore be assessed for their effectiveness in achieving their intended climate change goals.

However, there is a lack of clear and workable criteria and methods for assessing the actual climate-smartness of these interventions. In addition, often, there is limited inclusion of stakeholders' perspectives and, therefore, little buy-in resulting in a lack of wide scale adoption. This chapter proposes a participatory approach that—unlike many other assessments—involves stakeholders at every stage: from indicator selection, through indicator weighing, to actual intervention evaluation. Its application in Lushoto District, Tanzania, demonstrates that participatory assessment of the climate-smartness of agriculture interventions can be used to provide valuable indication supporting CSA groundwork. The protocol presented ensures the selection of locally relevant indicators and the inclusion of farmers' experiences through participatory monitoring of the interventions' local impact. We recommend its use for eliciting insights on the effectiveness of the on-farm components of CSA initiatives beyond this study. These insights can then inform necessary adjustments of such programmes.

The approach is easy to adapt to different types of interventions in a variety of contexts. We believe, however, that the protocol would be easier to implement with farmers after the adjustment of the quantitative scales used to rank indicators and value interventions according to these indicators. We suggest the use of qualitative descriptions of these scales for future applications.

Our framework deals with two pillars of CSA only, namely food security and adaptation. With its standard indicators and long-term and off-farm impacts, the mitigation potential of the interventions does not lend itself to such participatory approaches. We, therefore, recommend complementing the participatory assessments in terms of food security and adaption with science-led GHG emissions estimations. These could be a combination if ex- and in-situ measurements and modelling approaches. Such complementary studies would add value to the overall assessment of climate-smartness of tested interventions.

References

Altieri MA (1999) Applying agroecology to enhance the productivity of peasant farming systems in Latin America. Environ Dev Sustain 1:197–217

Amani H (2005) Making agriculture impact on poverty in Tanzania: the case on non-traditional export crops. Paper presented at A policy dialogue for accelerating growth and poverty reduction in Tanzania, Economic and Social Research Foundation, Dar es Salaam, 12 May 2005

Brown D (2009) Good practice guidelines for indicator development and reporting. Paper presented at the Third World Forum on statistics, knowledge and policy, Busan, Korea, 27–30 Oct 2009

Chigwa FC, Eik LO, Kifaro G et al (2015) Alternative goat kid-rearing systems for improved performance and milk sharing between humans and offspring in climate change mitigation. Sustainable Intensification to Advance Food Security and Enhance Climate Resilience in Africa. Springer, Heidelberg, pp 331–341

Cromwell E, Kambewa P, Mwanza R et al (2001) Impact assessment using participatory approaches: starter pack and sustainable agriculture in Malawi. Overseas Development Institute, Agricultural Research & Extension Network, London

Descheemaeker K, Amede T, Haileslassie A et al (2011) Analysis of gaps and possible interventions for improving water productivity in crop livestock systems of Ethiopia. Exp Agric 47:21–38

Dose H (2007) Securing household income among small-scale farmers in Kakamega District: possibilities and limitations of diversification. Transformation in the Process of Globalisation, GIGA Research Programme, Institute of Global and Area Studies (GIGA), Hamburg

Food and Agriculture Organization (2013) Climate smart agriculture sourcebook. FAO, Rome. http://www.fao.org/docrep/018/i3325e/i3325e.pdf. Date accessed 16/02/2018

Food and Agriculture Organization (2014) Knowledge on climate smart agriculture. FAO, Rome. http://www.fao.org/3/a-i4064e.pdf. Date accessed 16/02/2018

Hayati D, Ranjbar Z, Karami E (2010) Measuring agricultural sustainability. Biodiversity, biofuels, agroforestry and conservation agriculture. Springer, Heidelberg

Herrero M, Thornton PK, Notenbaert AM et al (2010) Smart investments in sustainable food production: revisiting mixed crop–livestock systems. Science 327:822–825

Hoang LA, Castella J-C, Novosad P (2006) Social networks and information access: implications for agricultural extension in a rice farming community in northern Vietnam. Agric Hum Values 23:513–527

Horrigan L, Lawrence RS, Walker P (2002) How sustainable agriculture can address the environmental and human health harms of industrial agriculture. Environ Health Perspect 110:445–456

Kamanga B, Waddington S, Robertson M et al (2010) Risk analysis of maize–legume crop combinations with smallholder farmers varying in resource endowment in central Malawi. Exp Agric 46:1–21

Kangalawe R, Liwenga E, Majule A (2007) The dynamics of poverty alleviation strategies in the changing environments of the semiarid areas of Sukumaland, Tanzania. Research report submitted to Research on Poverty Alleviation, Dar es Salaam

Kimaru-Muchai S, Mucheru-Muna M, Mugwe J et al (2013) Communication channels used in dissemination of soil fertility management practices in the central highlands of Kenya. B. Vanlauwe, P. van Asten, G. Blomme (Eds.), Agro-Ecological Intensification of Agricultural Systems in the African Highlands, Routledge, London (2013), pp. 283–307

Lebacq T, Baret PV, Stilmant D (2013) Sustainability indicators for livestock farming: a review. Agron Sustain Dev 33:311–327

Lobell DB, Schlenker W, Costa-Roberts J (2011) Climate trends and global crop production since 1980. Science 333:616–620

López-Ridaura S, Masera O, Astier M (2002) Evaluating the sustainability of complex socio-environmental systems: the MESMIS framework. Ecol Indic 2:135–148

Lusigi WJ (1995) Measuring sustainability in tropical rangelands: a case study from northern Kenya. Defining and measuring sustainability: the biogeophysical foundations. The World Bank, Washington, DC, pp 277–307

Lyamchai C, Yanda P, Sayula G et al (2011) Summary of baseline household survey results: Lushoto, Tanzania. CGIAR Research Program on Climate Change, Agriculture and Food Security (CCAFS). Copenhagen. https://ccafs.cgiar.org/es/node/48157#.WoLtP0x2vIU

Meul M, Van Passel S, Fremaut D et al (2012) Higher sustainability performance of intensive grazing versus zero-grazing dairy systems. Agron Sustain Dev 32:629–638

Mishra AK, Sandretto CL (2002) Stability of farm income and the role of nonfarm income in US agriculture. Rev Agric Econ 24:208–221

Mittal S, Bajwa J (2015) Identifying indicators to certify the climate smart villages. In: Accessed 16/02/2018

Mnenwa R, Maliti E (2010) A comparative analysis of poverty incidence in farming systems of Tanzania. Special Paper 10/4. Research on Poverty Alleviation, Ottawa

Msambichaka L, Mashindano O, Luvanda E et al (2009) Analysis of the performance of the agriculture sector and its contribution to economic growth and poverty reduction. Draft submitted to the Ministry of Finance and Economic Affairs, Dar es Salaam

Murthy IK, Dutta S, Vinisha V et al (2016) Impact of agroforestry sytems on ecological and socio-economic systems: a review. Glob J Sci Front Res: H Environ Earth Sci 16(5), Version 1.0, no. 3:15–27

Mwandosya MJ, Nyenzi BS, Lubanga M (1998) The assessment of vulnerability and adaptation to climate change impacts in Tanzania. Centre for Energy, Environment, Science and Technology, Dar es Salaam

Nambiar K, Gupta A, Fu Q et al (2001) Biophysical, chemical and socio-economic indicators for assessing agricultural sustainability in the Chinese coastal zone. Agric Ecosyst Environ 87:209–214

Notenbaert A, Massawa S, Herrero M (2010) Mapping risk and vulnerability hotspots in the COMESA region. Technical Report. ReSAKSS Working Paper No 32

Nyasimi M (2017) Adoption and dissemination pathways for climate-smart agriculture technologies and practices for climate-resilient livelihoods in Lushoto, Northeast Tanzania. J Clim 2017(5):63. https://doi.org/10.3390/cli5030063

Odini S (2014) Access to and use of agricultural information by small scale women farmers in support of efforts to attain food security in Vihiga County, Kenya. J Emerg Trends Econ Manag Sci 5:100

Organisation for Economic Co-operation and Development (2001) Measuring productivity: measurement of aggregate and industry-level productivity growth. OECD Manual, OECD, Paris

Quinney M, Bonilla-Findji O, Jarvis A (2016) CSA programming and indicator tool: 3 steps for increasing programming effectiveness and outcome tracking of CSA interventions. CCAFS, Copenhagen

Rasul G, Thapa GB (2003) Sustainability analysis of ecological and conventional agricultural systems in Bangladesh. World Dev 31:1721–1741

Rasul G, Thapa GB (2004) Sustainability of ecological and conventional agricultural systems in Bangladesh: an assessment based on environmental. Econ Soc Perspect Agric Syst 79:327–351

Rosenstock TS, Lamanna C, Chesterman S et al (2016) The scientific basis of climate-smart agriculture: a systematic review protocol. CCAFS Working Paper no. 138. Copenhagen, Denmark

Rowhani P, Lobell DB, Linderman M et al (2011) Climate variability and crop production in Tanzania. Agric For Meteorol 151:449–460

Saaty TL (1980) The analytic hierarchy process. McGraw-Hill, New York

Sarris A, Savastano S, Christiaensen L (2006) The role of agriculture in reducing poverty in Tanzania: a household perspective from rural Kilimanjaro and Ruvuma. FAO Commodity and Trade Policy Research Working Paper, 19

Sen A (1999) Development as Freedom (DAF). Oxford University Press, Oxford

Shikuku K, Mwongera C, Winowiecki L et al (2016) Understanding farmers' indicators in climate-smart agriculture prioritization in the Southern Agricultural Growth Corridor of Tanzania (SAGCOT). International Center for Tropical Agriculture (CIAT), Cali 43 p

Smith A, Thorne P, Snapp S (2015) Measuring sustainable intensification in smallholder agroecosystems: a review. Glob Food Secur 12(2017):127–138

Snapp SS, Blackie MJ, Gilbert RA et al (2010) Biodiversity can support a greener revolution in Africa. Proc Natl Acad Sci 107:20840–20845

Tumbo S, Mutabazi K, Kimambo A et al (2011) Costing and planning of adaptation to climate change in animal agriculture in Tanzania. International Institute for Environment and Development, London

United Republic of Tanzania (2014) Agriculture climate resilience plan 2014–2019. Ministry of Agriculture, Dar es Salaam

United Republic of Tanzania (2015) Tanzania climate-smart agriculture programme. Ministry of Agriculture, Dar es Salaam

United Republic of Tanzania (2016) Agricultural sector development programme phase two. Ministry of Agriculture, Dar es Salaam

Valet S, Ozier-Lafontaine H (2014) Ecosystem services of multispecific and multistratified cropping systems. Sustainable Agriculture Reviews 14. Springer, Heidelberg

Van Cauwenbergh N, Biala K, Bielders C et al (2007) SAFE—a hierarchical framework for assessing the sustainability of agricultural systems. Agric Ecosyst Environ 120:229–242

Van Passel S, Meul M (2012) Multilevel and multi-user sustainability assessment of farming systems. Environ Impact Assess Rev 32:170–180

Verchot LV, Van Noordwijk M, Kandji S et al (2007) Climate change: linking adaptation and mitigation through agroforestry. Mitig Adapt Strateg Glob Chang 12:901–918

Vignola R, Harvey CA, Bautista-Solis P et al (2015) Ecosystem-based adaptation for smallholder farmers: definitions, opportunities and constraints. Agric Ecosyst Environ 211:126–132

Waithaka M, Thornton PK, Herrero M et al (2006) Bio-economic evaluation of farmers' perceptions of viable farms in western Kenya. Agric Syst 90:243–271

Wanzala M (2010) Comprehensive Africa agriculture development programme review: renewing the commitment to African agriculture. Final Report. NEPAD Planning and Coordinating Agency, Johannesburg

Williams TO, Mul M, Cofie OO (2015) Climate smart agriculture in the African context. Background Paper. Feeding Africa Conference 21–23 Oct 2015

World Bank (2016) Climate-smart agriculture indicators: world bank group report number 105162-GLB. World Bank, Washington, DC

Yegbemey RN, Yabi JA, Dossa CSG et al (2014) Novel participatory indicators of sustainability reveal weaknesses of maize cropping in Benin. Agron Sustain Dev 34:909–920

Zhu Y, Chen H, Fan J et al (2000) Genetic diversity and disease control in rice. Nature 406:718–722

15

Adopting Stress-Tolerant Varieties and its Welfare Effects

Chris M. Mwungu, Caroline Mwongera, Kelvin M. Shikuku, Mariola Acosta, Edidah L. Ampaire, Leigh Ann Winowiecki, and Peter Läderach

15.1 Introduction

In most developing countries agriculture plays a significant role in enhancing food security among smallholder farmers. It is regarded as a significant economic activity that can reduce absolute and relative poverty among smallholder farmers in Sub-Saharan Africa (SSA) (Odame et al. 2013). However, both presently and in the future, the agricultural sector is increasingly threatened by the adverse impacts of climate risks. As a result of climate change, inconsistent and unstable agricultural yields will ultimately increase the risk of food and nutritional insecurity among the vulnerable populations in SSA. It is expected that climate change will ultimately lead to increased nutritional disorders, diseases, hunger and socio-economic instability in Africa (Msowoya et al. 2016). Since most families in rural SSA provide own farm labour in agriculture (Dieterich et al. 2016), poorly fed families may provide low quality labour, which can also affect production. With continuous deterioration in production over seasons, the standards of living for farmers in rural Africa will be compromised. In the case of Uganda, a decrease of a 2–4% in Gross Domestic Product is foreseen, if sufficient measures to combat climate change are not taken into consideration (Markandya et al. 2015).

C. M. Mwungu (✉) · C. Mwongera · K. M. Shikuku
International Centre for Tropical Agriculture (CIAT), Africa Regional Office, Nairobi, Kenya
e-mail: c.mwungu@cgiar.org

M. Acosta · E. L. Ampaire
International Institute of Tropical Agriculture (IITA), Kampala, Uganda

L. A. Winowiecki
World Agroforestry Centre (ICRAF), Nairobi, Kenya

P. Läderach
International Center for Tropical Agriculture (CIAT), Asia Regional Office c/o Agricultural Genetics Institute, Hanoi, Vietnam

Climate-smart agriculture (CSA) technologies, such as stress-tolerant varieties have the potential to increase productivity and reduce poverty levels of smallholder farmers (Food and Agriculture Organization (FAO) 2013). In addition, stress-tolerant varieties may reduce the risk of pests and diseases that are accelerated by climate change (Jellis 2009; Nyasimi et al. 2017). Among the challenges experienced by farmers in northern Uganda are a high prevalence of crop diseases and an increasing occurrence of inter- and intra-seasonal dry spells (Mwongera et al. 2014). Therefore, stress-tolerant varieties can reduce the cost of production and lower the economic risk of investing in agriculture. Although trade-offs are possible, adoption of stress-tolerant varieties can contribute to the three pillars of CSA by increasing production and enhancing the resilience of farming systems (Shiferaw et al. 2014). Furthermore, stress-tolerant varieties enhance the optimal use of available household resources and are, therefore, central to sustainable economic development (Khatri–Chhetri et al. 2017).

We carried out studies in 2015 to prioritise context-specific CSA practices for Nwoya District (Shikuku et al. 2015). The use of improved stress-tolerant varieties was ranked highest among the shortlisted CSA practices by stakeholders. However, the adoption of the stress-tolerant varieties was still low in the District, partly due to past experience of other improved varieties as well as a lack of financial resources. The most prevalent challenges to agriculture production, linked to climate stresses, were: the high prevalence of pests and diseases, unpredictable rainfall patterns, soil erosion, droughts and floods. Other practices that were selected as relevant to address these matters included: maize legume intercrop, agroforestry, silvo-pastoral systems and crop rotation. Few studies have assessed the impacts of climate change and climate-smart agriculture options on farm income, labour demand, food security and nutrition, thus empirical evidence is still insufficient. Existing studies include, Makate et al. (2016), which reported that households became more food secure and resilient to climate change on the adoption of crop diversification. Also, Manda et al. (2016), which argued that the adoption of improved varieties only increases the cost of production; but, when blended with a maize–legume intercrop, household crop income increased. And Brüssow et al. (2017), which found that the adoption of CSA technologies by farmers in Tanzania increased household food security in terms of diversity and stability. In this study, we assess the welfare effects of adopting stress-tolerant varieties in Nwoya District, using per capita crop income as a proxy to measure farmers' welfare. The study considered stress-tolerant varieties of maize, beans, cassava and groundnuts. To fill important gaps in the evidence, this study asked the following research questions: (i) what are the drivers for adoption of stress-tolerant varieties? (ii) What is the impact of adopting stress-tolerant varieties on households' welfare?

15.2 Data and Methods

The study used a household survey data set collected in Nwoya District, Uganda in October 2014. The District covers a geographical area of 4736.2 square kilometres (km^2) and has an average population density of 36.99/km^2. Over the course of the

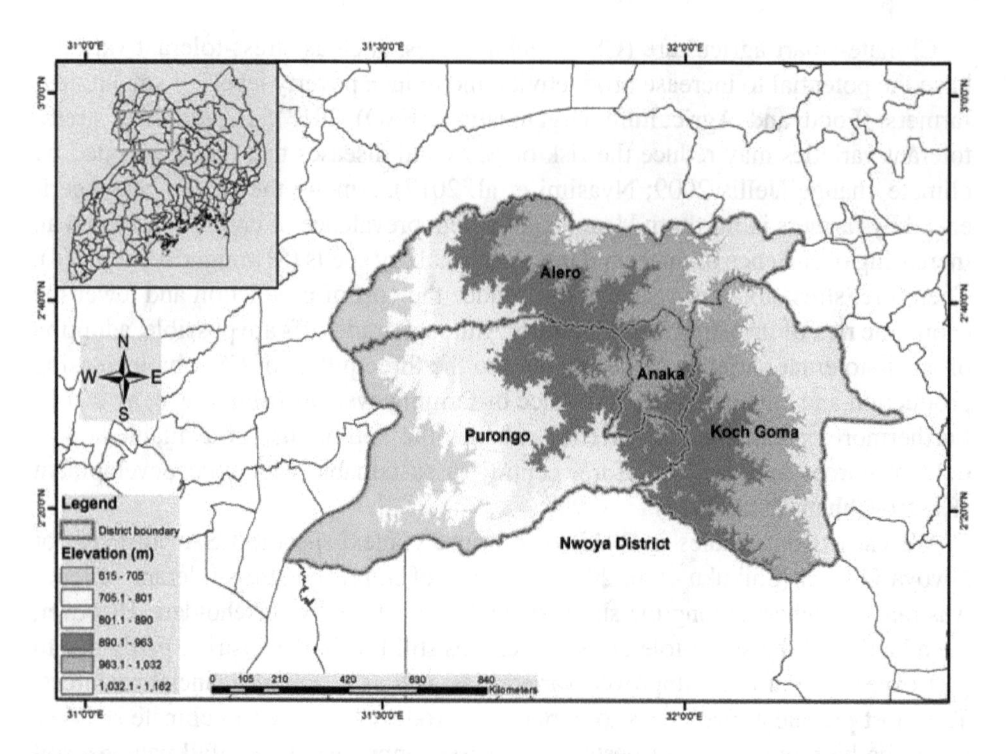

Fig. 15.1 Map of the study area—Nwoya District in Northern Uganda

year, the temperature varies from 18 to 36 °C. The region has a wet season that begins in March and ends in November each year. Planting of annual crops normally begins in April, while harvesting starts in July depending on the crop. Figure 15.1 presents a map of the study area. A detailed description of the study area, sample size, target population, sampling procedure and accessibility of the data is provided by Mwungu et al. (2017). Data were collected from 585 farm households in Nwoya District via one-to-one interviews. The survey questionnaire captured information on socio-demographics, dwelling characteristics, assets ownership, food availability, access to financial services, adoption of CSA technologies and practices, membership of agricultural groups, sources of agricultural information, farming activities and production for different crops at plot level, personal values, and farmers' perceptions of climatic changes. Empirical analysis began by assessing the determinants of adopting stress tolerant varieties using a binary choice logistic regression model. This analysis constituted the first step in the propensity score matching (PSM) technique of impact assessments. Within a regression framework, a binary choice model is specified as:

$$W_{it}^{*} = \beta x_i + \varepsilon_i$$
$$W_{it} = \begin{cases} 1, \text{ if } W_{it}^{*} > 0 \\ 0, \text{otherwise} \end{cases} \tag{15.1}$$

where W_{it}^{*} is a latent unobserved variable whose counterpart, W_{it}, is observed in dichotomous form only; where $W_{it} = 1$ represents households that adopted a CSA technology and $W_{it} = 0$ represents households that did not adopt; β refers to a vector of coefficients estimated by the model. The signs and magnitude of the marginal effects are important in explaining the effect of the independent variables on the adoption of improved varieties or simply the propensity to adopt. A vector of independent variables is represented by x while ε is the error term. The choice of the independent variables was informed by literature on the adoption of agricultural innovations (see, for example, Manda et al. 2016). This first step generated propensity scores, that is, the estimated probabilities of households to adopt stress tolerant varieties based on the observed covariates, x.

In the second step, average treatment effects (ATE) were estimated based on a matched sample of adopters and non-adopters, which was obtained using the propensity scores generated in the first step. Using this approach, we controlled for unobserved heterogeneity due to self-selection into adopter and non-adopter groups. The probability of self-selection bias in non-experimental studies might imply that the adopters are systematically different from non-adopters and, if this is not adequately controlled for, the estimated impacts may be biased and the conclusion misleading. The fundamental assumption in PSM is that outcomes are independent of treatment assignment and conditioned on explanatory variables. Furthermore, it is assumed that there is sufficient overlap in propensity scores so that both adopters and non-adopters have an equal probability greater than zero and less than one of adopting improved varieties. These two underlying assumptions are, respectively, referred to as the 'conditional independence assumption' and the 'common support assumption'. Causal effect, therefore, refers to the difference between the observed and counterfactual household welfare. Following Becker and Caliendo (2007), average treatment on the treated (ATT) can be calculated by:

$$ATT = E\left(y_{1i}(1)\middle|T=1\right) - E\left(y_{0i}(0)\middle|T=0\right) \tag{15.2}$$

where ATT is the observed per capita net crop income for improved varieties adopters, y_i is the observed average household welfare for adopters of improved varieties, while y_o is the estimated household welfare if the adopters had not adopted improved varieties.

15.2.1 Description of Variables

Table 15.1 provides a detailed description of the variables used in the study. The choice of the variables used in this study was informed by literature from past studies on the adoption and impact of agricultural technologies (Bonabana-Wabbi 2002; Joshua Udoh and Titus Omonona 2008; Simtowe et al. 2012; Asayehegn et al. 2017).

Table 15.1 Measurement and description of variables

Variable description	Measurement
Outcome variable	
Per capita net crop income (USD)	Net crop income divided by household size
Household characteristics variables	
Household size (count)	Number of persons in the household
Sex of the household head (0/1)	Sex of the principal decision-maker in the household
Age of the household head (years)	Age of the principal decision-maker
Literacy index (index)	Ratio between household members with post-primary education and household size
Number of years residence in the village (years)	Number of years a farmer has lived in the village
Household wealth characteristics	
Asset index (index)	Index of number and type of assets owned to determine the well-being of the household
Institutional and access related variables	
Access to government extension (0/1)	Dummy for access to government extension officers
Access to farmer organisation (0/1)	Dummy for access to farmer organisation
Access to NGO information (0/1)	Dummy for access to NGO information
Access to demo plots (0/1)	Dummy for access to demo plots information
Access to credit services (0/1)	Access to credit services
Member of an agricultural group (0/1)	Membership of an agricultural group
Gender roles and personal values	
Training on personal values (0/1)	Training on personal values
Perceptions of climate change and associated risks	
Noticed change in climate (0/1)	Noticed a change in climate
Perceiving likely change in climate change in future (0/1)	Perception that climate will change in future
Perceiving reduced current rainfall (0/1)	Perception that rainfall amount received has reduced
Climatic shocks	
Experienced floods (0/1)	Experienced floods
Experienced drought (0/1)	Experienced drought
Food security variable	
Number of hunger months in a year (count)	Number of hunger months in a year

15.3 Results of Econometric Analysis

15.3.1 Determinants for Adoption of Improved Varieties

The logistic model was initially fitted in the first stage of the PSM to assess the factors that influence adoption of improved varieties and determine the propensity to adopt for each household. The dependent variable was a binary stress tolerant varieties adoption. To establish the reliability of the estimates from the logit model, a variance inflation factor (VIF) test for multicollinearity and Hosmer–Lemeshow (HL) test for goodness of fit were conducted. The VIF test ruled out serious multicollinearity and the HL test showed that the logit model was properly specified. Additionally, the log likelihood ratio obtained was −608.5264, which was statistically significant at 1%, while the pseudo-R^2 value of the model was 0.1421. This indicated overall significance of the logistic model and a good fit for the data. As shown in Table 15.2, the decision to adopt stress tolerant varieties was positively influenced by household size, gender of the household head, access to agricultural information from NGOs, perception of future changes in climate, number of years' residence in the village and the asset index. This indicated that for every unit increase in any of the variables, the probability of adopting improved varieties increases by the corresponding marginal effects.

These results are in harmony with other past studies on theoretical and empirical literature about agricultural technology adoption. For instance, farmers who had access to NGO information were 10.33% more likely to adopt stress-tolerant varieties than their counterparts. This is partly because access to information reduces uncertainty about new technologies as farmers become aware of the new technology and how to use it effectively. These findings are in agreement with (Bonabana-Wabbi 2002) which reported that farmers who had access to agricultural information had a higher probability of adopting integrated pest management technologies in Uganda. However, against our expectation, farmers who had access to demo plots information were not more likely to adopt improved varieties. We hypothesise the reason for this finding is based on the context of the study site. Communities in northern Uganda suffered conflict and were displaced in camps and have only resettled back in their farms within the last decade. Approaches relying on trust and social networks are, therefore, more likely to influence learning and the adoption of stress tolerant varieties. In this case, we see that learning through NGOs—most of which have been in the community for long periods and have built good relationships with the farmers—is likely to be more effective compared to demonstration plots, which are often set up for short periods. In addition, CSA technologies are context-specific and so might be the approaches used to promote CSA. In Nwoya District, for example, households as well as villages tend to be geographically quite far from each other. In such cases, farmers (in a previous and related study) indicate that distance to the plot was the main reason why they were not actively participating in the demonstration plots (Shikuku et al. 2015). Such farmers often demanded the reimbursement of transport and refreshments costs during training, without which they were unwilling to actively learn.

Household size had a positive effect on the adoption of stress tolerant varieties. This is plausible because a greater number of household members means there are more people available to provide the intensive labour that comes with the adoption of new technologies. This is in agreement with Adepoju and Obayelu (2013) who reported that household size was an important factor in determining the type of livelihood strategies adopted. The significance of the

Table 15.2 Determinants of adoption of improved varieties

Variable	ME	SE	*p*-value
Household characteristics variables			
Household size (count)	0.0130[a]	0.0055	0.018
Sex of the household head (0/1)	−0.1248	0.0058	0.332
Natural log of age of the household head (years)	0.0449	0.0450	0.366
Literacy index (index)	0.1131	0.0908	0.213
Natural log of number of years in the village (years)	0.6283[a]	0.0300	0.041
Household wealth characteristics			
Asset index (index)	0.1955[a]	0.5058	0.018
Institutional and access related variables			
Access to government extension (0/1)	0.0642	0.0331	0.053
Access to farmer organisation (0/1)	−0.0532	0.0304	0.080
Access to NGO information (0/1)	0.1032[b]	0.0326	0.002
Access to demonstration plots information (0/1)	−0.1402[b]	0.0406	0.001
Access to credit (0/1)	−0.0556	0.0305	0.068
Member of an agricultural group (0/1)	0.0620	0.0493	0.209
Gender roles and personal values			
Training on personal values (0/1)	0.0186	0.0289	0.522
Perceptions of climate change and associated risks			
Noticed change in climate change (0/1)	−0.0809	0.0895	0.366
Perceiving likely change in climate change in future (0/1)	0.0931[b]	0.0280	0.001
Perceived reduced current rainfall (0/1)	0.0324	0.0289	0.262
Climatic shocks			
Experienced floods (0/1)	0.0515	0.0301	0.088
Experienced drought (0/1)	0.0143	0.0321	0.656
Food security variable			
Number of hunger months (count)	−0.0107	0.0154	0.485
Sub-county fixed effects			
Alero	0.2840[b]	0.0390	0.000
Anaka	0.0668	0.6209	0.282
Koch Goma	0.4524[b]	0.0382	0.000
Purongo	–	–	–
Constant	−2.8014[b]	0.7342	0.000
Log likelihood	−608.5264		
LR chi2 (22)	201.55		
Prob>chi2	0.000		
Pseudo-R^2	0.1421		

ME (marginal effects), *SE* standard error, *p-value* probability value
[a]Significance at 5%
[b]Significance at 1%

number of years of residence in the village meant that a farmer who stayed in the village for more than 1 year was 7.19% more likely to adopt new seed varieties. This could be attributed to strong social networks along with a greater number of years' farming experience in the village. Simtowe et al. (2012) also reported that farmers who'd lived in their village for a longer time were more likely to be exposed to the availability of improved pigeon pea varieties, unlike their counterparts, because of the social capital in information sharing. Asset index was used as a proxy for estimating the wealth of the farmers. Farmers with more assets are likely to have more money, equipment and materials that will aid easy access to new technologies. The results in Table 15.2 show that a 1% increase in the asset index increases the probability of adopting new varieties by 19.55%. This is in line with Tesfaye et al. (2016), where the authors reported that asset ownership was positively correlated with the adoption of improved wheat varieties in rural Ethiopia. Lastly, the significance of the variable 'noticed change in climate change' indicates that farmers who had noticed change were 10.98% more likely to adopt improved varieties. We can argue that such farmers know about the negative impacts of climate change and would, therefore, prefer to adopt technologies that will increase production and make them food secure, unlike their counterparts. Asayehegn et al. (2017) similarly argue that farmers who were aware of climate change were more willing to implement climate adaptation measures to mitigate themselves from the dangers.

15.3.2 Estimating the Impact of Improved Varieties Adoption Decision

In the second step of PSM, we applied three different matching algorithms: nearest neighbour matching, kernel matching, and radius matching. The PSM model was used to determine the impact of the different CSA technologies on household welfare. After matching, ATE was computed. The propensity scores for both adopters and non-adopters ranged from 0 to 1. The reduced magnitude of Pseudo-R^2 as well as the statistical insignificance of the *p*-values associated with the likelihood test, justified the choice of PSM model for our data. In addition, as shown in Table 15.3, there was a substantial reduction in bias after matching which is important in examining balancing powers of estimation. The reduction in the value and the insignificance of Pseudo-R^2 after matching indicated that there were no significant differences in the values of the independent variables for the adopters and non-adopters of stress-tolerant varieties after matching. Likewise, the *p*-values of the likelihood ratio test were insignificant after matching. Lastly, the mean and median bias were all below 20% justifying the choice of PSM model in this study.

Table 15.3 Statistical tests to justify matching

Matching method	Pseudo-R^2	Likelihood ratio Chi^2	p>Chi^2	Mean bias	Median bias
Before matching	0.1421	168.75	0.000	28.0	24.0
Radius matching	0.002	2.54	1.000	2.2	1.4
Kernel Based matching	0.014	17.16	0.309	5.9	6.4
Nearest neighbour matching	0.003	3.32	0.999	2.6	2.4

Table 15.4 ATT for household welfare

Matching algorithm	Treated (ATT)	Controls (ATT)	Difference	SE	t-stat	ATE
RM	1102.87	753.15	349.72	097.23	3.60	358.41
KBM	1102.87	808.89	293.98	122.94	2.39	319.77
NNM	1102.87	899.04	203.83	110.27	1.85	258.04

Note: The amount is stated in USD
RM radius matching, *KM* kernel based matching, *NNM* nearest neighbour matching, *SE* standard error, *t-stat* t-statistics

The estimates of the average adoption effects from all the three matching algorithms are presented in Table 15.4. The results showed that the adoption of stress tolerant varieties has a positive impact on household welfare. Results from all the three matching algorithms were consistent. As shown in Table 15.4, per capita crop income was higher for the adopters than the matched non-adopters. From the findings—assuming that the two groups were matched on the equality of their propensity score—we can infer that the difference in the household per capita income results from adopting stress tolerant varieties.

15.3.3 Sensitivity Tests for Estimated Average Treatment Effects

Statistically, it is important to test the reliability of the estimated values of ATT and ATE (Becker and Caliendo 2007). This helps the researcher to examine the sensitivity of the estimated treatment effect to small deviations in the propensity scores. Doing so also acts as a check on the quality of the comparison group. Due to the limitations of observed data, such as bias creation, sensitivity analysis helps in checking if the unobserved variation has a significant effect on the estimated values of ATT and ATE. As shown in Table 15.5, the significance level is unaffected even if gamma values are increased by threefold. This clearly shows that the estimated values of ATT and ATE will not change to any external deviation.

Table 15.5 Sensitivity analysis with Rosenbaum bounds

Gamma	sig+	sig–
1	0.093194	0.093194
1.25	0.757402	0.000394
1.5	0.990693	2.20E-07
1.75	0.999918	3.70E-11
2	1	2.80E-15
2.25	1	0
2.5	1	0
2.75	1	0
3	1	0

15.4 Implications for Development

This study assessed the drivers behind the adoption of stress-tolerant varieties and their impact on farmers' welfare. The results showed that household size, access to agricultural information from NGOs, perception of future changes in climate, number of years' resident in the village and asset index all have a positive influence on the propensity to adopt stress-tolerant varieties. Further results show that stress-tolerant varieties have the potential of increasing net crop income within a range of USD 500–864 per hectare per year, corresponding to an 18–32% increase. Our empirical results suggest the need to implement a bundled solution in scaling up the adoption of stress-tolerant varieties. Specifically, a bundled solution that includes the strengthened capacity of households to own farm assets and increased access to agricultural and weather information (relying on pathways reinforced by trust and social networks) can be effective for adaptation to climatic risks in northern Uganda. The findings support the view that, in a similar way, CSA interventions are context-specific as are the pathways for scaling up the adoption of these interventions. Overall, in harmony with existing literature, the adoption of stress-tolerant varieties as CSA technologies can be a corridor to improving the welfare of farm households in northern Uganda.

Acknowledgements This work was carried out by the International Center for Tropical Agriculture as part of the CGIAR Research Program on Climate Change, Agriculture and Food Security. The project, Increasing Food Security and Farming System Resilience in East Africa through Wide-Scale Adoption of Climate-Smart Agricultural Practices, is funded with support from the International Fund for Agriculture Development.

References

Adepoju AO, Obayelu OA (2013) Livelihood diversification and welfare of rural households in Ondo State. Nigeria J Dev Agricult Econ 5(12):482–489

Asayehegn GK, Temple L, Sanchez B et al (2017) Perception of climate change and farm level adaptation choices in central Kenya. Cahiers Agricult 26(2):1–11

Becker SO, Caliendo M (2007) Mhbounds-sensitivity analysis for average treatment effects. http://ftp.iza.org/dp2542.pdf. Accessed 23 Feb 2018

Bonabana-Wabbi J (2002) Assessing factors affecting adoption of agricultural technologiese: the case of Integrated Pest Management (IPM) in Kumi District, Eastern Uganda (Masters dissertation, Virginia Tech). https://vtechworks.lib.vt.edu/handle/10919/36266. Accessed 23 Feb 2018

Brüssow K, Faße A, Grote U (2017) Adopting climate-smart strategies and their implications for food security. http://www.tropentag.de/2015/abstracts/full/584.pdf. Accessed 23 Feb 2018

Dieterich C, Huang A, Thomas AH (2016) Women's opportunities and challenges in sub-Saharan African job markets. https://www.imf.org/external/pubs/ft/wp/2016/wp16118.pdf. Accessed 23 Feb 2018

Food and Agriculture Organization (2013) Sourcebook on climate smart agriculture, forestry and fisheries. FAO, Rome. http://www.fao.org/climatechange/374910c425f2caa2f5e6f3b9162d39c8507fa3.pdf. Accessed 23 Feb 2018

Jellis GJ (2009) Crop plant resistance to biotic and abiotic factors: current potential and future demands. In: Proceedings of the 3rd International Symposium on Plant Protection and Plant Health in Europe, Julius Kühn-Institut, Berlin-Dahlem, Germany, 14–16 May 2009, pp 15–21

Joshua Udoh E, Titus Omonona B (2008) Improved rice variety adoption and its welfare impact on rural farming households in Akwa Ibom State of Nigeria. J New Seeds 9(2):156–173

Khatri–Chhetri A, Aggarwal PK, Joshi PK et al (2017) Farmers' prioritization of climate-smart agriculture (CSA) technologies. Agric Syst 151:184–191

Makate C, Wang R, Makate M et al (2016) Crop diversification and livelihoods of smallholder farmers in Zimbabwe: adaptive management for environmental change. Springerplus 5(1):1–18

Manda J, Alene AD, Gardebroek C et al (2016) Adoption and impacts of sustainable agricultural practices on maize yields and incomes: evidence from rural Zambia. J Agric Econ 67(1):130–153

Markandya A, Cabot C, Beucher O (2015) Economic assessment of the impacts of climate change in Uganda. https://cdkn.org/wp-content/uploads/2015/12/Uganda_CC-economics_Final-Report2.pdf. Accessed 23 Feb 2018

Msowoya K, Madani K, Davtalab R et al (2016) Climate change impacts on maize production in the warm heart of Africa. Water Resour Manag 30(14):5299–5312

Mwongera C, Shikuku KM, Twyman J, Winowiecki L, Ampaire A, Koningstein M, Twomlow S (2014) Rapid rural appraisal report of Northern Uganda. International Center for Tropical Agriculture (CIAT), CGIAR Research Program on Climate Change, Agriculture and Food Security (CCAFS). https://cgspace.cgiar.org/rest/bitstreams/32075/retrieve. Accessed 23 Feb 2018

Mwungu CM, Mwongera C, Shikuku KM et al (2017) Survey data of intra-household decision making and smallholder agricultural production in northern Uganda and southern Tanzania. Data Brief 14:302–306

Nyasimi M, Kimeli P, Sayula G et al (2017) Adoption and dissemination pathways for climate-smart agriculture technologies and practices for climate-resilient livelihoods in Lushoto, north-east Tanzania. Climate 5(3):63

Odame H, Kimenye L, Kabutha C et al (2013) Why the low adoption of agricultural technology in eastern and central Africa? In: Selected paper prepared for presentation at the Association for Strengthening Agricultural Research in Eastern and Central Africa conference, Entebbe, Uganda, 21 Oct 2011, pp 21–23

Shiferaw B, Tesfaye K, Kassie M et al (2014) Managing vulnerability to drought and enhancing livelihood resilience in sub-Saharan Africa: technological, institutional and policy options. Weather Clim Extremes 3:67–79

Shikuku K, Mwongera C, Winowiecki L (2015) Understanding farmers' indicators in climate-smart agriculture prioritization in Nwoya District, northern Uganda. Centro Internacional de Agricultura Tropical, Cali (Publicación CIAT No. 412). https://cgspace.cgiar.org/rest/bitstreams/65373/retrieve. Accessed 23 Feb 2018

Simtowe F, Muange E, Munyua B et al (2012) Technology awareness and adoption: the case of improved pigeon pea varieties in Kenya. In: Selected paper prepared for presentation at the International Association of Agricultural Economists Triennial Conference, Foz do Iguaçu, Brazil, 18–24 Aug, 2012 pp 18–24

Tesfaye S, Bedada B, Mesay Y (2016) Impact of improved wheat technology adoption on productivity and income in Ethiopia. Afr Crop Sci J 24(s1):127–135

16

CSA and the Impact of Tradition and Religion

Julia Davies, Dian Spear, Angela Chappel, Nivedita Joshi, Cecile Togarepi, and Irene Kunamwene

16.1 Introduction

Rural communities in the semi-arid areas of southern Africa are particularly vulnerable to climate change because they depend predominantly on rain-fed agriculture to support their livelihoods. In addition, a number of non-climatic issues—including poverty, inequality, education deficits and poor governance—render communities in these areas even more susceptible to climate-related problems. Climate-smart agriculture (CSA) has the potential to increase the resilience of these vulnerable communities because it integrates environmental management and climate-change adaptation with social and economic sustainability (Chioreso and Munyayi 2015).

The implementation of CSA, however, has proven difficult in southern Africa. Previous studies have shown that key barriers include inadequate policy and insufficient access to finances, technology, land and human resources (Barnard et al. 2015; Sibanda et al. 2017; Williams et al. 2015). Less is understood, however, about how cultural barriers—norms, values, historical legacies, religious and traditional beliefs and social identities—affect the adoption of CSA (Thomalla et al. 2015). This study considers the role played by devotion (religious faith and belief) and respect for tradition (preservation of time-honoured customs) (Schwartz 1992) in

J. Davies (✉) · D. Spear · I. Kunamwene
African Climate and Development Initiative (ACDI), University of Cape Town, Rondebosch, South Africa
e-mail: julia.davies@uct.ac.za

A. Chappel · N. Joshi
Environmental and Geographical Sciences, University of Cape Town, Cape Town, South Africa

C. Togarepi
Department of Agricultural Economics and Extension, Old Administration Block, University of Namibia Ogongo Campus, Ogongo, Namibia

CSA adoption in Namibia. Research involved a review of existing literature and the collection of empirical data through 60 semi-structured interviews. These interviews were conducted with farmers in the semi-arid north-central region[1] of Namibia in July 2017 as part of the Adaptation at Scale in Semi-Arid Regions (ASSAR)[2] research project.

The agricultural sector contributes to only 3.7% of the country's gross domestic product, but small-scale and subsistence crop and livestock farming remains an important aspect of livelihood security in rural areas, where more than half of the national population resides (MAWF 2015). Farming, moreover, is fundamental to the cultural identity of Namibian people, particularly in the Oshiwambo culture. Increasing the resilience of Namibia's agricultural sector is thus key not only for alleviating poverty and food insecurity but also for preserving local socio-cultural identities.

In recent years the government of Namibia has developed policies and plans to enhance agricultural growth, improve natural resource management and upscale climate-change interventions.[3] In 2015 a draft national CSA programme was adopted (MET 2015). Farmers, with support from the government, non-governmental organisations and national research institutions, have begun implementing CSA practices. These include initiatives such as drip irrigation, planting early-maturing mahangu (pearl millet), engaging in small-scale rice farming, using draught-animal power and farm tools (instead of tractors) for ripping fields, and selecting more hardy, drought-tolerant cattle breeds.

Some farmers, however, have been slow to adopt such approaches. Our study found that cultural factors—especially religious belief, reliance on traditional knowledge and the symbolic significance of certain agricultural practices—have played a role in the low uptake of CSA in Namibia. We argue, however, that these barriers can be turned into opportunities: By working with rather than against religious and traditional value systems, extension workers could promote the adoption of CSA and thereby help to reduce the impacts of climate change and variability. These benefits could be achieved through one or more of the following avenues: (i) positioning religious and traditional leaders as climate change champions; (ii) integrating scientific information with traditional knowledge; and (iii) framing CSA in such a way that it complements rather than conflicts with religious beliefs or traditional practices.

[1] Interviews were conducted in the Onesi Constituency, which falls within the Omusati Region. The specific study sites were the Okathitukeengombe, Oshihau and Omaenene villages.

[2] The ASSAR project (2014–2018) aims to deepen the understanding of climate vulnerability and adaptation in semi-arid regions of Africa and Asia, where millions of people are highly vulnerable to climate-related impacts and risks. See http://www.assar.uct.ac.za/

[3] Some key policies relevant for CSA implementation and scale-out in Namibia include: Namibia Vision 2030; National Development Plan 4 (2012/13–2016/17); National Agricultural Policy (2015); National Disaster Risk Management Policy (2009); National Drought Policy and Strategy (1997) and National Climate Change Policy (2011) (as cited in MET 2015).

16.2 Cultural Barriers to CSA Adoption

16.2.1 Religious Faith and Belief

Religious faith has proved a hindrance to the embrace of scientific climate forecasts. In Namibia, seasonal climate forecasts (SCFs) are produced by the national meteorological services and then disseminated to farmers via agricultural extension officers and radio broadcasts, the latter of which reach 90% of the population. In addition, a new online platform (http://www.lisa.com.na) allows farmers to interact with experts via text message. Some farmers in northern Namibia, however, do not take advantage of this information. Reasons for the low levels of uptake of SCF by farmers in Africa include problems of downscaling, a limited capacity among farmers to understand forecast data, and mismatches between the information provided by SCF and what is perceived as useful on the ground (Luseno et al. 2016; Singh et al. 2017; Ziervogel and Calder 2003; Ziervogel and Opere 2010). Another major reason, however, is the Christian belief, prevalent in northern Namibia, that rainfall and crop productivity cannot be predicted but are solely dependent on the will of God (also see Angula et al. 2016; Selato 2017; Spear et al. 2015). As explained by an interview respondent, "I cannot tell if the drought is going to worsen or not—that is God's work" (Farmer 1). Another explained that "only God knows what is in the future" (Farmer 2). Strong religious beliefs can make people accept their circumstances rather than use forecasts to inform their practices. From this perspective, any anomalies in climate are perceived as punishment for people displeasing God in some way: "maybe God is angry because of the things people are doing—that is why we are not getting enough rainfall" (Farmer 3).

16.2.2 Symbolic Significance of Agricultural Practice

While livestock rearing contributes to livelihood security for farmers in southern Africa in non-drought times, it can increase farmers' vulnerability during extended periods of low rainfall. Overstocking places more pressure on food and water resources and may cause increased land degradation. Drought can also cause animals to die, a major financial loss for farmers. One farmer explained, "we used to use animals to plough the field but those animals died in the drought last year and the year before" (Farmer 10).

But even farmers who have previously experienced losses still choose to keep livestock in the face of drought. The reasons for this reluctance are many: low market prices, collapse of markets, a lack of access to markets and a poor understanding of the reasons to sell (Speranza 2010; Togarepi et al. 2016). In some cases, however, a failure to sell is linked to the symbolic significance of livestock (Doran et al. 1979; Hegga et al. 2016; Stroebel et al. 2008). In Namibia, cattle are often perceived as a direct measure of affluence, status, prestige and security. Livestock rearing is used as an informal insurance system, and some farmers see cattle as more valuable than money itself. One farmer in northern Namibia explained, "I believe that cows help to bail me out of my problems—that's why I will continue to keep them" (Farmer 6). Another said, "I believe that livestock is a part of our culture and by owning them, I am definitely better than someone who doesn't have them. If I sell my cows, I feel like I am cheating on my culture" (Farmer 7). The cultural identity of Ovambo men is closely tied to the ownership of livestock: "A man is his cattle" (Farmer 8); "My parents told me that as a man you should own livestock. One of the definitions of a man is having livestock. I fully agree with them" (Farmer 9). These strong cultural attachments make farmers reluctant to sell stock, even when forecasts call for severe drought.

16.2.3 Traditional Agricultural Knowledge

The lack of uptake of SCF is sometimes due to a tendency to favour traditional weather knowledge over scientific climate forecasts (Newsham and Thomas 2011; Mogotsi et al. 2011; Selato 2017; Jiri et al. 2016). In Namibia, farmers plant their crops according to traditional calendar dates (December 15–February 15) even if the SCF indicates an earlier or later onset of the rainfall season. A farmer describes her strategy: "I plough the first set of crops at the end of December … the second set mid-January … and the third set at the end January/beginning of February. … It should rain by the time of the second or third ploughing session" (Farmer 4). Traditional forecasting methods are also still widely used in southern Africa. In Botswana, for example, farmers observe the flowering of trees, the position of stars and the persistence of "pregnant" clouds to determine how much rain the season is likely to bring (Mogotsi et al. 2011; Selato 2017). Similarly, a stakeholder from Namibia explained that "according to indigenous knowledge, we can predict 'yes, it [drought] might worsen'" (Farmer 5). While traditional forecasting methods have indeed proven valuable, they are also becoming less accurate due to climate change (Angula et al. 2016), rendering SCF more important than ever.

Uptake of new farming technologies—including new crops or cultivars—can also be stymied by adherence to traditional norms. In northern Namibia many farmers refuse to adopt new practices even if they recognize that the climate is changing: "We haven't changed the crops that we grow. We don't change them at all even if it is dry or there is good rainfall" (Farmer 15). As with traditional forecasting methods, farmers are reluctant to change practices that have been passed down through generations: "We will keep farming the same way because in the Oshiwambo culture we don't like to change tradition" (Farmer 16). One farmer explained, "I fear new practices won't work and my yield will be even worse" (Farmer 17). In north-central Namibia, farmers have a strong cultural attachment to mahangu. While their preferred variety of millet is generally hardy and well-adapted to the semi-arid conditions of the region, productivity has declined over the last two decades (CPP

2012). The Namibian government has introduced more resilient seed varieties, but many farmers have rejected them because of their slightly different appearance and taste. One explained, "We will keep farming like this forever. It is tradition so it has to continue for generations. Change to what? Mahangu is our main meal, so there is no way we can change it" (Farmer 18). Some farmers are open to trying new crop varieties, but as a supplement rather than a replacement for traditional versions. Asked about adopting new crops, one respondent answered, "maybe, but we must still grow mahangu because it is part of our culture" (Farmer 16).

It is important to note that the embrace of tradition has a generational component. Farmer 3 explained, "I am not willing to use new practices because I am old and maybe I won't carry them out correctly," and farmer 2 said, "I am very old now but if I was young and energetic I would try new farming practices." Whereas older farmers are unlikely to market their livestock even in the face of drought, younger farmers often are more flexible. One man explained, "I'm an elder and I can't stay without livestock since it is a part of my culture" (Farmer 12). But a younger farmer claimed that "the culture is there, but it will not stop me from selling my livestock" (Farmer 11).

16.3 Working with Religious and Traditional Belief Systems to Enable Adaptation

Although religious and traditional beliefs sometimes prevent farmers from making more adaptive decisions, we argue that these cultural factors should not be viewed simply as barriers. These belief systems—precisely because they play such an important role in agricultural decision-making—should be viewed as an opportunity through which to catalyse the dissemination of CSA. We identify three possible avenues through which to do this.

16.3.1 Positioning Religious and Traditional Leaders as 'Champions'

The climate change literature increasingly acknowledges the role of "champions" or "lead farmers" in encouraging adaptation (Conservation Agriculture Task Force for Zimababwe 2008; Davies and Ziervogel 2017; Roberts 2008). However, there have been fewer attempts to recruit religious or traditional elites to play this role, despite the fact that many farmers tend to have greater confidence in information that comes from such figures. In Botswana, for example, the tradition of *Letsema* means that farmers must wait for permission from the village chief before they begin planting or harvesting. Even if a seasonal forecast indicates an earlier onset of rainfall, farmers will wait for word from the chief (Selato 2017). In this context, it would be essential for the chief to champion adaptation. Chishakwe et al. (2012) found that fostering relationships of trust with traditional leaders was essential for establishing local ownership of community-based adaptation projects in the Mayuni Conservancy in Namibia. Similarly, case studies from Malawi and Zambia highlight the role of traditional leaders in building adaptive capacity in their local communities (Reid et al. 2010).

Religious leaders could play a similar role. The staunch religious nature of much of the Ovambo society means that church leaders have substantial influence upon agricultural decision-making. They therefore could play an influential role in promoting CSA. Champions may also take the form of a church organisation rather than an individual. The Southern African Faith Communities' Environment Institute (safcei.org) is a regional multi-faith network that promotes religious education and teaching about the environment and climate change. In Zimbabwe, Foundations for Farming (www.fffzimbabwe.org) uses religious narratives to promote soil and water conservation practices such as no-till, mulching and crop rotation (Kassam et al. 2014). Similarly, the Green Anglicans in Swaziland aim "to fulfil God's call to be Earthkeepers and to care for Creation" (www.greenanglicans.org). Although such programmes can be effective, it is important that the approaches they promote are relevant to the local environmental and socio-economic contexts.

16.3.2 Integrating Traditional and Scientific Knowledge

Western science, though vital to creating resilience to climate change, still has much to learn from traditional bodies of ecological knowledge (Berkes et al. 2000; Mazzocchi 2006). For example, while SCF can provide information about climate change on broader spatial and temporal scales, traditional forecasting methods may help to counter downscaling issues associated with climate models, as they provide information that is more locally relevant. SCF could therefore be used to complement traditional understandings of risk and enable farmers to make more informed decisions (Ambani and Percy 2014; Singh et al. 2017; Ziervogel and Opere 2010).

Such integration is difficult and requires robust engagement between communities, experts and government (Kniveton et al. 2014; Singh et al. 2017; Thomalla et al. 2015). In Namibia the system of "indigenous land units" (ILU) has been used for decades to help farmers classify local environmental conditions and thus determine how specific areas should be used (Verlinden and Dayot 2005). This system has become less viable not only because of climate change but also because increasing densities of both human and livestock populations have added greater pressure to the environment. Despite its deficiencies, "for better or for worse, the land unit system is what farmers use to make farming decisions" (Newsham and Thomas 2011). Those wishing to successfully promote CSA approaches will need to take into account traditional methods such as ILU. A key component of this is working through traditional or religious leaders, who can advise on what changes may be practical and acceptable in the local context.

16.3.3 Changing the Framing of CSA

Corner et al. (2014) highlight how, in countries such as Uganda, climate change is increasingly being communicated in ways that resonate with religious or indigenous values and beliefs. In Namibia such an approach could be particularly important when working with indigenous San communities, whose cultural identities continue to be linked strongly to the semi-nomadic, hunter-gatherer lifestyles of their ancestors. Targeted attempts by government to promote farming among the San have had limited success (Dieckmann et al. 2014). As a result, the San are today the most marginalized population group in Namibia, and many have become increasingly dependent on government aid and piecework. If adaptation projects are not designed and implemented in ways that consider their cultural traditions and beliefs, then the San are likely to become increasingly vulnerable to climate change (Dieckmann et al. 2013).

Framing information in ways that honor traditional beliefs can improve the uptake of CSA approaches. In Namibia, for example, livestock plays an important role in wedding ceremonies, funerals, communal feasts and other social and cultural events (Ziervogel 2016). Increased livestock mortality because of drought limits people's capacity to engage in such crucial activities. As a result, advice to sell livestock before a predicted drought could be framed not only as a sound financial and ecological decision but also as a way to preserve a farmer's ability to participate in traditional culture. New practices must be promoted in ways that appeal to the cultural vulnerability of communities, because "people can take extraordinary measures to protect that which they view as sacred" (Sachdeva 2016). In addition, motivating agricultural adaptation by promoting the economic benefits of CSA might prove useful among poor farming communities. Evidence from Kenya has shown that farmers who use traditional forecasting methods in conjunction with SCF see greater returns than those who use only traditional methods (Ambani and Percy 2014). Working with these value systems could enable farmers to take up practices that are more forward-looking, even if they are not sold to farmers as such.

16.4 Implications for Development

The values and belief systems of local communities have played a significant role in the uptake of CSA in southern Africa. In Namibia, it is clear that religion and tradition have prevented some farmers from taking steps to become more climate resilient. We argue, however, that it is important to work *with* religious and traditional value systems. Because these systems play such a pivotal role in agricultural decision-making, they provide a key opportunity through which to promote the dissemination and uptake of climate change information in general and CSA in particular.

Mobilizing these approaches, however, will be difficult. Future research should consider empirically testing the application of the three avenues—positioning traditional leaders as agricultural champions, integrating traditional and scientific knowledge, and reframing CSA—suggested here. For example, it may be important for researchers to consider why, to date, there has been limited evidence of efforts to promote collaboration between agricultural extension services and religious or traditional leaders. Or, in cases where religious groups do promote CSA, studies could perhaps determine the degree to which this is done in conjunction with sound technical advice and appropriate technologies that are readily available to the congregation. In addition, future research agendas might benefit from testing novel extension approaches in neighbouring districts, revising the type of training provided to extension workers or, in cases where extension services are inaccessible, consider how individual "lead" farmers within a community may be trained in adaptation techniques and encouraged to disseminate these innovations to the broader community.

In carrying out such studies, it is essential to emphasize that there is no single solution, and future research should therefore promote flexibility and an awareness of local cultural, environmental and socio-economic contexts. Different types of advice, or alternative framing devices, may need to be adopted when communicating information to diverse cultural groups, or to older versus younger farmers. For instance, encouraging rational experimentation and innovation is likely to be more appropriate among young farmers who are more open to new ideas, whereas older farmers would perhaps respond better to advice that considers the importance of maintaining livestock as a source of wealth and prestige. In this way, religion and tradition can play a role in easing the transition to new information and practices. Paying proper respect to closely held traditional beliefs can help improve the likelihood of CSA measures being adopted, and therefore contribute to reducing the impacts of climate change and variability on the agricultural sector in southern Africa.

References

Ambani M, Percy F (2014) Facing uncertainty: the value of climate information for adaptation, risk reduction and resilience in Africa. CARE International, Nairobi. Retrieved from www.careclimatechange.org/adaptation-initiatives/alp

Angula MN, Ntombela KP, Samuels MI, Swarts M, Cupido C, Haimbili NE, Menjono-Katjizeu ME & Hoabes M (2016) Understanding pastoralist's knowledge of climate change and variability in Arid Namibia and South Africa. In Proceedings of centenary conference of the Society of South African Geographers: 25–28 Sept 2016, Stellenbosch. Stellenbosch: Research Gate

Barnard J, Manyire H, Tambi E, Bangali S (2015) Barriers to scaling up/out climate smart agriculture and strategies to enhance adoption in Africa. Forum for Agricultural Research in Africa (FARA), Accra

Berkes F, Colding J, Folke C (2000) Rediscovery of traditional ecological knowledge as adaptive management. Ecol Appl 10(5):1251–1262

Chioreso E, Munyayi R (2015) Climate smart agriculture. Desert Research Foundation of Namibia, Windhoek. Retrieved from http://www.fao.org/docrep/018/i3325e/i3325e00.htm

Chishakwe N, Murray L, Chambwera M (2012) Building climate change adaptation on community experiences. Pubs.Iied.Org. Retrieved from http://pubs.iied.org/pdfs/10030IIED.pdf

Conservation Agriculture Task Force for Zimababwe, C (2008) The conservation agriculture toolbox for Zimbabwe. Development. FAO. Retrieved from https://www.fsnnetwork.org/conservation-agriculture-toolbox-zimbabwe

Corner A, Markowitz E, Pidgeon N (2014) Public engagement with climate change: the role of human values. Wiley Interdiscip Rev Clim Chang 5(3):411–422. https://doi.org/10.1002/wcc.269

Country Pilot Partnership for Integrated Sustainable Land Management (CPP) Programme (2012) Improved Pearl Millet Production in an arid environemnt for the improvement of livelihoods and reversal of land degradation: a case study from Omusati Region, Namibia. Ministry of Environment and Tourism; the Ministry of Agriculture, Water and Forestry; and the United Nations Development Programme, Outapi

Davies J, Ziervogel G (2017) "Learning by Doing" – lessons from the co-production of three South African municipal climate change adaptation plans. In: Fünfgeld H, Maloney S, Granberg M (eds) Local action on climate change: opportunities and constraints. Routledge, Stockholm, pp 53–71

Dieckmann U, Odendaal W, Tarr J, Schreij A (2013) Indigenous peoples and climate change in Africa: report on case studies of Namibia's Topnaar and Hai//om communities. Legal Assistance Centre, Windhoek

Dieckmann U, Thiem M, Dirkx E, Hays J (2014) "Scraping the Pot": San in Namibia two decades after independence. Legal Assistance Centre and Desert Research Foundation of Namibia, Windhoek

Doran MH, Low ARC, Kemp RL (1979) Cattle as a store of wealth in Swaziland: implications for livestock development and overgrazing in eastern and southern Africa. Am J Agric Econ 61(1):41–47 https://doi.org/10.2307/1239498

Government of the Republic of Namibia (2009) National Disaster Risk Management Policy. Office of the Prime Minister, Directorate: Disaster Risk Management. Windhoek

Government of the Republic of Namibia (1997) National Drought Policy and Strategy. National Drought Task Force, Windhoek

Hegga S, Ziervogel G, Angula M, Spear D, Nyamwanza A, Ndeunyema E, Kunamwene I, Togarep C, Morchain D (2016) Vulnerability and risk assessment in Omusati Region in Namibia: fostering people-centred adaptation to climate change. Adaptation at Scale in Semi-Arid Regions (ASSAR)

Jiri O, Mafongoya PL, Mubaya C, Mafongoya O (2016) Seasonal climate prediction and adaptation using indigenous knowledge systems in agriculture systems in Southern Africa: a review. J Agric Sci 8(5):156. https://doi.org/10.5539/jas.v8n5p156

Kassam A, Derpsch R, Friedrich T (2014) Global achievements in soil and water conservation: the case of conservation agriculture. Int Soil Water Conserv Res 2(1):5–13. https://doi.org/10.1016/S2095-6339(15)30009-5

Kniveton D, Visman E, Tall A, Diop M, Ewbank R (2014) Dealing with uncertainty: integrating local and scientific knowledge of the climate and weather. Disasters 39(1):35–53. https://doi.org/10.1111/disa.12108

Luseno W, McPeak JG, Barrett CB, Little PD, Gebru G (2016) Assessing the value of climate forecast information for Pastoralists: evidence from southern Ethiopia and northern Kenya. Int Res Inst Clim Prediction, (February 2003). https://doi.org/10.1016/S0305-750X(03)00113-X

MAWF (2015) Namibia agriculture policy. Government Gazette of the Republic of Namibia. Ministry of Agriculture, Water and Forestry, Windhoek

Mazzocchi F (2006) Western science and traditional knowledge. EMBO Reports (European Molecular Biology Organization) 7(5):463–466. https://doi.org/10.1038/sj.embor.7400693

Ministry of Environment and Tourism (MET) (2011) National Climate Change Policy for Namibia. United Nations Development Programme, Namibia

Ministry of Environment and Tourism (2015) Republic of Namibia Country climate smart agriculture programme 2015–2030. Ministry of Environment and Tourism and Ministry of Agriculture, Water and Forestry, Windhoek

Mogotsi K, Moroka AB, Sitang O, Chibua R (2011) Seasonal precipitation forecasts: agro-ecological knowledge among rural Kalahari communities. Afr J Agric Res 6(4):916–922. https://doi.org/10.5897/AJAR10.756

Newsham AJ, Thomas DSG (2011) Knowing, farming and climate change adaptation in North-Central Namibia. Glob Environ Chang 21(2):761–770. https://doi.org/10.1016/j.gloenvcha.2010.12.003

Reid H, Huq S, Murray L (2010) Community champions: adapting to climate challenges. International Institute for Environment and Development, London

Roberts D (2008) Thinking globally, acting locally – institutionalizing climate change at the local government level in Durban, South Africa. Environ Urban 20(2):521–537. https://doi.org/10.1177/0956247808096126

Sachdeva S (2016) Religious identity, beliefs, and views about climate change. Oxford Res Encycl Clim Sci (1). https://doi.org/10.1093/acrefore/9780190228620.013.335

Schwartz SH (1992) Universals in the content and structure of values: theoretical advances and empirical tests in 20 countries. Adv Exp Soc Psychol 25(C):1–65. https://doi.org/10.1016/S0065-2601(08)60281-6

Selato J (2017) Credibility and scale as barriers to the uptake and use of seasonal climate forecasts in Bobirwa Sub-District, Botswana. Department of Environmental and Geographical Sciences University of Cape Town 1, Cape Town

Sibanda LM, Mwamakamba SN, Mentz M, Mthunzi T (eds) (2017) Policies and practices for climate-smart agriculture in Sub-Saharan Africa: a comparative assessment of challenges and opportunities across 15 countries. Food, Agriculture and Natural Resource Policy Analysis Network (FANRPAN), Pretoria

Singh C, Daron J, Bazaz A, Ziervogel G, Spear D, Krishnaswamy J, Zaroug M, Kituyi E (2017) The utility of weather and climate information for adaptation decision-making: current uses and future prospects in Africa and India. Clim Dev:1–17. https://doi.org/10.1080/17565529.2017.1318744

Spear D, Baudoin MA, Hegga S, Zaroug M, Okeyo AE Haimbili E (2015) Vulnerability and adaptation to climate change in the semi-arid Regions of Southern Africa, 111

Speranza CI (2010) Drought coping and adaptation strategies: understanding adaptations to climate change in agro-pastoral livestock production in makueni district, Kenya. Eur J Dev Res 22(5):623–642. https://doi.org/10.1057/ejdr.2010.39

Stroebel A, Swanepoel FJC, Nthakheni ND, Nesamvuni AE, Taylor G (2008) Benefits obtained from cattle by smallholder farmers: a case study of Limpopo Province, South Africa. Aust J Exp Agric 48(7):825–828. https://doi.org/10.1071/EA08058

Thomalla F, Smith R, Schipper ELF (2015) Cultural aspects of risk to environmental changes and hazards a review of perspectives. In: Disaster's impact on livelihood and cultural survival, (1), 3–18. https://doi.org/10.1201/b18233-3

Togarepi C, Thomas B, Kankono M (2016) Cattle marketing constraints and opportunities in north-central communal areas of Namibia, Ohangwena region. Livest Res Rural Dev 28:7

Verlinden A, Dayot B (2005) A comparison between indigenous environmental knowledge and a conventional vegetation analysis in north central Namibia. J Arid Environ 62(1):143–175. https://doi.org/10.1016/j.jaridenv.2004.11.004

Williams T, Mul M, Cofie O, Kinyangi J (2015) Climate smart agriculture in the African context. Cgspace.Cgiar.Org (October). Retrieved from https://cgspace.cgiar.org/bitstream/handle/10568/68944/Climate_Smart_Agriculture_in_the_African_Context[1].pdf?sequence=1

Ziervogel G (2016) What Africa's drought responses teaches us about climate change hotspots. Water Wheel 15(5):31–33

Ziervogel G, Calder R (2003) Climate variability and rural livelihoods: assessing the impact of seasonal climate forecasts in Lesotho. Society:403–417. https://doi.org/10.1111/j.0004-0894.2003.00190.x

Ziervogel G, Opere A (2010) Integrating meteorological and indigenous knowledge-based seasonal climate forecasts for the agricultural sector. International Development Research Centre, Ottawa, Canada. Climate Change Adaptation in Africa Learning Paper Series. Retrieved from http://web.idrc.ca/uploads/user-S/12882908321CCAA_seasonal_forecasting.pdf

Part IV
Value Chains and Climate-Resilience

17

Adoption of Agricultural Innovations and Entrepreneurship

Carlos Luis Barzola Iza, Domenico Dentoni, Martina Mordini, Prossy Isubikalu, Judith Beatrice Auma Oduol, and Onno Omta

17.1 Introduction

This chapter examines entrepreneurship as part of the broad debate surrounding when and why farmers adopt agricultural innovations, especially in the context of multi-stakeholder platforms (MSPs) (Kilelu et al. 2013; Schut et al. 2015) and similar organizations seeking to scale climate-smart agriculture (CSA) practices. Farmer entrepreneurship generally refers to a process of recombining agricultural resources innovatively to create opportunities for value creation (Shane and Venkataraman 2000; Lans et al. 2013). Entrepreneurial farmers may be among the first in a community to experiment with new practices, mobilize a previously underutilized resource, or use fresh information to build a new market for agricultural products. Such skills will be particularly crucial in the coming decades, as climate change forces farmers to adapt their agricultural systems. Recent investigations suggest that entrepreneurship may help agents respond to environmental shocks or adapt to rapidly changing market and conditions (Naudé 2010; York and Venkataraman 2010; Bruton et al. 2008; Khavul and Bruton 2013; Bruton et al. 2013). This literature,

C. L. Barzola Iza (✉)
Business Management and Organisation (BMO), Wageningen University,
Hollandseweg 1, Wageningen, 6706 KN, The Netherlands

Facultad de Ciencias de la Vida (FCV), Escuela Superior Politecnica del Litoral (ESPOL),
Campus Gustavo Galindo, Km 30.5 Via perimetral, Guayaquil, Ecuador
e-mail: carlos.barzolaiza@wur.nl

D. Dentoni · M. Mordini · O. Omta
Business Management and Organisation (BMO), Wageningen University,
Hollandseweg 1, Wageningen, 6706 KN, The Netherlands

P. Isubikalu
College of Agricultural and Environmental Science, Makerere University, Kampala, Uganda

J. B. Auma Oduol
World Agroforestry Center (ICRAF), Nairobi, Kenya

however, is largely lacking in empirical evidence. Similarly, the burgeoning literature that analyzes the adoption of CSA-related innovations (Lipper et al. 2014; Zilberman et al. 2018) has not yet considered farmer entrepreneurship as a significant influence on such processes.

Our study seeks to fill this gap in the literature by collecting evidence on the relationship between entrepreneurial mindsets and an openness to agricultural innovation. To explore why certain farmers innovate and adapt, this chapter (i) proposes an adapted measurement model for farmer entrepreneurship, and (ii) investigates quantitatively the influence of farmer entrepreneurship and farm characteristics on three types of innovation: "product innovation," which here refers to the use of new farm inputs, transformation of farm output into new products, or production according to a new quality standard; "process innovation," which involves adopting new farm practices, embracing new ways of farm organization, or putting new information into use; and "market innovation," which entails opening a new market channel.

We ground our study in interviews with farmers associated with Ugandan MSPs involved in coffee and honey value chains. Recently, MSPs have attempted to introduce CSA practices not only in coffee and honey but also in other sectors (Bomuhangi et al. 2016; Sabiiti et al. 2016). This study has implications for agents in MSPs—farmers and their representatives, researchers, value-chain partners and policy-makers—or similar multi-actor organizations who wish to promote the adoption of CSA or other novel practices. By understanding and encouraging farmer entrepreneurship, MSPs may be able to stimulate a broad range of agricultural innovation.

17.2 Methods

For this study, a survey questionnaire was completed by 152 farmers in four sub-counties (Mukoto, Namabya, Bukhofu and Namboko) of Uganda's Manafwa district. The questionnaire covered the farm's characteristics and the farmers' entrepreneurial orientation. Farm characteristics included demographics (age, gender, education), farm size and access to resources (both tangible resources such as credit, fertilizers, etc. and intangible resources such as intellectual capital). To assess farmer entrepreneurship, we selected four key dimensions of entrepreneurial orientation: innovativeness, risk-taking, proactiveness and entrepreneurial intentions. Finally, farmers' innovation was measured in line with the empirical analyses by (Wu and Pretty 2004) on product innovation, (Yang 2013) on process innovation, and (Johne 1999) on market innovation (see Table 17.1). All questionnaire items were organized on a 5-point Likert scale.

As a first step, a confirmatory factor analysis (CFA) was performed. This allows one to assess whether a measurement model for a latent or intangible variable (such as entrepreneurial orientation) is appropriately reflected by a questionnaire (Harrington 2009). The following indices help in this assessment: chi-squared test, the root mean square error of approximation (RMSEA), goodness of fit index (GFI) and the comparative fix index (CFI). The CFA confirmed that slightly adapted measures of innovativeness, proactiveness and entrepreneurial intentions suited the

Table 17.1 Operationalization of concepts: entrepreneurial competences and farmer characteristics

Measure	Literature	Questionnaire item
Entrepreneurial competences		
Innovativeness	"Entrepreneurial orientation: a psychological model of success among southern African small business owners" (Krauss et al. 2005), "Adapting the measurement of youth entrepreneurship potential to the context of Mindanao, Philippines" Lai et al. (2017a, b)	I always like to search for the latest information and technology
		I like to try new technology in my farm
		If there is an improvement in my coffee/honey product, I am willing to change where I sell it
		I am willing to include new high-yielding varieties/more bee hives in my farm, to satisfy more customers
Risk-taking	"Entrepreneurial orientation: a psychological model of success among southern African small business owners" (Krauss et al. 2005)"Adapting the measurement of youth entrepreneurship potential to the context of Mindanao, Philippines" Lai et al. (2017a, b)	I would keep my current varieties/bee hives in the farm, rather than substituting them with others that I do not know
		I prefer avoiding doing an investment in my farm, if I do not know the benefits that I will get
		I do not want to enlarge my farm, because I do not want to incur more costs
		If someone suggests me to include more high-yielding varieties/bee hives in my farm, I will do it and I take great risk (chances for very high profits)
Proactiveness	"Entrepreneurial orientation: a psychological model of success among southern African small business owners" (Krauss et al. 2005), "Adapting the measurement of youth entrepreneurship potential to the context of Mindanao, Philippines" Lai et al. (2017a, b)	I am willing to start practices that other farms do not do yet
		If asked to adopt another type of farming technology, I am one of the first farmers to use it
		For my job, I perform above and beyond expectations, but there is always something more to be done or improved
		I do not mind failing if I learn something different from another coffee/honey farming practice

(continued)

Table 17.1 (continued)

Measure	Literature	Questionnaire item
Intentions	"Social structure, reasonable gain, and entrepreneurship in Africa" (George et al. 2016)	I intend to start a new coffee-honey-related business in the next 3 years (i.e. trading, processing)
		I intend to include a new technology to increase the yield of my coffee/honey productions in the next 3 years
		I intend to expand the contacts with other actors in my value chain in the next 3 years
		With my credit and savings, I intend to enlarge my farm with only coffee/honey production in the next 3 years
Farmers innovation		
Product innovation	"Social connectedness in marginal rural China: The case of farmer innovation circles in Zhidan, north Shaanxi" (Wu and Pretty 2004)	I have improved the use of my production practices in my coffee/honey farm to improve the quality of my coffee/ honey, in the past 5 years
Process innovation	"An empirical research on farmer innovation in agriculture industrial clusters" (Yang 2013)	I have improved my production practices, because other fellow farmers suggested it to me, in the past 5 years
Process innovation	"An empirical research on farmer innovation in agriculture industrial clusters" (Yang 2013)	I have improved my production practices, because other actors in my value chain suggested it to me, in the past 5 years
Market innovation	"Successful market innovation" (Johne 1999)	I have changed where I sell my coffee/honey production in the past 5 years

Ugandan context. However, the measures of risk-taking were not sufficiently well fitted to this context. After minor adaptations to the initial measurement model,[1] the index values were the following[2]: GFI = 0.941, AGFI = 0.9, RMSEA = 0.055, CFI = 0.933; chi square p-value = 0.05. This shows that the measurement model of entrepreneurial orientation fitted well with the Ugandan context.

After performing the CFA, linear regressions were used to analyze the impact of entrepreneurial orientation on product, process and market innovations, in interaction with farm characteristics. Multiple regression models were run using different interaction terms (e.g., the combined effect of innovativeness and education level, or the combined effect of proactiveness and age) in order to: (i) understand whether, when considering different control and interaction variables, the effects on agricultural innovations were stable; and (ii) assess whether the effect of entrepreneurial orientation and farm characteristics vary in their impact on innovation in general as well as on specific types of innovation.

17.3 Findings

Figure 17.1 shows the key tested relationships among the variables of interest: entrepreneurial orientation, farm characteristics and farmer innovations. In the first tested regression model (when farm characteristics and entrepreneurial orientation are considered together with interaction terms), it was found that only education

[1] Different combinations have been created between the different first-order latent constructs. While running the analysis, problems emerge if risk-taking is included amongst the latent constructs. If risk-taking is included the values are represented as follows: GFI = 0.855, AGFI = 0.803, RMSEA = 0.084, CFI = 0.661, and the chi square is significant (p value = 0.000). Problems also arises whether a CFA is performed for the first-order latent construct risk-taking, when taken alone, thus without any combinations with innovativeness, proactiveness and intentions. At the same time, CFA was conducted for each of the first-order latent constructs, which did not register any issues: innovativeness, proactiveness and intentions. The correlation values of each variable with the latent construct were high and the model fit was good as well. Furthermore, one questionnaire item for the measurement of innovativeness and one questionnaire item (out of four total in each) for proactiveness were excluded. It has been proven that even with three items for dimension, the questionnaire can still maintain statistical authenticity (Cook et al. 1981). If risk-taking should not be included in the questionnaire to measure entrepreneurial competences—even with a one-item reduction each for innovativeness and proactiveness—the questionnaire is still statistically authentic.

[2] A value of the RMSEA of about 0.05 or less would indicate a close fit of the model in relation to the degrees of freedom. The requirement of exact fit corresponds to RMSEA = 0.0. A value of about 0.08 or less for the RMSEA would indicate a reasonable error of approximation, and one would not want to employ a model with a RMSEA greater than 0.1. GFI is less than or equal to 1. A value of 1 indicates a perfect fit. It is acceptable when GFI >0.9 The AGFI (adjusted goodness of fit index) considers the degrees of freedom available for testing the model. The AGFI is bounded above by one, which indicates a perfect fit. It is not, however, bounded below by zero, as the GFI is. It is acceptable when AGFI >0.9. CFI falls in the range from 0 to 1. CFI values close to 1 indicate a very good fit.

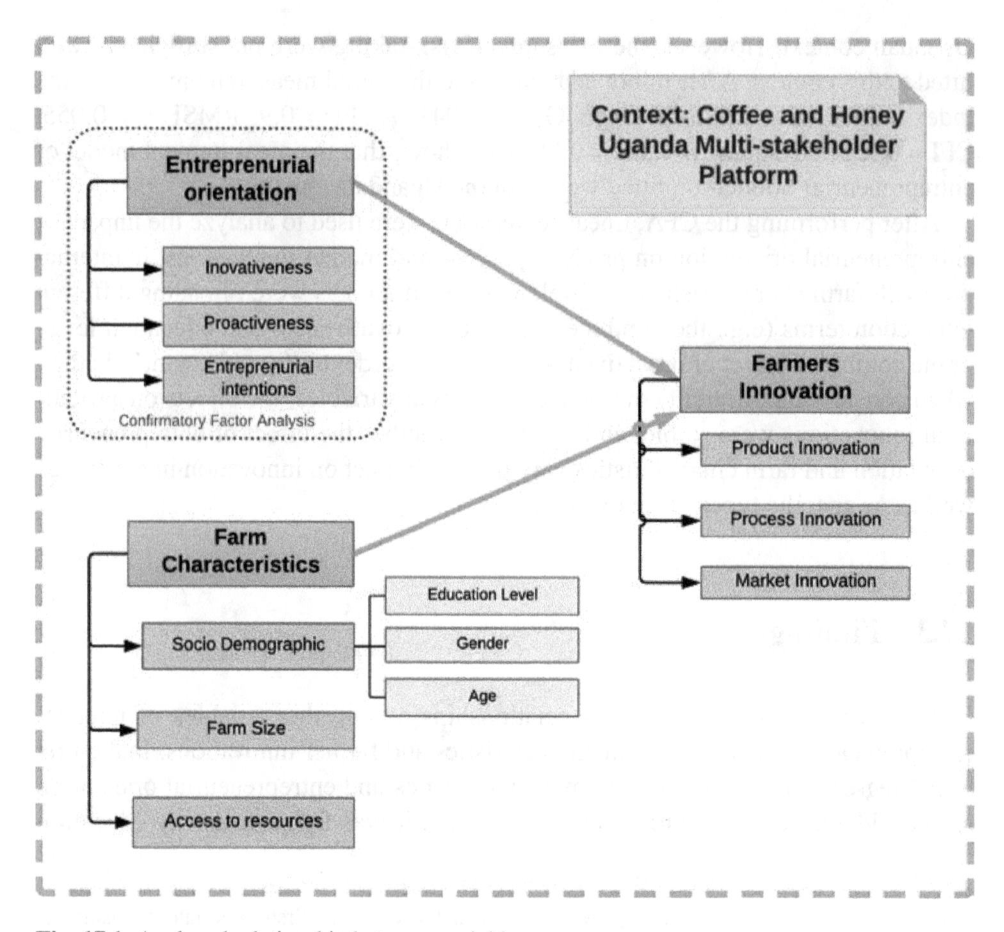

Fig. 17.1 Analyzed relationship between variables

level, farm size and access to resources have a significant effect on all forms of agricultural innovation (product, process and market). Entrepreneurial orientation variables showed no significant impact. At this stage of the analysis all variables were included in the tested model, which was too many variables given the relatively limited sample (n = 152). To decrease the number of variables, we separately included each dimension of entrepreneurial orientation (innovativeness, proactiveness, risk-taking and entrepreneurial intentions) together with all of the farm characteristics. After running this second set of regression models, it was found that education has a significant impact on process innovation. Similarly, when entrepreneurial proactiveness or entrepreneurial intentions are taken into account, a higher education level has a significant effect on process innovation (see Appendix 1). Simply put, farmers who engage in process innovation in their coffee and honey value chains do so not because of their proactiveness, innovativeness or intentions, but rather because of their higher education levels.

When we exclude education from the regression models, however, entrepreneurial orientation showed a positive effect on farmers' innovations. In particular, when farmers have higher entrepreneurial innovativeness, smaller farms have a more pos-

Table 17.2 Model summary and significance of interaction between entrepreneurial innovativeness, farm size and process innovation

Model		Unstandardized coefficients		Standardized coefficients	t	Sig	Collinearity statistics	
		B	Std. error	Beta			Tolerance	VIF
1	Constant	1.860	0.966		1.926	0.056		
	Entr. innovativeness	0.121	0.067	0.146	1.797	0.074	0.999	1.001
	Farm size	0.118	0.199	0.048	0.590	0.556	0.999	1.001
2	Constant	−4.292	2.773		−1.548	0.124		
	Entr. innovativeness	0.573	0.202	0.688	2.830	0.005	0.107	9.304
	Farm size	5.276	2.192	2.145	2.406	0.017	0.008	124.936
	Farm size – entr. innovativeness	−0.379	0.160	−2.162	−2.362	**0.019**	0.008	131.673

Table 17.3 Model summary and significance of interaction between entrepreneurial proactiveness, farm size and process innovation

Model		Unstandardized coefficients		Standardized coefficients	t	Sig	Collinearity statistics	
		B	Std. error	Beta			Tolerance	VIF
1	Constant	10.321	2.308		4.472	0.000		
	Farm size	0.730	0.578	0.102	1.263	0.209	1.000	1.000
	Entr. proactiveness	0.220	0.169	0.105	1.301	0.195	1.000	1.000
2	Constant	23.497	5.909		3.977	0.000		
	Farm size	−9.628	4.325	−1.350	−2.226	0.028	0.017	57.809
	Entr. proactiveness	−0.821	0.462	−0.393	−1.777	0.078	0.130	7.702
	Farm size – entr. proactiveness	0.818	0.339	1.547	2.416	**0.017**	0.016	64.441

itive effect on all forms of agricultural innovation. Specifically, if the interaction between farm size and entrepreneurial innovativeness increases, process innovation decreases (see Table 17.2).

This means that farmers with a smaller farm size and higher innovativeness were the most likely to engage in all forms of agricultural innovation. Furthermore, the interaction between farmer proactiveness and farm size has a positive effect on all forms of farmer innovation. Specifically, if the interaction effect among farm size and entrepreneurial proactiveness increases, the computed variable of process innovation increases (Sig. 0.18) (see Table 17.3). This means that proactive farmers with a larger farm size are the most likely to engage in all forms of agricultural innovations. Third, contrary to our initial hypothesis, farm size has a negative effect on innovation. In particular, if the size of the farm increases, process innovation decreases (Appendix 2). This means that larger farms (with more than 20 beehives or more than 1 acre of coffee-cultivated land) are generally less inclined than smaller farms to innovate their processes. Finally, when access to farm input resources increases, process innovation increases as well, when entrepreneurial competencies are also considered (Appendix 3).

No other variable—not entrepreneurial orientation or farm characteristics—influences process innovations as much as a farmer's education level, a finding that has been demonstrated frequently in many other contexts (e.g., Thangata and Alavalapati 2003). In Uganda, however, relatively few farmers attain high levels of education, so it is important to evaluate the findings with education level taken out of the equation. Results suggested that farmers with smaller farm size and higher access to resources have significantly higher levels of agricultural innovation. This finding on farm size is somewhat surprising because it contrasts with a wide literature suggesting that larger farms engage more often in innovations (Adesina and Baidu-Forson 1995; Weir and Knight 2004). To better understand these results, we further analyzed the interaction effect of these farm characteristics and entrepreneurial orientation on innovation. We found:

- The higher farmers' innovativeness is, the stronger the negative effect of farm size on their innovations. This may suggest that smaller farmers would be the most reactive in taking up new product, process and market innovations when they become more innovative.
- The higher farmers' proactiveness is, the stronger is the positive effect of farm size on their innovations. This may suggest that larger farmers would be the most reactive in taking up new product, process and market innovations when they become more proactive.

These results confirm that factors such as education levels, farm size and access to resources are key factors shaping the triggering and scaling of agricultural innovations, including those related to CSA practices.

A couple of methodological cautions are in order. First, by testing the measurement model through the CFA, we found that the measures of risk-taking as a dimension of entrepreneurial orientation did not fit the data in the Ugandan context. This means that—in contrast to Lai et al. (2017a) in the Philippines—the farmers in this Ugandan survey did not understand how the questionnaire items on risk-taking together corresponded to one concept. More significantly, this signals that risk-taking may not be a suitable or desirable dimension of entrepreneurial orientation in a farm context afflicted by market, social and environmental shocks. Given the limited sample size, though, it is worth conducting further tests on the risk-taking in other context before recommending to definitely drop this dimension in similar study contexts.

Second, in our database, multicollinearity among variables is high (meaning that the dimensions of entrepreneurial orientation are highly correlated with each other and with some farm characteristics) and sample size is relatively small (n = 152). This created statistical problems that forced us to build multiple smaller regression models to analyze all the variables of interest. If future research allows the collection of a larger sample, multivariate statistics may offer a valid alternative to the use of linear regressions. Furthermore, to avoid desirability bias (i.e., farmers trying to give perceived-as-desirable answers to the interviewer), future research may combine survey questions with direct participant observation of farmers' innovative and proactive actions over time, as well as their interactions with stakeholders in their environments.

17.4 Implications for Development

Our findings suggest that there are opportunities for development efforts to promote innovation among farmers. It is important to note that current psychological theories see entrepreneurial orientation not as a personality trait (Rauch and Frese 2007) fixed early in life but as a mindset that can develop over time—that can, in other words, be learned (Baum et al. 2014; Campos et al. 2017). At least two learnable dimensions of farmers' entrepreneurial orientation—proactiveness and innovativeness—may play a role in the adoption and scaling of agricultural innovations when tailored to certain types of farmers (York and Venkataraman 2010; Bruton et al. 2013). The development of a proactive and innovative mindset can be encouraged through workshops and other training activities for farmers. MSPs can thereby act as spaces for engaging in entrepreneurship training and thus supporting the development of entrepreneurial ecosystems (Bruton and Ahlstrom 2003; Dentoni and Klerkx 2015; Seuneke et al. 2013). In particular, smallholder farmers may benefit from tailored trainings on innovativeness, while larger farmers would benefit from capacity-building activities focusing on proactiveness. Therefore, if confirmed on studies at a larger scale, these findings suggest that training focused on shifting the mindsets of farmers can lay the groundwork for agricultural innovation.

Appendix 1

Coefficients – process innovation

Model		Unstandardized coefficients		Standardized coefficients	t	Sig	Collinearity statistics	
		B	Std. error	Beta			Tolerance	VIF
1	Constant	1.967	0.765		2.571	0.011		
	Education level	0.280	0.127	0.179	2.202	0.029	0.966	1.035
	Entr. [roactiveness	0.076	0.058	0.106	1.301	0.195	0.966	1.035

Model		B	Std. error	Beta	t	Sig	Tolerance	VIF
2	Constant	4.898	2.810		1.743	0.083		
	Education level	–0.978	1.167	–0.624	–0.838	0.404	0.011	87.238
	Entr. proactiveness	–0.152	0.218	–0.211	–0.696	0.488	0.069	14.447
	Education level – entr. proactiveness	0.097	0.090	0.920	1.084	**0.280**	0.009	113.171

Coefficients – process innovation

Model		Unstandardized coefficients		Standardized coefficients	t	Sig	Collinearity statistics	
		B	Std. error	Beta			Tolerance	VIF
1	Constant	3.091	0.822		3.760	0.000		
	Education level	0.322	0.131	0.205	2.462	0.015	0.927	1.079
	Entr. intentions	–0.015	0.048	–0.026	–0.307	0.759	0.927	1.079
2	Constant	2.318	2.769		0.837	0.404		
	Education level	0.637	1.087	0.407	0.586	0.559	0.013	74.265
	Entr. intentions	0.030	0.160	0.052	0.186	0.852	0.085	11.800
	Education Level – entr. Intentions	–0.018	0.062	–0.236	–0.293	**0.770**	0.010	100.066

Dependent Variable: I have improved my production practices because other actors in my value chain suggested it to me in the past 5 years

Appendix 2

Coefficients – process innovation

Model		Unstandardized coefficients		Standardized coefficients	t	Sig	Collinearity statistics	
		B	Std. error	Beta			Tolerance	VIF
1	Constant	10.321	2.308		4.472	0.000		
	Farm size	0.730	0.578	0.102	1.263	0.209	1.000	1.000
	Entr. proactiveness	0.220	0.169	0.105	1.301	0.195	1.000	1.000
2	Constant	23.497	5.909		3.977	0.000		
	Farm size	−9.628	4.325	−1.350	−2.226	0.028	0.017	57.809
	Entr. proactiveness	–0.821	0.462	–0.393	−1.777	0.078	0.130	7.702
	Farm size – entr. proactiveness	0.818	0.339	1.547	2.416	**0.017**	0.016	64.441

Dependent Variable: innovation

Appendix 3

Coefficients – process innovation

Model		Unstandardized coefficients		Standardized coefficients			Collinearity statistics	
		B	Std. error	Beta	t	Sig	Tolerance	VIF
1	Constant	2.812	0.826		3.405	0.001		
	Access to resources	0.075	0.025	0.250	3.064	0.003	0.949	1.054
	Entr. intentions	–0.015	0.047	–0.027	–0.327	0.744	0.949	1.054
2	Constant	3.460	2.928		1.182	0.239		
	Access to resources	0.031	0.194	0.102	0.158	0.874	0.015	65.504
	Entr. intentions	–0.054	0.175	–0.094	–0.310	0.757	0.069	14.510
	Access to resources – Entr. Intentions	0.003	0.011	0.177	0.231	**0.818**	0.011	92.249

Dependent Variable: I have improved my production practices because other actors in my value chain suggested it to me in the past 5 years

References

Adesina AA, Baidu-Forson J (1995) Farmers' perceptions and adoption of new agricultural technology: evidence from analysis in Burkina Faso and Guinea, West Africa. Agric Econ 13(1):1–9

Baum JR, Frese M, Baron RA (2014) The psychology of entrepreneurship. Psychology Press, Mahwah

Bomuhangi A, Namaalwa J, Nabanoga G (2016) Natural resources policy environment in Uganda, implication for gendered adaptation to climate changes. Environ Sci Ind J 12(10):117

Bruton GD, Ahlstrom D (2003) An institutional view of China's venture capital industry: explaining the differences between China and the West. J Bus Ventur 18(2):233–259

Bruton GD, Ahlstrom D, Obloj K (2008) Entrepreneurship in emerging economies: the research go in the future. Entrep Theory Pract 32(January):1–14

Bruton GD, Ketchen DJ, Ireland RD (2013) Entrepreneurship as a solution to poverty. J Bus Ventur 28(6):683–689. Available at:. https://doi.org/10.1016/j.jbusvent.2013.05.002

Campos F et al (2017) Teaching personal initiative beats traditional training in boosting small business in West Africa. Science 357(6357):1287–1290

Cook, J. D., Hepworth, S. J., Wall, T. D., & Warr, P. B. (1981). The experience of work: A compendium and review of 249 measures and their use: Academic Press London

Dentoni D, Klerkx L (2015) Co-managing public research in Australian fisheries through convergence–divergence processes. Mar Policy 60:259–271

George G et al (2016) Social structure, reasonable gain, and entrepreneurship in Africa. Strateg Manag J 37(6):1118–1131

Harrington D (2009) Confirmatory factor analysis. Oxford University Press, Oxford

Johne A (1999) Successful market innovation. Eur J Innov Manag 2(1):6–11

Khavul S, Bruton GD (2013) Harnessing Innovation for Change: Sustainability and Poverty in Developing Countries. J Manag Stud 50:285–306. https://doi.org/10.1111/j.1467-6486.2012.01067

Kilelu CW, Klerkx L, Leeuwis C (2013) Unravelling the role of innovation platforms in supporting co-evolution of innovation: contributions and tensions in a smallholder dairy development programme. Agric Syst 118:65–77 Available at: http://linkinghub.elsevier.com/retrieve/pii/S0308521X1300036X [Accessed 17 July 2014]

Krauss SI et al (2005) Entrepreneurial orientation: a psychological model of success among southern African small business owners. Eur J Work Organ Psy 14(3):315–344

Lai C Chan C, et al. (2017a) 5 measuring youth entrepreneurship attributes: the case of an out-of-school youth training program in Mindanao, Philippines. Enabling agri-entrepreneurship and innovation: empirical evidence and solutions for conflict regions and transitioning economies, p 72

Lai C, Dentoni D et al (2017b) Adapting the measurement of youth entrepreneurship potential in a marginalised context: the case of Mindanao, Philippines. J Int Bus Entrep Dev 10(3):273–297

Lans T, Seuneke P, Klerkx L (2013) Agricultural entrepreneurship. In Encyclopedia of creativity, invention, innovation and entrepreneurship (pp. 44–49). Springer, New York, NY

Lipper L et al (2014) Climate-smart agriculture for food security. Nat Clim Chang 4(12):1068–1072

Naudé W (2010) Entrepreneurship, developing countries, and development economics: new approaches and insights. Small Bus Econ 34(1):1

Rauch A, Frese M (2007) Let's put the person back into entrepreneurship research: A meta-analysis on the relationship between business owners' personality traits, business creation, and success. Eur J Work Organ Psy 16(4):353–385

Sabiiti G et al (2016) Empirical relationships between banana yields and climate variability over Uganda. J Environ Agric Sci 7:3–13

Schut M et al (2015) Innovation platforms: experiences with their institutional embedding in agricultural research for development. Exp Agric 44:37 Available at: internal-pdf://84.152.52.160/Schut-2015-INNOVATION PLATFORMS_ EXPERIENCES W.pdf

Seuneke P, Lans T, Wiskerke JSC (2013) Moving beyond entrepreneurial skills: key factors driving entrepreneurial learning in multifunctional agriculture. J Rural Stud 32:208–219. Available at:. https://doi.org/10.1016/j.jrurstud.2013.06.001

Shane S, Venkataraman S (2000) The promise of entrepreneurship as a field of research. Acad Manag Rev 25(1):217–226

Thangata PH, Alavalapati JRR (2003) Agroforestry adoption in southern Malawi: the case of mixed intercropping of Gliricidia sepium and maize. Agric Syst 78(1):57–71

Weir S, Knight J (2004) Externality effects of education: dynamics of the adoption and diffusion of an innovation in rural Ethiopia. Econ Dev Cult Chang 53(1):93–113

Wu B, Pretty J (2004) Social connectedness in marginal rural China: the case of farmer innovation circles in Zhidan, north Shaanxi. Agric Hum Values 21(1):81–92

Yang, Li (2013) An empirical research on farmer innovation in agriculture industrial clusters. International conference on the modern development of humanities and social science. Hong Kong.

York JG, Venkataraman S (2010) The entrepreneur-environment nexus: uncertainty, innovation, and allocation. J Bus Ventur 25(5):449–463. Available at:. https://doi.org/10.1016/j.jbusvent.2009.07.007

Zilberman D et al (2018) Climate smart agriculture. Building resilience to climate change. Volume 52. B. Branca, LL, McCarthy N, Zilberman D, Solomon A, Giacomo, ed., Natural resource management and policy. Available at: http://link.springer.com/10.1007/978-3-319-61194-5

18

Shea Butter Production: A Climate-Smart Solution for the Vulnerable

James Hammond, Mark van Wijk, Tim Pagella, Pietro Carpena, Tom Skirrow, and Victoria Dauncey

18.1 Introduction

People suffering extreme poverty are typically the most vulnerable to system shocks, including those caused by climate change (FAO 2016). Finding climate-smart interventions that help the most vulnerable people is difficult because those people are typically less educated, have fewer resources to draw upon, and are less able to tolerate risk and adopt new practices (Ahmed et al. 2007).

This chapter examines marginal smallholder farmers in the Sahel, in Eastern Province, Northern Ghana, to explore whether shea butter production might offer a climate-smart solution to help the most vulnerable. Shea trees are highly abundant across the Sahel region. While the tree is culturally familiar and valued across the dry lands of West Africa (Carpena et al. 2016) it has yet to be domesticated (Hall et al. 1996). The fruits of the shea trees can be eaten, and the sun-dried kernels can be boiled down to produce a vegetable fat known as shea butter, which is used in both the food and cosmetics industries. The processing of the shea nut is laborious and considered to be a lowly form of work, which means it is often undertaken by women (who tend to be more vulnerable than men) and by poorer households. Shea butter has been widely promoted as a rural development intervention, as it is a freely accessible resource with a clear and reliable market value (Elias and Carney 2007; Hatskevich et al. 2011; Pouliot and Elias 2013). The trees also serve as a defence

J. Hammond (✉) · T. Pagella
World Agroforestry Centre (ICRAF), Nairobi, Kenya

School of Environment, Natural Resources and Geography, Bangor University, Bangor, UK
e-mail: J.Hammond@cgiar.org

M. van Wijk
International Livestock research Institute (ILRI), Nairobi, Kenya

P. Carpena · T. Skirrow · V. Dauncey
TREE AID, Bristol, UK

against encroaching desertification, and preserving them helps in both the mitigation and adaptation to climate change (Mbow et al. 2014).

Shea butter production could be considered outside the definition of climate-smart agriculture (CSA)—after all, it involves the gathering of products from non-cultivated trees and therefore strictly speaking does not qualify as agriculture. From a broader perspective, however, shea nut production supports the objectives of CSA, the three pillars of which are widely defined as increasing food security, increasing adaptive capacity, and mitigation of greenhouse gas emissions (FAO 2013; Neufeldt et al. 2013; Campbell et al. 2014; Lipper et al. 2014). Shea production has a positive impact on adaptive capacity: by enhancing the economic value of shea trees, it encourages the retention of trees in the landscape which will continue to provide buffering ecosystem services, promoting water and soil retention and guarding against desertification (Sinare et al. 2016). A healthy shea nut industry therefore promotes landscape-scale adaptation to climate change. The industry appears to have a neutral impact on greenhouse gas emissions, and a positive impact on food security, as is detailed below. We argue that increased shea butter production, supported by better business infrastructure leads to increased incomes and makes households more resilient to negative shocks.

The non-governmental organization (NGO) TREE AID led a 5-year programme[1] (2012–2017) with several goals: (i) increase income of communities involved in sourcing and processing shea nuts through improvements in product quality and quantity; (ii) increase women's empowerment by building organizational capacity and commercial infrastructure including business groups, warehouses and credit schemes; (iii) diversify the buyers' base to allow long-term and stable incomes for the producers; (iv) protect ecosystems and promote climate resilience through the reduction of the environmental impact of shea nut sourcing and production. TREE AID's efforts included helping producers form "union" organisations focused on regional marketing, services and value addition. It worked to build buffers against market fluctuations by securing minimum price guarantees from national and international buyers of shea butter. TREE AID also provided training on improved methods for shea butter processing, including the use of hand tools and electric machinery. For this study, we evaluated the TREE AID programme using the Rural Household Multi-Indicator Survey (RHoMIS), a carefully designed, low-cost, flexible household survey tool for efficient characterisation of farm systems in communities suffering from poverty and food insecurity (Hammond et al. 2017; Rosenstock et al. 2017). This study used RHoMIS to test whether the TREE AID shea butter programme helped increase the resilience of the extremely poor in Northern Ghana.

[1] This project was implemented with funding from Comic Relief.

18.2 Methods

This study focused on a population in the Upper East and Upper West regions of Northern Ghana, in the Lambussie Karni, Kassena Nankana East and Kassena Nankana West districts. Interviews with 223 households were conducted in March 2017. Informants were selected randomly from 26 villages within the project area, where informants were either project beneficiaries (101 households) or members of a control group (122 households) of non-beneficiaries. The villages were selected on the basis of their already established relationships with partner organisations. Beneficiaries in the project were self-selecting, and so it can be assumed they had more interest in shea compared to the general population. The control group, comprising households identified as future project beneficiaries, were chosen because they are directly comparable to the beneficiary households.

The RHoMIS tool uses a modular, rapid (40–60 min) digital survey to derive standardised indicators on agricultural practices, livelihoods, food security and dietary diversity, as well as gender roles (Hammond et al. 2017). A survey module was developed to collect information on use of non-timber forest products (NTFPs) and woody environmental resources. The indicator used was food availability (Frelat et al. 2016), which converts all household income and agricultural produce into a calorie per person score. Food availability was chosen in preference to cash incomes as it also takes account of self-produced and consumed items and thus provides a more comprehensive perspective on the livelihoods of the very poor (Ritzema et al. 2017). Other rapid and well-tested indicators were also gathered: experience of hunger, quantified using the Household Food Insecurity of Access Scale (HFIAS) (Coates et al. 2007); dietary diversity, assessed using the Household Dietary Diversity Score method (Swindale and Bilinsky 2006); and food groups, gathered using the Minimum Dietary Diversity for Women guide (MDD-W) (FAO and FHI 360 2016). The Progress out of Poverty Indicator (PPI) was used to cross-check the household income figures gathered from direct questioning (IPA 2015). The use of these standard indicators allows evaluation of the project impacts in a wider frame of reference, comparison to other locations, and evaluation of changes over time should a further RHoMIS study be done at a later date.

Households were classified into three poverty classes based on their food availability scores. Households with access to less than 2500 kcal per male adult equivalent (MAE) person per day were classed as "below the calorie line." Households above the calorie line but with a total value of activities (i.e., actual cash income plus the value of consumed agricultural produce) below US$1.90 were classed as "below the poverty line." Households with total value of activities above US$1.90 were classed as "above the poverty line." Welfare indicators have been presented as medians per household group, and incomes have been presented as trimmed means, where 5% of the observations at either extreme of the scale were dropped to reduce the effect of outliers. The Kruskal-Wallis test for significance was used when comparing between beneficiary and non-beneficiary households within paired poverty classes, and unless otherwise stated all significance was attributed at the 0.95 level.

18.3 Results

18.3.1 Household Livelihoods and Farm Characteristics

The majority of the population was very poor and suffered from food insecurity. The median income per person per day was $0.09, or $144 per household per year. The PPI predicted that 51% of households were below the $1.90 poverty line, although from reported household income we calculated that 99% of households were below that poverty line. Median household population was eight persons, and median land owned was 2 ha per household, with 1.6 ha cultivated in the last year. Crops sales accounted for the majority of household income ($96 per year), followed by environmental resources, including woody resources and non-timber forest products ($33 per year). Livestock sales and off-farm income were low, returning median values of zero, although some households did derive income from these sources. Livestock animals were, however, widely kept, with 80% of the population keeping some form of livestock. The main crops grown were ground nut (85% of households), maize (82%), millet (58%), rice (53%) and sorghum (25%). The main livestock were goats (65%), chickens (48%), sheep (39%) and cattle (28%). The NTFPs reported were shea, baobab and mango, with shea by far the most widely used. Shea was gathered by 72% of the study population, baobab by 19% and mango by 8%. The environmental resources were fuelwood (65% of the population) and charcoal (5%).

Using the food availability indicator, we calculated that the median amount of kcal available per person (adult male equivalent) per day was 3023; but that 42% of the population had less than 2500 kcal available per day. Households reported on average 3 months during which it was difficult to source enough food, with the worst period being May through August. Using the HFIAS indicator, 81% of households were categorised as severely food insecure during the lean season, 9% moderately food insecure, 3% mildly food insecure and 7% food secure. Dietary diversity was low during the lean season, with a median score of 3 food groups eaten at least weekly. Outside the lean season the dietary diversity score was considerably better, with a median score of 7.

Very few households were considered above the poverty line, either among beneficiaries or non-beneficiaries (see Fig. 18.1). There were, however, more households in the poorest category (below the calorie line) among the non-beneficiary group than the beneficiaries ($p < 0.05$). The plausible reason for this, looking at the sources of calories and income illustrated in Fig. 18.1, is the major role played by NTFPs. The mean amount of income derived from NTFPs is greater amongst project beneficiaries. Amongst both beneficiaries and non-beneficiaries the importance (as a proportion of calorie provision) of NTFPs and woody resources is greater for poorer households.

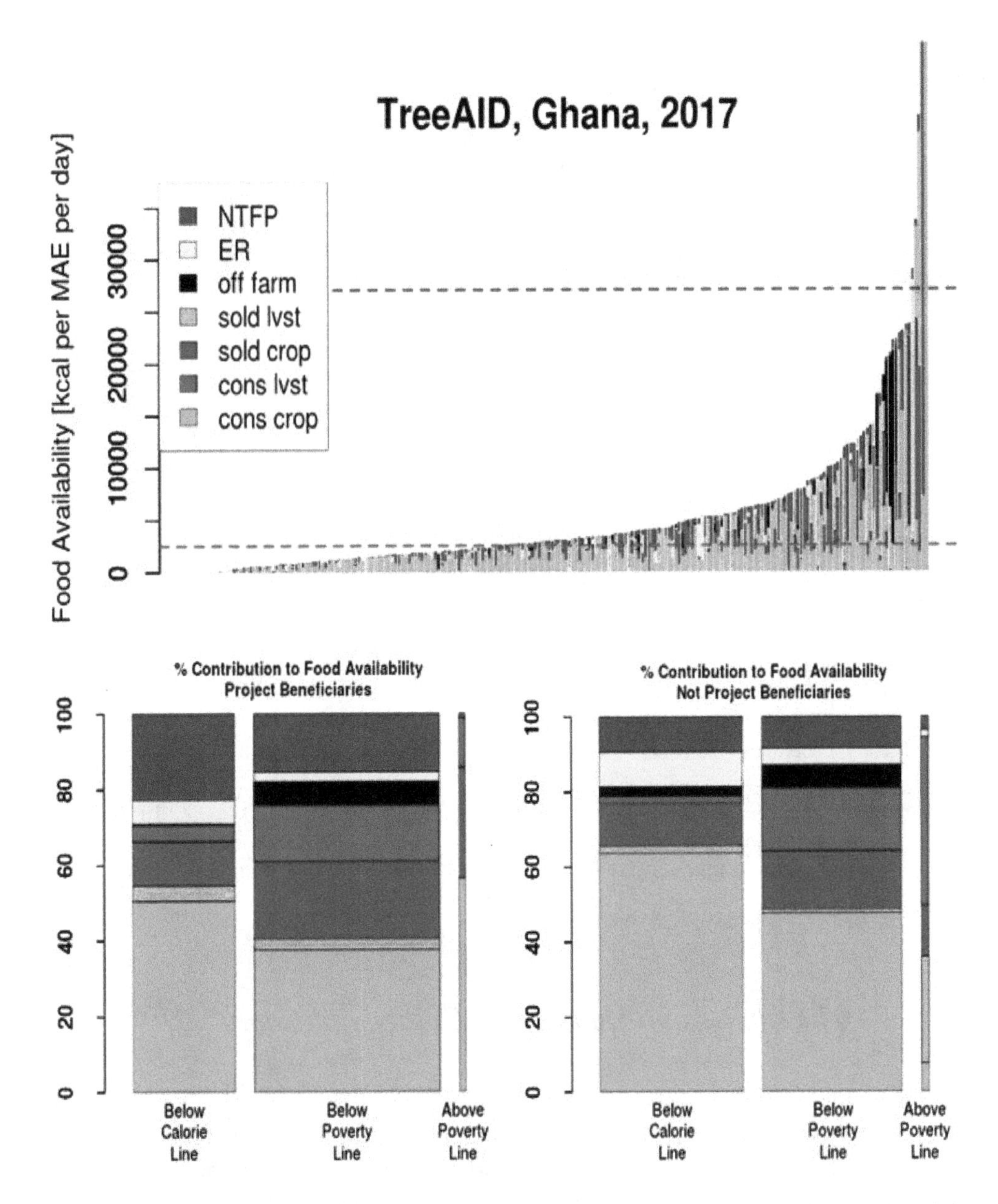

Fig. 18.1 Household livelihoods displayed as potential food availability, kcal per male adult equivalent person per day. The upper panel shows the amount of calories potentially derived from different income (or food) sources, and each column represents an individual household. The horizontal dashed lines represent thresholds used to divide the population. The red dashed line represents minimum calorie requirement per day (2500 kcal per MAE), and the blue line represents the poverty line threshold of $1.90 per person per day. The lower panel shows the mean proportion of income derived from each income source for households in three poverty classes: those below the calorie line, those below the poverty line and those above the poverty line; the width of the column represents the number of households in that category. The population is also divided into beneficiaries of the project and non-beneficiaries. The livelihood sources are represented in the legend in the upper right corner, with the following abbreviations: NTFP non-timber forest products, ER environmental resources, lvst livestock, cons consumed

Table 18.1 Household welfare indicators, by beneficiary and non-beneficiary (control) households, and by poverty class. Food availability is shown as kilocalories per MAE; the proportion of households suffering from severe food insecurity is determined using HFIAS; the dietary diversity score is determined using the household dietary diversity score method (HDDS) and the ten food categories from the MDD-W indicator; and PPI is used to predict the likelihood of households being in poverty using the $1.90 poverty line. All values shown are median averages, and statistical significance was established using the Kruskal-Wallis rank sum test, comparing between beneficiary and control households within the same poverty class. Differences significant at $p < 0.05$ are highlighted in bold, and differences at $p < 0.1$ are italicised

Project beneficiary	Poverty class	Food availability (kcal/MAE)	Income $/pers/ day	% hh severley food insecure	Dietary diversity score (lean season)	Hungry months	PPI predicted % under poverty line
Control	all	**2558**	**0.05**	84	3	3	**51**
Beneficiary	all	**3885**	**0.14**	76	3	3	**35**
Control	below calorie line	1307	**0.01**	87	3	3	62
Beneficiary	below calorie line	1277	**0.04**	83	2	3	51
Control	below pov line	*4756*	0.17	83	3	3	35
Beneficiary	below pov line	*5548*	0.24	73	4	3	35
Control	above pov line	43795	1.92	67	4	1	10
Beneficiary	above pov line	128014	1.15	50	3	2	17

18.3.2 The Impacts on Household Welfare Indicators

When looking at the whole population, significant effects on household welfare indicators were found (see Table 18.1). Beneficiary households had higher potential calorie availability, higher cash incomes, and better progress out of poverty scores. Furthermore, the reduction in the number of households classified as severely food insecure (using the HFIAS indicator) scored a low but non-significant *p* value of 0.12, implying, in combination with the above-mentioned significant effects, positive project outcomes on the beneficiary population.

The poorest households, below the calorie line, showed an increase in actual cash incomes from US$0.01 per person per day to US$0.04 per person per day. Increase in the food availability score for beneficiary households below the poverty line was found to be significant only at the $p < 0.1$ level.

18.3.3 Shea Derived Incomes

Table 18.2 shows a breakdown of incomes derived from shea and firewood, as well as proportions of the populations engaged in each activity. The project achieved a statistically significant increase of income from sales of shea at the whole population level. The increased income was due to sales of shea butter, and not shea nuts or fruits. Furthermore, the beneficiary population derived less income from sale of fuelwood compared to the control. The total number of households using shea was also higher in the beneficiary population ($p < 0.1$), the total number selling shea butter was higher, and the total number selling fuelwood was lower.

When considering households of different poverty classes, those below the calorie line showed the most marked changes: average income from shea butter was almost ten times higher among the beneficiary population, and more than twice the proportion of households took part in shea butter selling. A similar pattern was observed amongst the households below the poverty line, although effects were at the $p < 0.1$ level, perhaps reflecting the greater variation in income sources among households in that poverty class. An unexpected observation was that the beneficiary households above the poverty line showed less income and engagement with shea than non-beneficiary households, although there were so few households in that class that the finding cannot be considered robust.

18.4 What Factors Led to the Success of This Project?

The reasons for the beneficiaries' higher incomes from shea are multiple. The survey data shows that the quantity of shea fruit gathered per household did not significantly differ between beneficiaries and non-beneficiaries (mean 130 kg/year), but the amount converted into shea butter did. Beneficiary households yielded on average 37 kg/year of shea butter compared to 13 kg/year for non-beneficiaries ($p < 0.01$). The high difference in average shea butter production may be in part due to the higher number of beneficiaries who produced shea butter compared to non-beneficiaries, as well as more efficient production techniques, including access to tools and machines that reduced the drudgery of the process. Also, the ability to store shea nuts or butter may have reduced wastage. There was no significant evidence that beneficiaries sold more nuts or fruits compared to non-beneficiaries, nor was there significant evidence that the project achieved higher sale prices for shea butter for beneficiaries (median price 1.5 $/kg). It therefore appears that the project created a greater "market pull" by facilitating easier and more efficient processing, and by establishing sales groups.

The different usage of fuelwood may be an important clue as to the production of shea butter. Non-beneficiaries collected the same amount of fuelwood as beneficiaries but sold more of it as fuelwood. It may be, therefore, that beneficiaries burned the fuelwood they gathered in their production of shea butter. This is strongly implied

Table 18.2 The use of shea products and firewood by households, sub-divided into beneficiary and non-beneficiary (control) groups, and also separated by poverty class. Incomes presented are trimmed means, in US$ per household per year. Statistical significance was established using the Kruskal-Wallis rank sum test, comparing between beneficiary and control households within the same poverty class. Differences significant at $p < 0.05$ are highlighted in bold, and differences at $p < 0.1$ are italicised

			Shea		Shea butter		Shea seed		Shea fruit		Fuelwood	
Project beneficiary	Poverty class	number hh	Income $/yr	% hh selling	Income $/yr	% hh selling	Income $/yr	% hh selling	Income $/yr	% hh selling	Income $/yr	% hh selling
Control	all	122	**28**	*60*	**12**	**29**	4	16	3	21	**9**	**37**
Beneficiary	all	101	**57**	*70*	**40**	**46**	3	19	3	18	**3**	**20**
Control	below calorie line	60	**15**	*45*	**4**	**18**	0	7	3	22	9	38
Beneficiary	below calorie line	35	**49**	*63*	**42**	**40**	2	17	2	14	6	29
Control	below pov line	59	40	73	*19*	*36*	9	24	5	22	**11**	**34**
Beneficiary	below pov line	64	67	75	*45*	*50*	4	20	3	19	**3**	**16**
Control	above pov line	3	*95*	100	*73*	**100**	22	67	0	0	33	67
Beneficiary	above pov line	2	*17*	50	*0*	**0**	0	0	17	50	0	0

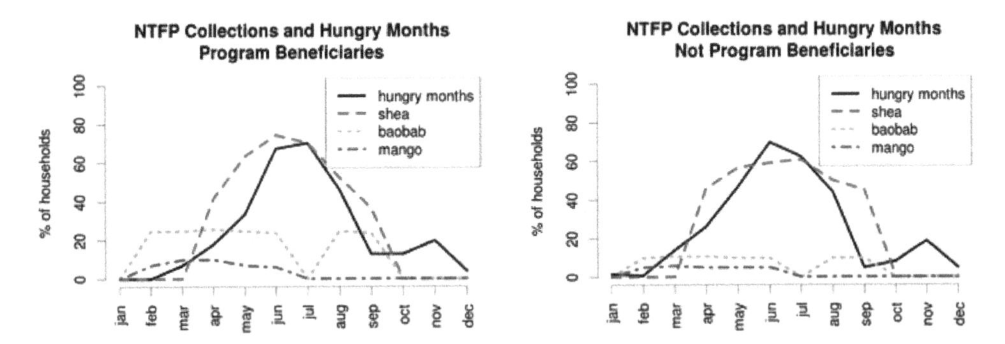

Fig. 18.2 The timings of reported "hungry months" during which food is in short supply, and of NTFP gathering. The collection of shea fruit seems to be well matched with the hungry season

from the survey data, and if true would be a clear case of adding extra value to already gathered environmental resources. It also implies that total greenhouse gas emissions were not increased through increased shea production, as non-beneficiaries gathered an equal amount of fuel wood but sold it instead of using it for shea production. Local informants believed that this fuel wood was not being sold to shea butter producers, but this was not established quantitatively. If it was in fact sold to producers, this fact could undermine our conclusion regarding emissions.

The households below the calorie line showed a much higher adoption rate of shea and shea butter sales amongst the beneficiary group compared to non-beneficiaries. This partly reflects the fact that shea butter is highly labour intensive and does not immediately generate a large amount of income. As a result, shea does not attract wealthier families who have opportunities elsewhere. The difficulty of shea production may be a blessing as well as a curse: it does not offer an easy path out of poverty, but due to the initial low cash investment and high labour cost, it may be a commodity which is well suited to improving incomes and food security for the very poor and vulnerable.

The timing of the shea fruit season also makes it a useful crop to combat food insecurity, and may explain in part the popularity of the crop. Figure 18.2 shows the timing of lean season and NTFP harvesting, as reported by beneficiary and non-beneficiary households. It is clear that shea harvesting coincides with the lean season, and that baobab and mango do not. Furthermore, it can be seen that the lean season starts a little later for project beneficiaries, possibly as an effect of the project interventions. The shea harvest seems to be particularly well timed to meet a local need.

Shea collection is also a strongly gendered activity, practiced mainly by women: 70% of households surveyed reported women gathering shea, with only 21% reporting men involved. Most importantly, the income is predominantly controlled by women, with 70% of households reporting female control of shea incomes and only 11% reporting male control. The gender breakdown of work and income control did not differ significantly between the beneficiary and control populations. The gendered nature of shea activities may also have helped the project gain traction in an environment where opportunities for women can be scarce and where attempts to increase the female share of household income can be a challenge (Johnson et al. 2016).

The project's implementing staff considered the construction of the warehouses for the union organizations to store shea products to be an important part in the project. The warehouses acted as a hub for the unions, a safe and pest-free storage area, and a place to access machinery to process shea. The warehouses also may have contributed to the female control of shea income because they were not gendered spaces: homesteads can have gender taboos associated with storage areas, making it difficult for women to extract full value from shea products. Study results did include suggestions for improving the warehouses. First, they were constructed late in the project; had they been constructed earlier the unions may have been more successful in negotiating guaranteed minimum prices. Another suggested improvement was a credit system whereby union members could receive some payment when depositing shea in the warehouses, to be set against the final payment they received when selling the shea butter. Unfortunately, this system could not be established due to logistical complications.

18.5 Implications for Development

This project once again demonstrated the usefulness of the RHoMIS tool. It permitted evaluation of the project at low cost, and the data gathered can now be pooled with that from other sites and used to build a body of evidence on routes to achieving resilience of small holder rural households. The use of a rapid and well-designed evaluation tool permitted a deeper understanding of the project impacts on household welfare than could otherwise have been achieved.

This study reveals the benefits of shea butter value chain work. The more successful interventions were training of households in shea butter extraction techniques and the formation of unions providing access to storage and machinery. The financial infrastructure proved more challenging to organize, with credit schemes and minimum price guarantees coming either too late or not at all. Despite these challenges, we have shown that the poorer sectors of society, and particularly females, benefited from the project in terms of income and food security.

A number of factors contributed to the success of this project and consideration of these may help improve other value-chain projects relating to climate-smart objectives.

- *Gender inclusive:* Supporting shea chains makes it easy to reach women, as shea is already a gendered (female-biased) product and not linked to land ownership.
- *Pro-poor:* Due to the high labour requirements and low initial cash investments, shea butter is a commodity well-suited to improving incomes and food security for the very poor and vulnerable sections of society. It unattractive to wealthier households, which may create more opportunities for the poor.
- *Culturally acceptable:* Shea was already culturally well accepted, and abundant, with little risk entailed in entering the market.

- *Timely:* The timing of the potential shea fruit income suited a local need: income during the lean season.
- *Adoptable:* The project interventions were simple and accessible to many households: hand tools, training, unions and access to storage space and machinery.

The project did reveal some challenges. The business training and value chain enhancement took longer to establish than was initially hoped, and price guarantees from buyers could not be secured. Earlier prioritisation of these activities may make them more successful in the future.

Evaluating the full environmental impact of the project was beyond the scope of the study, but there is no doubt that continued use of shea trees entails ecosystem benefits. One possible negative environmental consequence could be increased use of fuel wood for shea processing. We did not see evidence of this, but if it is found to be a problem, it could be managed by establishment of fuel lots. By preserving and encouraging the maintenance of trees in the landscape, shea production combats desertification and promotes preservation of soil and water resources. By providing both a source of food and opportunities for cash income, it contributes to healthier households and communities, making them more resilient in the face of environmental shocks. In this case, the facilitation of increased shea butter production and sales offered significant benefits to the most vulnerable smallholder farmers: decreasing the number of people in extreme poverty.

References

Ahmed A, Hill R, Smith L et al (2007) The world's most deprived: characteristics and causes of extreme poverty and hunger. International Food Policy Research Institute, Washington, DC. https://doi.org/10.2499/0896297705

Campbell B, Thornton P, Zougmoré R et al (2014) Sustainable intensification: what is its role in climate smart agriculture? Curr Opin Environ Sustain 8:39–43. https://doi.org/10.1016/j.cosust.2014.07.002

Carpena P, Kaboret B, Ouedraogo D et al (2016) Supporting the development of democratic and locally controlled small-scale enterprises based on non-timber forest products in Burkina Faso. Food Chain 6(2):77–91. https://doi.org/10.3362/2046-1887.2016.005

Coates J, Swindale A, Bilinsky P (2007) Household food insecurity access scale (HFIAS) for measurement of food access: indicator guide. Food and Nutrition Technical Assistance, Washington, DC

Elias M, Carney J (2007) African shea butter: a feminized subsidy from nature. Africa 77:37–62. https://doi.org/10.3366/afr.2007.77.1.37

FAO (2013) Climate-smart agriculture sourcebook. Available at: http://www.fao.org/docrep/018/i3325e/i3325e00.htm

FAO (2016) State of Food and Agriculture. Rome

FAO and FHI 360 (2016) Minimum dietary diversity for women: a guide to measurement. Available at: http://www.fao.org/3/a-i5486e.pdf

Frelat R, Lopez-Ridaura S, Giller K et al (2016) Drivers of household food availability in sub-Saharan Africa based on big data from small farms. Proc Natl Acad Sci 113(2):458–463. https://doi.org/10.1073/pnas.1518384112

Hall J, Aebischer D, Tomlinson H et al (1996) Vitellaria paradoxa: a monograph. School of Agricultural and Forest Sciences. University of Wales, Bangor

Hammond J, Fraval S, van Etten J et al (2017) The rural household multi-indicator survey (RHoMIS) for rapid characterisation of households to inform climate smart agriculture interventions: description and applications in East Africa and Central America. Agric Syst 151:225–233. https://doi.org/10.1016/j.agsy.2016.05.003

Hatskevich A, Jeníček V, Antwi Darkwah S (2011) Shea industry: a means of poverty reduction in northern Ghana. Agric Trop Subtrop 44(4):223–228

IPA (2015) Progress out of poverty index. Available at: http://www.progressoutofpoverty.org/

Johnson N, Kovarik C, Meinzen-Dick R et al (2016) Gender, assets, and agricultural development: lessons from eight projects. World Dev 83:295–311. https://doi.org/10.1016/j.worlddev.2016.01.009

Lipper L, Thornton P, Campbell B et al (2014) Climate-smart agriculture for food security. Nature Clim Change 4(12):1068–1072

Mbow C, Van Noordwijk M, Luedeling E, Neufeldt H, Minang PA, Kowero G (2014) Agroforestry solutions to address food security and climate change challenges in Africa. Curr Opin Environ Sustain 6:61–67. https://doi.org/10.1016/j.cosust.2013.10.014

Neufeldt H, Jahn M, Campbell B et al (2013) Beyond climate-smart agriculture: toward safe operating spaces for global food system. Agric Food Secur 2(1):12. https://doi.org/10.1186/2048-7010-2-12

Pouliot M, Elias M (2013) To process or not to process? Factors enabling and constraining shea butter production and income in Burkina Faso. Geoforum 50(Supp C):211–220. https://doi.org/10.1016/j.geoforum.2013.09.014

Ritzema R, Frelat R, Douxchamps S et al (2017) Is production intensification likely to make farm households food-adequate? A simple food availability analysis across smallholder farming systems from East and West Africa. Food Security 9(1):115–131. https://doi.org/10.1007/s12571-016-0638-y

Rosenstock T, Lamanna C, Chesterman S et al (2017) When less is more: innovations for tracking progress toward global targets. Curr Opin Environ Sustain 2:54–61. https://doi.org/10.1016/j.cosust.2017.02.010

Sinare H, Gordon L, Kautsky E (2016) Assessment of ecosystem services and benefits in village landscapes: a case study from Burkina Faso. Ecosyst Serv 21:141–152. https://doi.org/10.1016/j.ecoser.2016.08.004

Swindale A, Bilinsky P (2006) Household dietary diversity score (HDDS) for measurement of household food access: indicator guide. Food and Nutrition Technical Assistance Project. Academy for Educational Development, Washington, DC

19

Climate Change, CSA and the Private Sector

Kealy Sloan, Elizabeth Teague, Tiffany Talsma, Stephanie Daniels, Christian Bunn, Laurence Jassogne, and Mark Lundy

19.1 Introduction

Agricultural researchers understand that there are no one-size-fits-all solutions to production issues: variations in climate, soil, farmer experience and many other factors mean that any advice must be specifically tailored to the given context (Osorio-Cortes and Lundy 2018). It is less well understood, however, that private sector supply-chain actors exhibit just as much variability, and also require tailor-made solutions. Civil society and public-sector donors, when working with businesses often lump them under a generic heading and approach them in the same way. To effectively engage private-sector to make substantial contributions to the promotion of climate-smart agriculture (CSA), they must be understood and approached in more nuanced ways.

This paper assesses how private-sector actors from different parts of the supply chain view, understand, and engage with climate change and the promotion of CSA practices. The private sector is increasingly at the center of market systems approaches because of their ability to facilitate innovation, access to producers and continuity of initiative (Vorley et al. 2009, Lundy et al. 2003). Our analysis draws

K. Sloan · S. Daniels
Sustainable Food Lab, Hartland, VT, USA
e-mail: ksloan@sustainablefood.org; sdaniels@sustainablefood.org

E. Teague
Root Capital, Cambridge, MA, USA
e-mail: eteague@rootcapital.org

T. Talsma · C. Bunn · M. Lundy (✉)
International Center for Tropical Agriculture (CIAT), Cali, Colombia
e-mail: t.talsma@cgiar.org; bunn@cgiar.org; t.talsma@cgiar.org; m.lundy@cgiar.org

L. Jassogne
International Institute of Tropical Agriculture (IITA), Oyo, Nigeria
e-mail: L.Jassogne@cgiar.org

on semi-structured interviews and broader engagement with 42 private firms working in coffee, cocoa and other commodity crops ("Private Sector Consultation" 2018).[1] Our findings indicate that many food and beverage companies already support action on climate change, at least in general terms. Most, however, say that they need more guidance on climate risks and CSA solutions, in order to deepen and scale their engagement. This study indicates that efforts to encourage private supply-chain actors to embrace CSA should emphasise the following efforts: (i) offering granular, subnational-level climate-risk data that will allow companies to integrate CSA into their broader risk-management strategies; (ii) providing CSA information and resources that are tailored to companies' specific position within the supply-chain; and (iii) emphasising the business case for CSA to make CSA uptake viable for companies that are held accountable to revenue goals.

19.2 Provide Granular Data to Assist in Risk Management

Most food and beverage companies recognise that climate change both exacerbates business risks and threatens ongoing sustainability efforts. In spite of this, many are reluctant to act because of uncertainty about how and when their supply chains will be affected, what role they should play and how to coordinate a response that is a part of holistic sourcing and sustainability strategies. Even those companies that are already taking action require more information in order to engage more deeply and at scale.

All companies conduct risk management as a core commercial function, and our interviews showed most food and beverage companies now routinely include climate change as one aspect of risk assessment. Companies generally spoke of two categories of climate-change risk: operations risk, or risk to physical assets such as processing facilities; and supply-chain risk, or risk of supply disruption. Risk varies according to the companies' physical footprint and supply-chain concentration. For instance, the mainstream cocoa and chocolate industry is heavily exposed to supply-chain risk, because most of the world's cocoa comes from West Africa, a region already experiencing the effects of climate change. The industry recognises the immediate and long-term threat of climate change to both the livelihoods of farmers and to a stable supply, as well as the pressures on forest health that may result from these threats (Lundy 2017).

[1] The Learning Community for Supply Chain Resilience, funded by USAID's Feed the Future program, interviewed 18 coffee companies (roasters and traders), 11 cocoa and chocolate companies (brands and traders) and 13 grain and ingredient companies. The goal was to better understand how they think about climate-smart agriculture, the types of activities in which they engage, and the types of climate information they use and/or need. Coffee and cocoa companies feature prominently because of the vulnerability of their supply chains to climate change: Coffee and cocoa are tree crops with long productive life cycles, and most producers are smallholder farmers in low-income countries.

Despite widespread recognition of climate-change risks, most companies are at the early stages of developing strategies that explicitly address CSA. Interviews revealed most companies address select pillars of CSA but rarely all three in a cohesive manner. For example, corporate sourcing and sustainability programs usually seek to increase productivity (the first pillar of CSA) via training, inputs, credit and efforts to strengthen community-level institutions. Multinational companies often have policies focused on reducing greenhouse gas emissions (the second pillar) in facilities under their direct control. Yet companies rarely reported efforts related to adaptation (the third pillar), in large part because adaptation action requires climate data that is more detailed than what is commonly available (Private Sector Consultation 2018).

What does climate action in the supply chain look like?
Some food and beverage companies are already moving from risk assessment to action. The trader Olam, for example, committed to buy climate-smart cocoa, which secured market access for farmers, and to pay premiums for Rainforest Alliance-certified cocoa. Similarly, coffee companies like Coop Coffee, JDE, Keurig Green Mountain, Lavazza and Nestlé are promoting CSA across their supply chains through training programs such as the Initiative for Coffee & Climate from the NGO Hanns R. Neumann Stiftung, or finance initiatives like the Coffee Farmer Resilience Initiative and the Rust Relief Fund.

Most companies interviewed explained that, as they make their first steps toward deepening their engagement in CSA, they would like the research community to clarify the key differences between CSA practices and long-promoted "good agricultural practices". All companies interviewed positioned their interest in CSA as an extension of both ongoing risk-management practice and sustainability programs focused on socio-economic development, environmental conservation and supply security through good agricultural practices. The companies seek to make their existing efforts more climate-smart rather than implementing new, isolated programs (Private Sector Consultation 2018).

The companies also expressed interest in particular types of data that would help inform their climate strategy. With some exceptions, most companies sought (i) granular (i.e., generally subnational) climate-risk data to diagnose and monitor their supply-chain and operational risks; (ii) guidance on specific, practical technologies to build resilience; (iii) more robust quantification of the economic impacts of climate change across producing regions; and (iv) risk projections for companion and/or alternative crops in regions facing diversification or transition. The companies called for this information to be more accessible: They would like researchers to provide more user-friendly data, such as brief fact sheets available through a central portal rather than academic papers housed behind a paywall (Private Sector Consultation 2018).

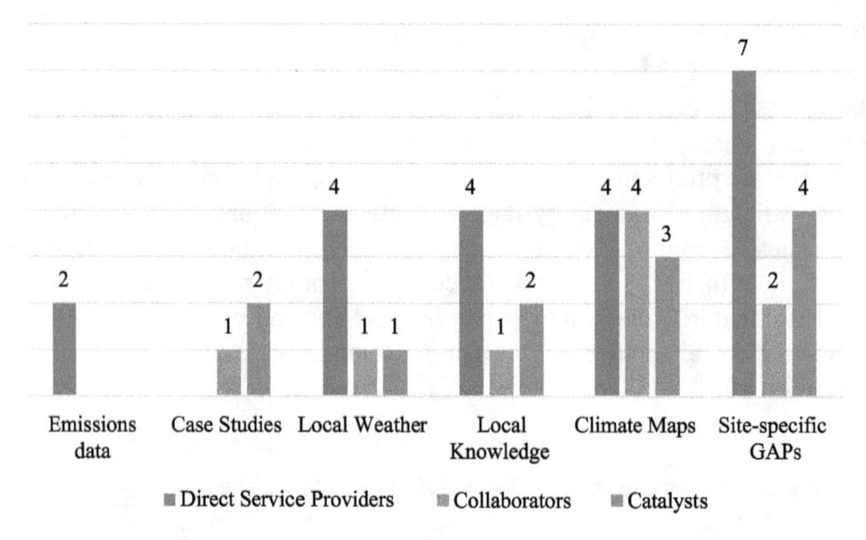

Fig. 19.1 Demand for climate change information by role (multiple choices allowed)

19.3 Tailor Information to Companies' Position in Supply Chain

Because food and beverage companies adopt different CSA strategies based largely on their position within the supply chain, researchers and policymakers should seek to tailor information and resources to suit individual needs.

Based on our research, we divided actors in the supply chain into three different categories[2]: (i) "direct service providers" (those providing services to smallholder farmers) such as ECOM Agroindustrial Corp. Ltd., (ii) "collaborators" (those working with direct service providers to deliver services to smallholder farmers), such as JDE, and (iii) "catalysts" (those working at a high level on climate issues with a light touch at the farm level), such as Tchibo. These actors get their climate information through a variety of different sources (Fig. 19.1).

Depending on their role, these actors see climate change through different lenses (see Table 19.1). **Direct service providers** were unlikely to distinguish between climate and sustainability efforts, but rather focused on holistic programs to increase productivity and make farming viable for today's farmers and attractive for the next generation. These companies were driven to action by farmer needs and were most interested in local knowledge and site-specific practices to help farmers adapt to climate change. **Collaborators** were more dependent on the direct service providers for information to shape their program design and often worked in partnership at a slightly higher level, looking to area-specific climate maps and case studies on successful programming to inform a broader strategy. **Catalysts** were more likely to be involved in broader, often industry-level conversations and interventions about

[2] Each category is followed by an example of a coffee company that fits that particular "role" in the typology.

Table 19.1 Access to and demand for climate information, by role within supply chain

Role	Access to and demand for climate information
Direct service providers	Currently have the most access to detailed farm-level data
	Seek more local information to supplement existing knowledge, such as changing local weather patterns and site-specific good agricultural practices (GAPs) that pair with their specific climate risks
Collaborators	Are dependent on the direct service providers for information to shape their program design and implementation
	Often work in collaborations at a slightly higher level, looking to area-specific climate maps and case studies on successful programming to inform a broader strategy
Catalysts	Rely on secondary sources of information from sector groups, such as sector platforms and trade groups, as well as desk research to answer particular questions
	Seek broad origin and risk mitigation information to inform global strategy
	May provide funding for research or services provision, may be visible as leaders in the sector, and may be interested in risk at origin, but rarely implement programs on the ground

climate change without directly intervening at the farm level. These companies seek multi-site risk mitigation and origin information to inform global strategy and collaborative solutions, often through sector platforms (Private Sector Consultation 2018).

Given the different needs of the actors with varying roles in the supply chain, researchers and policy makers should focus on providing the information most relevant to each. The varied demand for different types of information between different company roles can be seen in Fig. 19.1

19.4 Make the Business Case for CSA

Companies first and foremost are for-profit entities. They may see the need for longer-term solutions, but have to remain competitive and ensure they are meeting short-term financial goals as well as securing future supply. Whenever possible, researchers should emphasise return on investment and cost of inaction while connecting long-term climate projections to short-term productivity gains that both benefit the companies and build greater resilience in the agricultural system (Private Sector Consultation 2018).

For private companies, investment in CSA is driven primarily by efforts to secure a reliable supply and to avoid risks to their reputations. Supply security depends largely on the quantity sourced (those sourcing smaller quantities are less likely to feel this impact directly) and the sourcing region (the impacts of climate change are experienced more severely in some areas than in others). In the case of companies sourcing products of especially high quality, impacts can be pronounced even when volumes are low, if the regions that produce those goods are hard-hit by climate change. Reputational risks can range from severe to inconsequential, depending on the expectation by the consumer and/or the added value of a product grown according to climate-smart standards.

The degree to which companies choose to—or find themselves able to—invest in CSA depends on a range of factors. Among companies interviewed, those with dedicated sustainability staff embedded within procurement and sourcing departments often reported having an easier time incorporating CSA into their core sourcing strategies. Companies known for sustainability principles are often better able to prioritise such investment than their peers. In contrast, companies with shareholders who demand shorter-term profitability or quality results often have a more difficult time justifying the need for longer-term investment (Private Sector Consultation 2018). This is in line with recent findings on the determinants of corporate commitments to reduce deforestation as well (Lambin et al. 2017).

For most companies, private investment is a viable choice when contained within the company's own supply chain. As a lead firm, they are able to directly provide incentives to support CSA adoption amongst producers up the chain. However, when the benefits are less tangible or at risk for "leakage", blended finance models are well suited to these types of investment that deliver both public and private goods. This entails deliberate use of funds from capital providers with a range of financial and impact return expectation, from philanthropic capital with a negative rate of return, to those seeking capital preservation and below-market to market-rate returns (Private Sector Consultation 2018). Blended finance approaches can attract capital for investments addressing market failures or delivering significant social or environmental impact in emerging and frontier markets and enable more thoughtful longer-term investments in resilience by private sector actors.

Although many of the food and beverage companies surveyed already invest in CSA to some degree, they stressed the need for tangible, short-term business cases to justify ongoing investments in CSA. Companies must be able to capture the benefits of such investments via gains in volume or quality, increased supplier loyalty or deferred costs (Private Sector Consultation 2018).

19.5 Implications for Development

Our research highlights the need for the scientific community to provide more detailed, actionable information to incentivise companies' investments in CSA. Understanding the role each company plays in the supply chain—as direct service providers, collaborators or catalysts—can help define the type of information needed. Insights and approaches that effectively connect long-term climate projections with short-term productivity and weather variability are still needed to increase alignment between existing productivity focused approaches and effective CSA investments (Fig. 19.2)

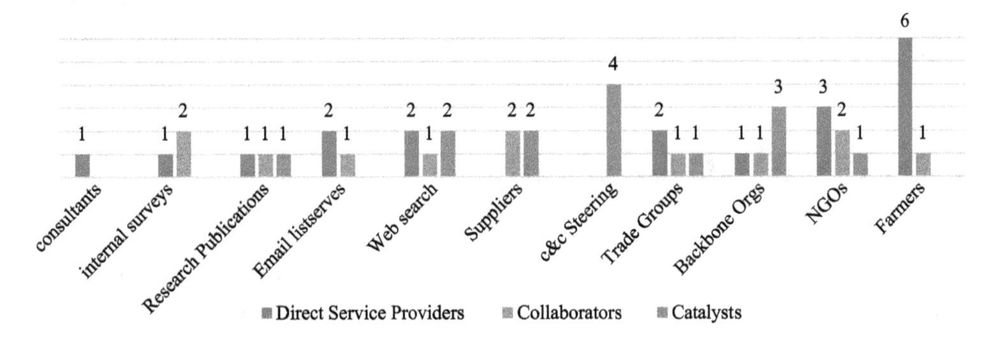

Fig. 19.2 Sources of climate-change information for various actors within the coffee industry (multiple choices allowed)

References

Lambin F, Gibbs HK, Heilmayr R et al (2017) The role of supply-chain initiatives in reducing deforestation. Nat Clim Chang 8:109–116. Available from:. https://doi.org/10.1038/s41558-017-0061-1

Lundy, M. (2017) Private Sector Development in Middle Income Countries. International Fund for Agricultural Development

Lundy M, Gottret MV, Cifuentes W, Ostertag CF, Best R (2003) Diseño de estrategias para aumentar la competitividad de cadenas productivas con productores de pequeña escala. International Center for Tropical Agriculture, CIAT, Rural Agroenterprise Development Project, Cali

Osorio-cortes, L. E., & Lundy, M. (2018). Scaling-Up Behaviour Change in Market Systems Development: A literature review. CGIAR Research Program on Policies, Institutions, and Markets (PIM), International Food Policy Research Institute (IFPRI).

Private Sector Consultation on Climate Smart Agriculture (2018) [online] Available at: http://sustainablefoodlab.org/wp-content/uploads/2018/01/Private-Sector-Consultation-on-Climate-Smart-Agriculture-FtF.pdf [Accessed 29 Mar. 2018].

Vorley, B., Lundy, M., and MacGregor, J. (2009) Business models that are inclusive of small farmers. Agroindustries for Development, Wallingford, UK: CABI for FAO and UNIDO, 186–222

20

CSA Value Chains

Caroline Mwongera, Andreea Nowak, An M. O. Notenbaert, Sebastian Grey, Jamleck Osiemo, Ivy Kinyua, Miguel Lizarazo, and Evan Girvetz

20.1 Introduction

Because of climate change, millions of people in sub-Saharan Africa are coping with rising temperatures (IPCC 2007; Jentsch et al. 2007; Engelbrecht et al. 2015) increases in the severity and frequency of droughts and (Jentsch et al. 2007; Allen et al. 2010; Ogalleh et al. 2012; Zhao and Dai 2015) floods (Mason et al. 1999; Frich et al. 2002; Douglas et al. 2008), rising pest and disease incidence, (Cheke and Tratalos 2007; Gregory et al. 2009) and soil degradation (Prospero and Lamb 2003; Brevik 2013). Some regions, such as southern Africa, will likely get drier during the winter season, while others (particularly at higher altitudes) may benefit as increased temperatures create new farming options (Christensen, J. H. et al. (2007)). Yields are likely to decrease substantially for cereal crops sensitive to heat and drought (wheat, maize, rice) but less so for crops with higher heat tolerance (such as millet) (Nelson et al. 2009; Leclerc et al. 2014). Overall, agricultural productivity and incomes have declined for smallholder farmers, pastoralists and fishermen, and are likely to decline further (FAO 2009; Gregory et al. 2009; Thulani and Phiri 2013; Junaidu et al. 2017).

Food security poses a growing challenge for much of the continent. To help people adapt to changing conditions, governments, the private sector and development partners

C. Mwongera (✉) · A. M. Notenbaert · S. Grey · J. Osiemo · I. Kinyua · E. Girvetz
International Center for Tropical Agriculture (CIAT), Nairobi, Kenya
e-mail: c.mwongera@cgiar.org; a.notenbaert@cgiar.org; S.Grey@cgiar.org; J.Osiemo@cgiar.org; I.Kinyua@cgiar.org; e.girvetz@cgiar.org

A. Nowak
World Agroforestry Center (ICRAF), Nairobi, Kenya
e-mail: a.nowak@cgiar.org

M. Lizarazo
International Center for Tropical Agriculture (CIAT), Cali, Colombia
e-mail: m.lizarazo@cgiar.org

have become interested in the uptake and scaling of climate-smart agriculture (CSA). Many of the studies to date have focused on the production end of the value chain—i.e., ways to help farmers grow more food. This limited focus neglects the importance of the harvesting, storage, processing and marketing stages. More researchers now are recognizing that food security is not just an issue of production but also of distribution, access and affordability (Ericksen 2008; Ingram 2011) CSA studies must follow suit.

This study argues that successful adaptation requires consideration of how climate change will affect all aspects of the value chain. It draws upon the county climate risk profiles (CRPs), a project of the International Center for Tropical Agriculture (CIAT) in collaboration with the Government of Kenya through the Ministry of Agriculture, Livestock and Fisheries and with funding through the World Bank. Addressing different stages of the value chain—input provision, on-farm production, harvesting, storage, processing and marketing—these CRPs assess actual and potential climate risks. The project's aim is to provide county governments and stakeholders with localized evidence of climate vulnerabilities and possible adaptation responses.

Each climate risk profile is framed around six key analytical stages: (i) overview of the agricultural context in the county; (ii) assessment of climate vulnerabilities across agricultural value-chain commodities; (iii) overview of on- and off-farm adaptation strategies specific to each selected value chain; (iv) analysis of available policies and programs to address climate change impacts on agriculture; (v) assessment of governance, institutional resources and capacity to incentivize uptake of adaptation strategies; and (vi) recommendations for addressing gaps that hinder effective institutional operation and collaboration. To date, profiles of 31 Kenyan counties have been developed.

This chapter presents a case study conducted in Nyandarua County. Our goal is to demonstrate the necessity of including value-chain perspectives in the design and scaling of CSA interventions.

20.2 Methodology

This paper draws on data collected and analyzed for Nyandarua County between June and September 2016. Nyandarua is located in the central area of the country and has a population of 596,268 (2009) over a land area of 3245 km^2. Temperatures range from 12 °C (July) to 25 °C (December), and annual rainfall ranges between a minimum of about 700 mm and a maximum of about 1700 mm spread over two seasons, mostly in the first wet season (January–June), but also in the second (short) wet season (September–December) (GOK 2014). The rainfall decreases from East to West. Agriculture is the main income-earning activity, employing 69% of the people, with crop production (estimated at 17 billion KES) and livestock keeping (7 billion KES) contributing 73% to the household incomes (MoALF 2016). Crop production in the county is mostly rain-fed, small-scale and for subsistence purposes. Malnutrition is a key challenge in the county, with 39% of the population estimated to be affected by food insecurity and 35% of children below 5 years stunted.

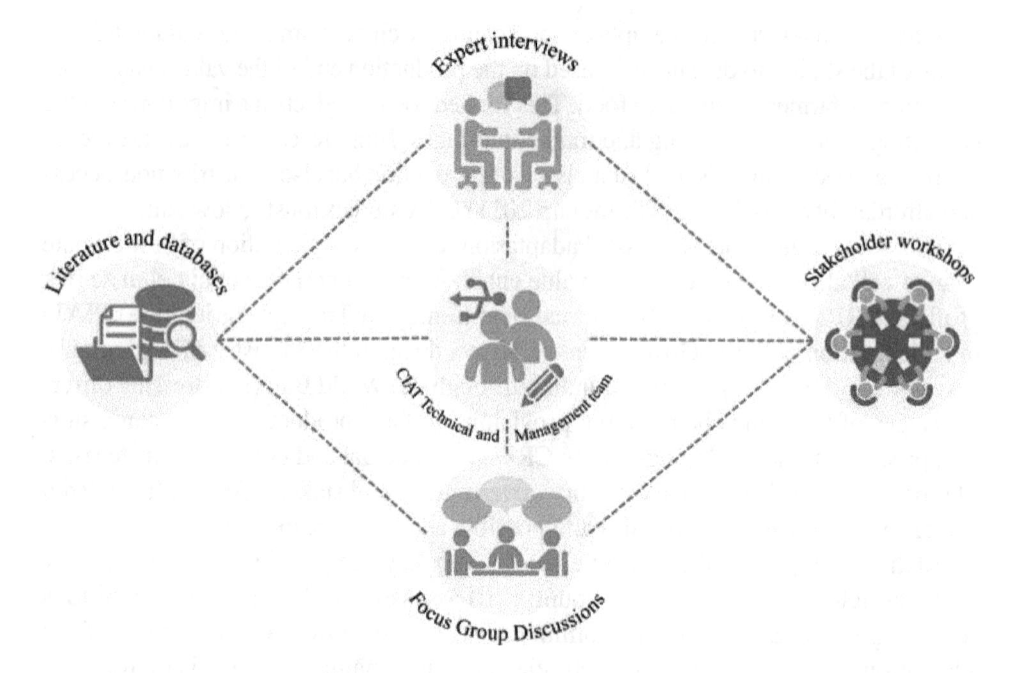

Fig. 20.1 Methodology of the Climate Risk Profiles showing the different approaches used for data collection

Creating the county climate risk profile for Nyandarua involved identifying major value-chain commodities, the key climate risks each faces and the adaptation options available. The study of each county relied on desktop research, climate-data analysis, farmers' focus groups, key informant interviews and a 3-day county stakeholder workshop attended by 30 farmers, service providers and representatives of governments, NGOs and farmer groups (Fig. 20.1). The focus groups brought together six to ten stakeholders representing each value chain. A total of 12 key informant interviews and six focus-group discussions were undertaken, with the goal of identifying stakeholder perceptions regarding: (i) activities along the value chain; (ii) current and potential climate-change impacts along the value chain, (iii) ongoing and potential adaptation options, and (iv) institutions, policies and programmes related to climate change adaptation in the county.

With input from the stakeholder workshop, we narrowed the list of agricultural commodities for analysis down to the four considered most important for food security and livelihoods: cow milk (dairy), poultry, peas and Irish potato. These were chosen based on contribution to food security, productivity, importance to the economy, resilience to current and future climate change, population engaged in the value chain and engagement of poor and marginalized groups. It emerged that at least 61% of the total population in the county are engaged in each of the four chosen value chains, involving all gender groups.

A mix of scientific and participatory approaches were used to identify which climate risks matter most for each commodity. The main climate hazards were identified based on the analysis of historical climate data (1981–2015) and climate projections (2021–2065) under RCPs 2.6 and 8.5. Climate indicators selected for the scientific assessments and initial presentation at the stakeholder workshop included moisture stress, drought stress, erosion risk, total precipitation, flooding and heat stress. During the stakeholder workshop, participants identified key value-chain activities, the two key climate risks for each value chain (from the six initially presented to them), magnitude of impact of the risk, underlying vulnerability factors (for specific groups of people) and who is most impacted (by geographical scope, age, gender and economic status). Participants also mapped currently available adaptation options and identified gaps in in the available options.

20.3 Results

As the main findings were consistent across all four value chains (dairy, poultry, peas, potato), this chapter presents results from the value chains of one crop (pea) and one type of livestock (dairy cows).

20.3.1 Effects of Climate Change on Value Chains

Based on the historic and future climate scenarios of the six indicators presented in the workshop, participants identified drought (represented by the number of consecutive days with moisture stress) and floods (represented by the magnitude of the wettest one-day event in mm/day) as the most relevant to the pea and dairy value chains. Historic climate analysis and participant perception agreed that both dry spells and extreme precipitation have been major hazards in the county. Future climate analyses for Nyandarua project significant increases in moisture stress in both seasons, as well as an increase in flood risk mostly in the second season (Figs. 20.2 and 20.3).

Based on stakeholder discussions, we also linked the perceived impacts of climate hazards to each stage in the value chain (Fig. 20.4). Drought affects all stages of the pea and dairy value chains, although in different ways. For example, while the effect of drought on pea inputs is largely moderate due to a limited availability of quality seed, the effect of drought on dairy inputs is severe, as it results in reduced breeding, poor quantity and quality of pasture and fodder, and increased costs in buying feed. In terms of the production stage, droughts severely affect both pea and dairy: peas suffer from low germination rates, hardened soils and increased incidence of pests and diseases; dairy cattle become emaciated and lose resistance to pests and diseases. In the dairy value chain, stakeholders perceive major to severe impacts from drought, which affects the harvesting, storage and processing stage. Drought also contributes to milk spoilage and increases operational costs in the collection and bulking of milk. Similarly, low levels of milk production can limit farmers' access to markets. Drought most adversely affects production activities in

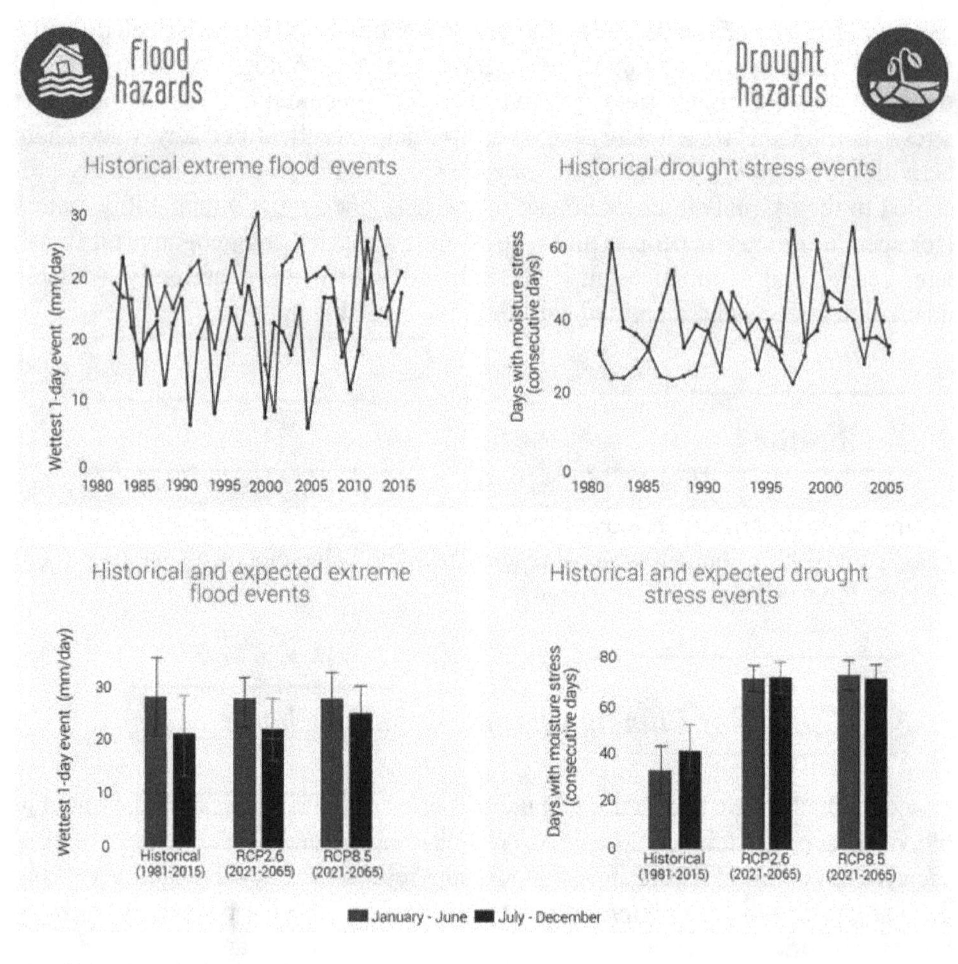

Fig. 20.2 Historical (1981–2015) and future projections (2021–2065) of flood and drought events in Nyandarua County, Kenya

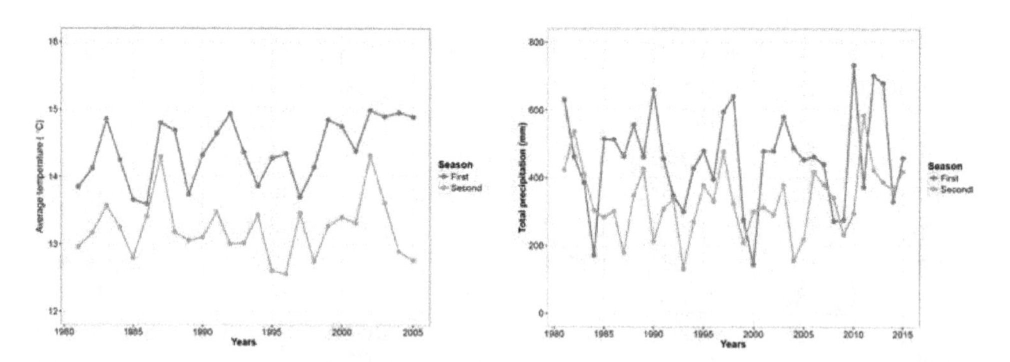

Fig. 20.3 Historical (1981–2015) average temperature and total precipitation in Nyandarua County, Kenya

Fig. 20.4 Drought impacts and adaptation options along the pea and dairy value chains in Nyandarua County, Kenya

pea: planting requires more time and labor due to hard soils; low germination increases the need for irrigation; and water stress leads to greater crop susceptibility to pest and diseases, low yields and poor quality produce.

In the dairy value chain, floods are perceived as having major to moderate negative impacts on provision of inputs, harvesting, storage and processing. In particular, excessive rainfall leads to destruction of roads, making inputs more expensive and increasing the cost of milk collection. Flooding also leads to damage of milk storage structures. In the pea value chain, impacts of floods on production were

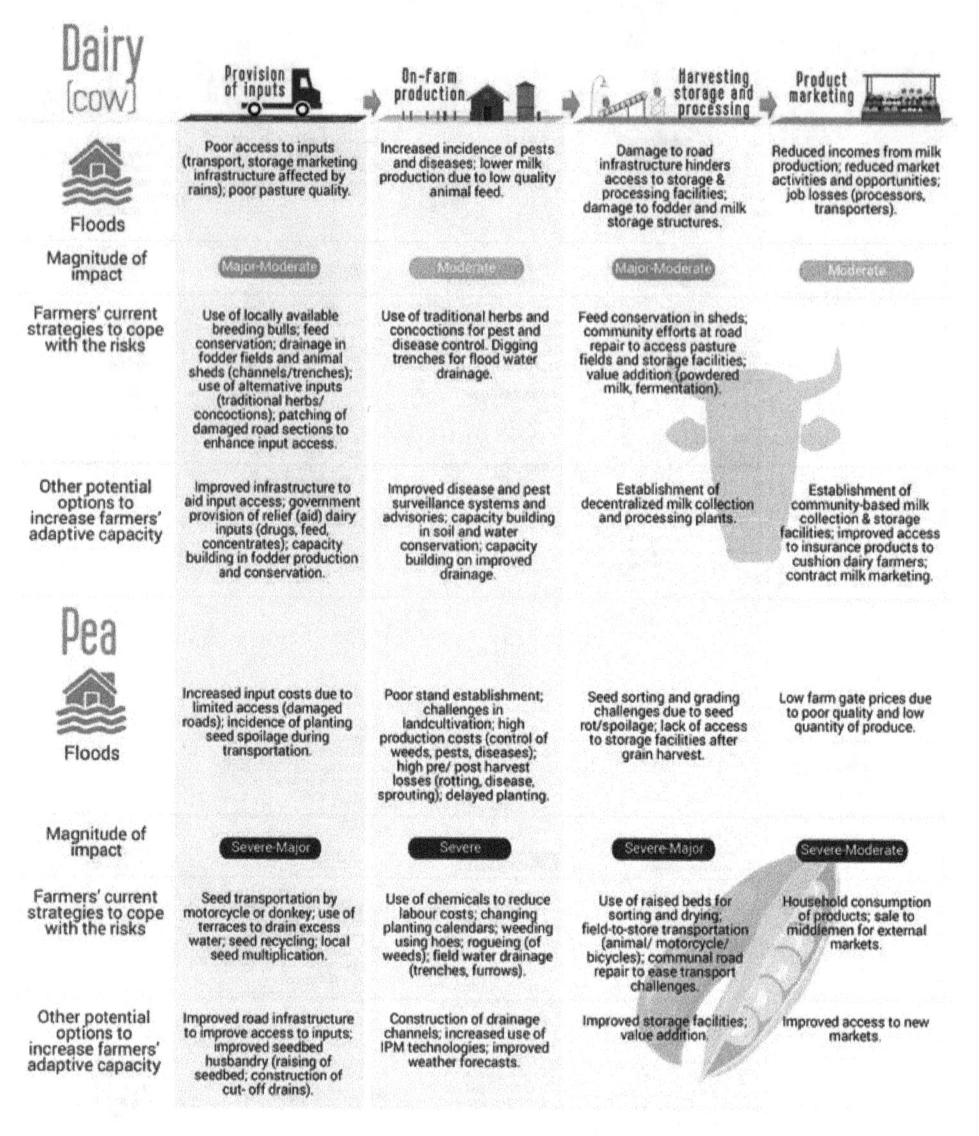

Fig. 20.5 Flood impacts and adaptation options along the pea and dairy value chains in Nyandarua County, Kenya

perceived as severe, leading to delayed planting, poor stand establishment, higher costs for labor and weed management, increased incidence of pests and diseases, and rotting of plants (Fig. 20.5). Apart from affecting on-farm production, floods also affect the transportation of inputs required for production, as roads may be damaged or become impassable. This damage to transport infrastructure can also hinder access to storage facilities, processing infrastructure and markets—consequences that often have knock-on effects for processors, agricultural buyers and their employees.

20.3.2 Options for Adapting Value Chains to Climate Change

These results show that climate hazards already negatively affect all activities along the chain. The impacts, however, vary by commodity and by stage of the chain, and therefore require different approaches in adaptation. This section examines the coping strategies used currently as well as the longer-term adaptation options.

Actors are already making some efforts to minimize the negative impacts of climate hazards and reduce climate risk, although our study indicates that their adaptation efforts are too heavily focused on production. For both peas and dairy, the number of adaptation options correlates to the perceived severity of the impacts. For example, a higher number of options were available for on-farm production in pea and at provision of inputs for dairy because of the impacts of drought (Fig. 20.4). For floods, the highest number of options adopted by actors is at the provision of input stage for both value chains (Fig. 20.5). Specifically, current adaptation strategies in dairy include feed conservation, fodder diversification (utilization of crop residues, herbs and shrubs for feed), use of herbal medicines, use of locally available breeding bulls, construction of drainage channels, local road repairs, sale of milk at farm gate, and value addition (milk fermentation). Strategies in the pea value chain include change of planting calendar, use of improved varieties, manure application, use of terraces, local seed multiplication, use of herbicides and pesticides, conservation agriculture, agroforestry, planting seedlings in raised beds and use of donkeys and motorcycle taxis for transportation (Figs. 20.4 and 20.5).

Our interviews identified the following potential priority actions in Nyandarua: (i) investing in climate-resilient infrastructures such as roads, irrigation systems, storage facilities and markets; (ii) engagement of the public and private sectors and financial and insurance services to support climate-resilient and inclusive agro-value chains; (iii) improve existing platforms and structures for climate adaptation along the value chain, such as standards, relief services, emergency funds, disease and pest surveillance, climate information services, early warning systems, land-use planning and zonation, agroforestry, soil and water conservation, value addition, collective marketing and climate responsive policies.

20.3.3 Impediments to Adaptation at the Local Level

The interviews in Nyandarua revealed a lack of understanding of climate change and the options available to adapt to it. Events such as reduction in crop cycle, rising temperatures and changes in length of the growing season were perceived as isolated or non-severe. There was also low awareness of potential adaptation options for managing risks. Similarly, there was a low understanding of the Kenyan government's climate-related policies and how they support adaptation at the local level. Most farmers in Nyandarua also fail to take advantage of the infrastructure and services (road networks, storage facilities, microfinance, and insurance) that might help them confront climate risks—either because they don't know about these options or because they can't afford them.

Overall, our results reveal the need to strengthen efforts to address climate change in Nyandarua County. It is noteworthy, that the focus is towards adaptation, and there is less attention to the mitigation potential of each adaptation option. CSA approaches have a weak presence, largely due to low institutional capacity and a weak policy environment. The institutions lack adequate guiding principles on climate change suited for the local context. Coordination among institutions also was noted as a challenge. Other institutional challenges included insufficient finances to enable wider project coverage, poor targeting of beneficiaries, poor monitoring and evaluation of the initiatives, and failure to properly engage stakeholders. The climate adaptation interventions that are undertaken have suffered from poor policies and weak implementation. Most significantly, for the purposes of this paper, the institutions focus primarily on the input acquisition and on-farm production stages, therefore missing the advantages of a value-chain approach.

20.4 Implications for Development

Our study indicates some strategies for addressing the policy and institutional challenges in Nyandarua County. First, the research project itself may have helped nudge the adaptation process forward. Recent research has recognized that stakeholder platforms can engage diverse actors and foster learning, coordination and fundraising (Wilson 2013; Wenger-Trayner et al. 2014; Ampaire et al. 2017). Further, such platforms can identify adaptation priorities and integrate them into development plans, directly influencing climate policy at the subnational level (Fleming et al. 2014). Our research supports these findings. In Nyandarua County the climate-risk profiling process brought stakeholders together, helped to identify the most vulnerable sections of the community in relation to each agricultural value chain and each hazard, and documented some of the ongoing projects aimed at mitigation. The CRPs can also be shared with stakeholders to help them better understand the climate-risk along different agricultural value chains as well as the best adaptation options. Engagement in the climate-risk profiling process helps experts evaluate CSA practices and determine which are most effective in helping the full length of the value chain adapt to the local context.

Policy can guide climate adaptation at many stages along the value chain. In Nyandarua, however, implementers at the local level appear to be not well informed of policy opportunities and barriers. Another key constraint there is the lack of local climate-change policies, as well as the lack of money and tools to implement national policies at local level. Agricultural development stakeholders and county government authorities plan to use information collected in the CRPs as part of the County Integrated Development Plans (CIDPs). For this to happen, stakeholders at every level must better understand the process of risk profiling and how it can help local farmers and the local economy. With local buy-in, risk profiling can be scaled out to the full range of agricultural commodities across the value chain. Climate risk profiling was a key input in the design of the US$250 million IDA-World Bank funded Kenya Climate-Smart Agriculture Project, for which Nyandarua is 1 of the 24 target counties.

Overall, our climate-risk analysis based on a value-chain approach showed that stakeholders are aware of the impacts of climate change along different stages of the value chain, and it revealed opportunities for adaptation in each of these stages. It also showed that value-chain analysis must reach beyond climate risks. Value chains are also vulnerable to pests and diseases; environmental degradation; changes in supply or demand; price fluctuations; logistical and infrastructural risks; financial, monetary, fiscal and tax policies; political risks; and security-related risks. Therefore, there is need for more comprehensive risk analysis in order to protect and build value chains.

References

Allen CD, Macalady AK, Chenchouni H, Bachelet D, McDowell N, Vennetier M, Kitzberger T, Rigling A, Breshears DD, Hogg ET, Gonzalez P (2010) A global overview of drought and heat-induced tree mortality reveals emerging climate change risks for forests. For Ecol Manag 259(4):660–684

Ampaire E, Acosta M, Mwongera C, Läderach P, Eitzinger A, Lamanna C, Mwungu C, Shikuku K, Twyman J, Winowiecki L (2017) Formulate equitable climate-smart agricultural policies. https://cgspace.cgiar.org/handle/10568/89093. Accessed 3 Mar 2018

Brevik EC (2013) The potential impact of climate change on soil properties and processes and corresponding influence on food security. Agriculture 3(3):398–417

Cheke RA, Tratalos JA (2007) Migration, patchiness, and population processes illustrated by two migrant pests. AIBS Bull 57(2):145–154

Christensen, J. H. et al. (2007) 'Regional Climate Projections', in Solomon, S. et al. (eds) Climate Change 2007: The Physical Science Basis. Report of the International Panel on Climate Change. Cambridge, United Kingdom and New York, NY, USA: Cambridge University Press, pp 847–940

Douglas I, Alam K, Maghenda M, Mcdonnell Y, McLean L, Campbell J (2008) Unjust waters: climate change, flooding and the urban poor in Africa. Environ Urban 20(1):187–205

Engelbrecht F, Adegoke J, Bopape MJ, Naidoo M, Garland R, Thatcher M, McGregor J, Katzfey J, Werner M, Ichoku C, Gatebe C (2015) Projections of rapidly rising surface temperatures over Africa under low mitigation. Environ Res Lett 10(8):085004

Ericksen PJ (2008) What is the vulnerability of a food system to global environmental change? Ecol Soc 13(2):14

FAO (2009) Climate change in Africa: The threat to agriculture. Food and Agricultural Organization of the United Nations, Regional Office for Africa. Accra, Ghana, p 5

Fleming A, Hobday AJ, Farmery A, Van Putten EI, Pecl GT, Green BS, Lim-Camacho L (2014) Climate change risks and adaptation options across Australian seafood supply chains–a preliminary assessment. Clim Risk Manag 1:39–50

Frich P, Alexander LV, Della-Marta PM, Gleason B, Haylock M, Tank AK, Peterson T (2002) Observed coherent changes in climatic extremes during the second half of the twentieth century. Clim Res 19(3):193–212

GOK (2014) Population and Housing Census 2009. Nairobi, Kenya. https://www.knbs.or.ke/2009-kenya-population-and-housing-census-analytical-reports/. Accessed 3 Mar 2018

Gregory PJ, Johnson SN, Newton AC, Ingram JS (2009) Integrating pests and pathogens into the climate change/food security debate. J Exp Bot 60(10):2827–2838

IPCC (2007) Climate Change 2007: the physical science basis. In: Solomon S, Qin D, Manning M, Chen Z, Marquis M, Averyt KB, Tignor M, Miller HL (eds) Contribution of Working Group I to the Fourth Assessment Report of the Intergovernmental Panel on Climate Change. Cambridge University Press, Cambridge, United Kingdom and New York, NY, USA, pp 996

Ingram J (2011) A food systems approach to researching food security and its interactions with global environmental change. Food Security 3(4):417–431

Jentsch A, Kreyling J, Beierkuhnlein C (2007) A new generation of climate-change experiments: events, not trends. Front Ecol Environ 5(7):365–374

Junaidu M, Ngaski AA, Abdullahi BS (2017) Prospect of Sub-Saharan African Agriculture Amid climate change: a review of relevant literatures. Int J Sustain Manag Inf Technol 3:20–27

Leclerc C, Mwongera C, Camberlin P, Moron V (2014) Cropping system dynamics, climate variability, and seed losses among East African smallholder farmers: a retrospective survey. Weather, Clim, Soc 6(3):354–370

Mason SJ, Waylen PR, Mimmack GM, Rajaratnam B, Harrison JM (1999) Changes in extreme rainfall events in South Africa. Clim Chang 41(2):249–257

MoALF (2016) Climate risk profile of Nyandarua, Kenya County Climate Risk Profiles Series. Nairobi, Kenya. https://cgspace.cgiar.org/rest/bitstreams/119946/retrieve. Accessed 3 Mar 2018

Nelson GC, Rosegrant MW, Koo J, Robertson R, Sulser T, Zhu T, Ringler C, Msangi S, Palazzo A, Batka M, Magalhaes M (2009) Climate change: impact on agriculture and costs of adaptation (Vol. 21). Intl Food Policy Res Inst. https://doi.org/10.2499/0896295354 Accessed 3 Mar 2018

Ogalleh SA, Vogl CR, Eitzinger J, Hauser M (2012) Local perceptions and responses to climate change and variability: the case of Laikipia District. Kenya Sustain 4(12):3302–3325

Prospero JM, Lamb PJ (2003) African droughts and dust transport to the Caribbean: climate change implications. Science 302(5647):1024–1027

Thulani D, Phiri K (2013) Rural livelihoods under stress: the impact of climate change on livelihoods in South Western Zimbabwe. Am Int J Contemp Res 3(5):11–25

Wenger-Trayner E, Fenton-O'Creevy M, Hutchinson S, Kubiak C, Wenger-Trayner B (2014) Learning in landscapes of practice: boundaries, identity, and knowledgeability in practice-based learning. Routledge, London, UK

Wilson GA (2013) Community resilience, policy corridors and the policy challenge. Land Use Policy 31:298–310

Zhao T, Dai A (2015) The magnitude and causes of global drought changes in the twenty-first century under a low–moderate emissions scenario. J Clim 28(11):4490–4512

21

Climate Change and Development of Nutrition-Sensitive Value Chain

Summer Allen and Alan de Brauw

21.1 Introduction

Food security has improved over the past quarter century in developing countries, with the number of undernourished people declining from 900 million in 2000 to 815 million in 2017 (FAO 2017). Yet while a larger proportion of the world's population can now access enough food in terms of caloric requirements, it is not necessarily nutritious. The 2016 Global Nutrition Report states that micronutrient deficiency remains stubbornly high, with obesity rates increasing rapidly in low- and middle-income countries (IFPRI 2016). For example, wasting (low weight for height, a sign of undernutrition) affected 52 million children under 5 in 2016, yet 41 million children were overweight the same year (FAO 2017).

In sub-Saharan Africa and South Asia, regions which rank highest in malnutrition rates, climate change poses a severe threat to food security; changes in temperature, precipitation patterns and disease environments are expected to reduce yields by levels as high as 2% per decade (GLOPAN 2015). Whilst heat and water stress will increase the incidence of pests and diseases, higher temperatures will also increase spoilage of fresh, nutritious foods, and climate events such as flooding will prevent their transport to market (Vermeulen et al. 2012). Climate change can also exacerbate nutritional deficiencies – increased CO_2 concentrations reduce the nutritional quality of crops, such as the protein content of grain crops and soybeans (Myers et al. 2017; Taub et al. 2008).

Most studies focus on the effects of climate change on agricultural productivity levels, but few have analysed the impact of climate change on household nutrition (e.g., Kabubo-Mariara et al. 2016; Springmann et al. 2016). For example it is estimated that globally, reduced fruit and vegetable consumption (caused by reduced

S. Allen (✉) · A. de Brauw
Markets, Trade and Institutions Division, International Food Policy Research Institute (IFPRI), Washington, DC, USA
e-mail: s.allen@cgiar.org; a.debrauw@cgiar.org

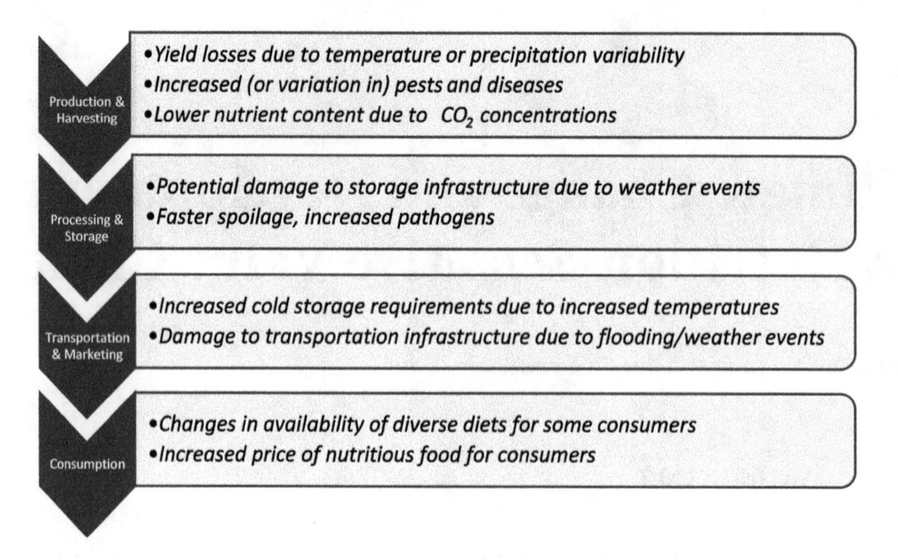

Fig. 21.1 Potential climate-related impacts to food value chains

crop availability and changes in consumption patterns) will result in 534,000 deaths (Springmann et al. 2016). Another study estimates that the number of malnourished children in developing countries is likely to increase by 8.8–10% due to climate change (Nelson et al. 2010).

By definition, food value chains include all actors and activities from producer to consumer, including: inputs into production, crop production, storage and processing, distribution and transportation, food retail and labeling, and consumption. The vulnerability of value chain activities to climate change could make production more expensive. Other factors could also affect costs, such as changes in energy or agricultural policies (Fig. 21.1). For example, rising temperatures and variable precipitation patterns will impact growing seasons, locations and water and nutrient demand, whilst also risking food safety, making storage and transportation even more critical (Fanzo et al. 2017).

Globally, agriculture and food systems need to adapt to meet the challenges of climate change if they are to support the diet of the growing global population. One promising option is the development of more nutrition-sensitive value chains that increase access to nutritious foods for local markets (e.g., Hawkes and Ruel 2012; Gelli et al. 2015). This approach relies on crop varieties that are tolerant to drought and heat, commodities with increased nutrient content, and reduced food losses.

This chapter provides a brief overview and examples of nutrition-sensitive value chains, and the research and findings thus far regarding how they can improve nutrition at the household level in Africa. The policy efforts supporting nutrition-focused agricultural practices in a changing climate will also be discussed.

21.2 Nutrition-Sensitive Value Chains

Achieving the second Sustainable Development Goal (to end hunger, achieve food security and improved nutrition, and promote sustainable agriculture) is challenging in a changing climate. Under most conditions, dietary choice does not align with what is optimal nutritionally (Allen and de Brauw 2017). There are, however, ways to improve the nutritional intake of consumers. Nutrition sensitive value chain interventions are a class of interventions that take place through a range of value chain actors to ensure more nutritious products reach consumers. Relative prices can shift either through improvements in the value chain or through regulation. Likewise, marketing campaigns and improved labelling can persuade the consumer to purchase more nutritious foods.

Decisions made on the supply side, for example regarding which foods to produce or with whom to trade, largely depend on the expected profit, which limits the nutritional composition of foods in the value chain. The nutrition-related or environmental consequences of value chain activity are rarely monetised. Producers are unlikely to shift production to more nutritious or environmentally sustainable foods if they will not result in increased profit. Previous research on nutrition-sensitive value chains, therefore looked for ways to ensure profitability, such as temperature-controlled supply chains for perishable foods, contracts that support the production of vegetables, and increased subsidies for infrastructure and inputs (Allen et al. 2016; Chege et al. 2015; Stifel and Minten 2017).

To create sustainable, climate-smart value chains consideration of synergies and trade-offs among economic, environmental and social objectives, including nutrition and health may be required (FAO 2013). For example animal source foods (meat, milk, eggs) are nutrient-dense, but producing them is both land and water intensive (Marlow et al. 2009). As the climate changes, these social, environmental, and economic trade-offs will shift with relative prices and the profitability of specific activities will change. Examples of this include the effect of variations in monsoon timing and strength in India (upon which both agriculture and energy depend), and the effects of biofuel policies in countries such as Ethiopia, that can lead to higher food prices and land use changes when compared to policies focused on agriculture and food security (Lobell et al. 2014).

Figure 21.2 illustrates the complex relationships that must be considered in nutrition-sensitive value chains: a delicate balance of economic (profit), social (including nutrition), and environmental impacts, including greenhouse gas emissions (figure adapted from FAO 2013). Though many of these social and environmental impacts are not monetised, pressure on resources (for example, through drought, floods and changes in soil productivity) and changes in consumer demand (for example, for products such as palm oil or animal source foods) can affect profitability and promote unsustainable production decisions.

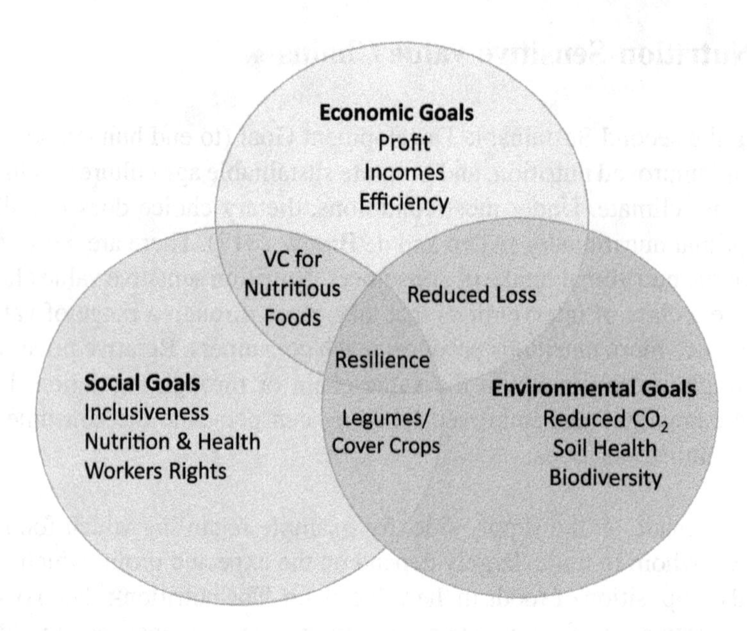

Fig. 21.2 Trade-offs and synergies for sustainable food chain development. (Source: Authors' Adaptation from FAO 2013)

21.3 Value Chains for Nutritious Foods: Lessons from the Field

Increasing the availability of nutritious foods is necessary to deal with substantial micronutrient malnutrition. In 2014, 32% of children below the age of 5 in sub-Saharan Africa were stunted (IFPRI 2016). Unfortunately, household-based interventions to improve the diet are often expensive (e.g., de Brauw et al. 2015). Such interventions are also complex. In Ethiopia, for example, increasing knowledge about nutrition was successful in improving dietary diversity, but only when households had sufficient market access (Hirvonen et al. 2017). Despite the challenges, value chain interventions have the potential to be cost-effective as they involve the private sector and, if successfully expanded, can reach a much larger set of beneficiaries than interventions focusing on individual households.

One intervention which would increase food security and diet diversity and that has potential for upscaling is the distribution, through input dealers, of nutritious crops that are more resilient to climate shocks. Beans are nutritious, but their productivity gains have lagged behind grain crops (Joshi and Rao 2017). In Malawi, the International Center for Tropical Agriculture (CIAT) and the Pan-Africa Bean Research Alliance (PABRA) have collaborated to study access to improved bean varieties. Though there is evidence of higher yields and dietary diversity for those who adopt the improved varieties, their initial use is associated with access to extension and mobile phones (Katungi et al. 2017).

Given the positive relationship between production diversity and dietary diversity shown in subsistence-oriented contexts, crop diversification could meet the dual needs for a more resilient crop mix (in the case of pests, disease, or extreme weather) and a more diverse diet at the household level. However, this relationship is not always positive. Once strong access to markets or increased technology adoption in agriculture is attained with concurrent increases in agricultural income, the relationship between production diversity and dietary diversity does not appear to be as strong (Koppmair, Kassie, and Qaim 2016). Farmers that are able to specialise do so because they both have higher incomes and they are able to mitigate the risk of specialisation.

Local supply chains that support more diverse diets can address these challenges. For example, in several countries homegrown school feeding programs that source school food from local producers have been implemented (WFP 2017). Recent work in Malawi has focused on testing whether such programs, when combined with behavior change communication (to improve nutrition, support local agriculture, and improve attendance at schools), can be effective in addressing malnutrition. The results are currently being finalised, but there is already evidence that these types of interventions can lead to improved dietary intake in preschool children and growth in their younger siblings (Gelli et al. 2017).

Finally, one of the most promising initiatives to increase the content of micronutrients in diets is biofortification, which involves breeding staple crops, including sweet potato, maize, beans and cassava, for higher micronutrient levels (Bouis et al. 2011). HarvestPlus has released biofortified crops and is supporting their inclusion into value chains and the market system in the Democratic Republic of Congo, Rwanda, Nigeria, Uganda and Zambia. For example, they are working with food processors to develop and market products using yellow cassava fortified with Vitamin A in Nigeria, and orange maize in Zambia (HarvestPlus 2017). Randomised control trials have demonstrated that biofortification can be effective in reducing the prevalence of inadequate micronutrient intake (Hotz et al. 2012a, b).

21.4 Nutrition-Sensitive Value Chains in a Changing Climate

In 2017 the World Economic Forum highlighted the need for inclusive, sustainable and efficient food systems that deliver nutritious food. Climate change significantly impacts malnutrition, both directly, through heat stress and water constraints, and indirectly, through loss in production. These impacts could be as large as changes in other socioeconomic indicators such as access to electricity and educational attainment (Davenport et al. 2017). The effect on stunting, however, could be partially mitigated by investments in education and electricity (Davenport et al. 2017). To build food systems that are resilient to climate change, it is critical that limited resources are used efficiently and losses reduced across the value chain (FAO 2013).

At the farm level there are opportunities for agroforestry to promote nutritious crops while ensuring more sustainable production in terms of soil health and carbon sequestration. The Initiative for the Adaptation of African Agriculture to Climate Change (AAA Initiative) notes opportunities for more integrated management of pastoral and forest systems that, in turn, can improve management of limited resources; agroforestry in particular offers the opportunity for producers to diversify their production (and income), maintain soil fertility and water resources, and provide carbon sequestration (AAA Initiative 2017). In addition to maintaining soil health and increasing pollination, forests also provide nutritious food (fruits, berries, mushrooms), cooking fuel and income opportunities (through the sale of forest products). In addition to maintaining soil health and increasing pollination, forests also provide nutritious food (fruits, berries, mushrooms), cooking fuel and income opportunities (through the sale of forest products); in southern Ethiopia proximity to a forest increased dietary diversity due to the increase in feed for livestock and resulting organic fertilizer for home gardens (Baudron et al. 2017).

It is critical that loss at the farm level is reduced by developing nutritious, resilient crop varieties that can tolerate climate variability. For example, researchers at the World Vegetable Center evaluated heat-tolerant and disease-resistant tomato varieties in Tanzania and found the rate of return to seed improvement to be as high as that reported for some staple crops (Schreinemachers et al. 2017). In the same region, access to improved pigeonpea varieties also increased income returns for farmers (Shiferaw et al. 2008).

Ultimately, to take full advantage, resilience at the farm level should be pursued together with activities such as increasing soil organic carbon and diversity of production and trade (FAO 2013). On farm, activities that promote nutrition-sensitive agriculture can improve soil organic carbon and incentivise crop diversity through strategic (and nutritious) cover crops, such as pigeonpea. Though rotation systems that use legumes and vegetables can increase rice yields in Africa, joint public-private sector strategies are needed to ensure quality legume seed production and distribution (Ojiewo et al. 2015).

After harvest, additional steps can be taken to reduce loss. Many nutritious foods (including fruit, vegetables, and dairy products) are perishable, therefore technology and good agricultural practices would increase the resilience of these value chains to climate shocks. Increased efficiency can also be achieved by providing storage and distribution infrastructure in value chains. This would also limit the impact of the production system on the environment, and vice versa (Gomez and Ricketts 2017).

Beyond adaption, opportunities also exist to introduce more nutritious crop varieties and increase uptake through location-specific interventions. The orange sweet potato from HarvestPlus and the International Potato Center (CIP) expanded to 14 sub-Saharan African countries through partnerships with Feed the Future and private companies, and exemplifies positive progress in interventions for nutritious foods. Indeed, by September 2016 orange sweet potato was estimated to have reached 2.89 million households (Low et al. 2017). However, the effectiveness of such interventions is dependent on consumer behavior and ultimately, consumer acceptance of the biofortified varieties. In the above example of the orange sweet potato, several consumer acceptance trials preceded large rollouts of specific varieties (e.g., Chowdhury et al. 2010). The importance of consumer behavior also extends to other nutritious crops; in Rwanda, the acceptance of high-iron beans was dependent upon location, income, and related nutritional information provided (Murekezi et al. 2017).

In addition, there are opportunities for public-private partnerships like the Feed the Future supported partnership in Ethiopia, to fortify wheat flour and provide iodised salt, with UNICEF and the Global Agriculture Information Network

(Gillespie et al. 2017). The advantage of a value chain approach that incorporates the private sector is that, if profitable, entrepreneurs will have an incentive to further develop them, but a focus on both consumer and producer is necessary to target (and affect) malnutrition (Allen and de Brauw 2017). Partnerships between the public and private sectors have been developed to increase access to fortified foods, which can be especially important in rural areas (Gomez and Ricketts 2017). Understanding economic mechanisms behind consumer choices as well as how those may change as a result of climate change or variability will be necessary to ensure that activities targeting more nutritious crop production and consumption are sustainable.

21.5 Implications for Development

A growing body of evidence demonstrates that climate change will strain current agricultural production systems, with negative consequences for food security. However, the ramifications of climate change and increased yield variability on nutrition are not so well documented. The most nutritious crops may be less desirable for producers as they often require more inputs and need to be stored quickly after harvest to mitigate against spoilage.

Value chain interventions are an attractive option because they can overcome constraints on the use of inputs and support the development of transport and storage facilities for healthier products. As the climate continues to change, it will be increasingly important to strengthen nutrition-sensitive value chains so that producers have inputs, markets, and price incentives for these products.

These interventions will need to be tailored to the constraints and opportunities of specific regions, and attention must be paid to any environmental trade-offs that might be required. A number of nutrition-sensitive chains could provide resilience to climate change, including chains for biofortified crops and varieties (such as beans) that are more tolerant to heat and moisture stress.

As noted, it will be important to consider social and environmental trade-offs when evaluating the cost-effectiveness of value chain interventions and related programs, including capacity development infrastructure. Finally, public-private partnerships that strengthen market linkages can also be developed to improve the nutritious content of food and account for its environmental footprint.

References

Allen S, de Brauw A (2017) Nutrition sensitive value chains: theory, progress, and open questions. Global Food Security. In press. Available from: https://doi.org/10.1016/j.gfs.2017.07.002

Allen S, de Brauw A, Gelli A (2016) Nutrition and sustainability: harnessing value chains to improve food systems. In: International Food Policy Research Institute. 2016 Global Food Policy Report. International Food Policy Research Institute, Washington DC, p 49–56

Baudron F, Chavarría JV, Remans R et al (2017) Indirect contributions of forests to dietary diversity in Southern Ethiopia. Ecol Soc 22(2):28. https://doi.org/10.5751/ES-09267-220228

Bouis HE, McClafferty B, Meenakshi JV et al (2011) Biofortification: a new tool to reduce micronutrient malnutrition. Food Nutr Bull 32:S31–S40 Available from: https://www.ncbi.nlm.nih.gov/pubmed/21717916

Chege C, Andesson C, Qaim M (2015) Impacts of supermarkets on farm household nutrition in Kenya. World Dev 72:394–407

Chowdhury S, Meenakshi JV, Tomlins KI et al (2010) Are consumers in developing countries willing to pay more for micronutrient-dense biofortified foods? Evidence from a field experiment in Uganda. Am J Agric Econ 93(1):83–97. https://doi.org/10.1093/ajae/aaq121

Davenport F, Grace K, Funk C et al (2017) Child health outcomes in sub-Saharan Africa: a comparison of changes in climate and socio-economic factors. Glob Environ Chang 46:72–87. https://doi.org/10.1016/j.gloenvcha.2017.04.009

De Brauw A, Eozenou P, Gilligan D et al (2015) Biofortification, crop adoption and health information: impact pathways in Mozambique and Uganda. HarvestPlus Working Paper #21. Available from: http://ebrary.ifpri.org/cdm/ref/collection/p15738coll2/id/129795

Fanzo J, McLaren R, Davis C et al (2017) Climate change and variability: what are the risks for nutrition, diets, and food systems? IFPRI Discussion Paper 01645: International Food Policy Research Institute, Washington, DC. Available from: http://ebrary.ifpri.org/cdm/ref/collection/p15738coll2/id/131228

FAO (2013) Climate-smart agriculture sourcebook. Food and Agriculture Organization of the United Nations (FAO), Rome. Available from: http://www.fao.org/docrep/018/i3325e/i3325e.pdf

FAO (2017) The state of food security and nutrition in the world: building resilience for peace and food security, Food and Agriculture Organization of the United Nations (FAO), Rome. Available from: http://www.fao.org/3/a-I7695e.pdf

Gelli A, Hawkes C, Donovan J et al (2015) Value chains and nutrition: a framework to support the identification, design and evaluation of interventions, IFPRI Discussion Paper 01413, International Food Policy Research Institute, Washington, DC. Available from: http://ebrary.ifpri.org/cdm/ref/collection/p15738coll2/id/128951

Gelli A, Margolies A, Santacroce M et al (2017) The NEEP-IE Study: rationale, design, and preliminary findings. Presentation at policy dialogue: improving food security, diets, and nutrition through multi-sectoral action: new evidence, challenges, and opportunities on 30 May 2017 in Lilongwe

Gillespie S, Van den Bold M, Stories of Change Study Team (2017) Stories of change in nutrition: an overview. Glob Food Sec 13:1–11. Available from. https://doi.org/10.1016/j.gfs.2017.02.004

GLOPAN (2015) Climate-smart food systems for enhanced nutrition. Policy Brief. Global Panel on Agriculture and Food Systems for Nutrition, London. Available from: https://glopan.org/sites/default/files/pictures/GloPan%20Climate%20Brief%20Final.pdf

Gomez M, Ricketts K (2017) Innovations in food distribution: food value chain transformations in developing countries and their implications for nutrition. In: Dutta S, Lanvin B, Wunsch-Vincent S (eds) The global innovation index. Cornell University, INSEAD, and WIPO, Ithaca

HarvestPlus (2017) Driving impact: harvest plus annual report 2016. HarvestPlus, Washington, DC. Available from: https://issuu.com/harvestplus/docs/harvestplus_annual_report_2016_fina

Hawkes C, Ruel MT (2012) Value chains for nutrition. In: Fan S, Pandya Lorch R (eds) Reshaping agriculture for nutrition and health. International Food Policy Research Institute, Washington, DC

Hirvonen K, Hoddinott J, Minten B et al (2017) Children's diets, nutrition knowledge, and access to markets. World Dev 95:303–315. Available from. https://doi.org/10.1016/j.worlddev.2017.02.031

Hotz C, Loechl C, de Brauw A et al (2012a) A large-scale intervention to introduce orange sweet potato in rural Mozambique increases vitamin A intakes among children and women. Br J Nutr 108(1):163–176. https://doi.org/10.1017/S0007114511005174

Hotz C, Loechl C, Lubowa A et al (2012b) Introduction of beta-carotene-rich orange sweet potato in rural Uganda resulted in increased vitamin A intakes among children and women and improved vitamin A status among children. J Nutr 142(10):1871–1880. https://doi.org/10.3945/jn.111.151829

IFPRI (2016) Global nutrition report 2016: from promise to impact: ending malnutrition by 2030. International Food Policy Research Institute, Washington, DC Available from: http://www.globalnutritionreport.org/the-report-2016/

AAA Initiative (2017) Addressing the challenges of climate change and food security, White Paper. The Initiative for the Adaptation of African Agriculture Climate Change. Available from: http://www.aaainitiative.org/sites/aaainitiative.org/files/AAA_livre_blanc_ENG.pdf

Joshi PK, Rao PP (2017) Global pulses scenario: status and outlook. Ann N Y Acad Sci 1392(1):6–17. https://doi.org/10.1111/nyas.13298

Kabubo-Mariara J, Mulwa RM, DiFalco S (2016) The impact of climate change on food calorie production and nutritional poverty: evidence from Kenya. Environment for Development Discussion Paper Series 16–26. The Environment for Development (EfD) initiative. Available from: http://www.rff.org/files/document/file/EfD-DP-16-26.pdf

Katungi E, Magreta R, Letaa E et al (2017) Adoption and impact of improved bean varieties on food security in Malawi. CIAT-PABRA Research Technical Report. Pan Africa Bean Research Alliance, Cali. Available from: https://cgspace.cgiar.org/handle/10568/82725

Koppmair S, Kassie M, Qaim M (2016) Farm production, market access and dietary diversity in Malawi. Public Health Nutr 20(2):325–335

Lobell DB, Naylor RL, Field CB (2014) Food, energy, and climate connections in a global economy. In: Naylor R (ed) The evolving sphere of food security. Oxford, New York

Low JW, Mwanga RO, Andrade M et al (2017) Tackling vitamin A deficiency with biofortified sweet potato in sub-Saharan Africa. Glob Food Sec 14:23–30

Marlow HJ, Hayes WK, Soret S et al (2009) Diet and the environment: does what you eat matter? Am J Clin Nutr 89(S5):1699S–1703S

Murekezi A, Oparinde A, Birol E (2017) Consumer market segments for biofortified iron beans in Rwanda: evidence from a hedonic testing study. Food Policy 66:35–49

Myers SM, Smith MR, Guth S et al (2017) Climate change and global food systems: potential impacts on food security and undernutrition. Annu Rev Publ Health 38:259–277. Available from. https://doi.org/10.1146/annurev-publhealth-031816-044356

Nelson GC, Rosegrant MW, Palazzo A et al (2010) Food security, farming, and climate change to 2050: scenarios, results, policy options. International Food Policy Research Institute, Washington, DC

Ojiewo C, Keatinge DJDH, Hughes J et al (2015) The role of vegetables and legumes in assuring food, nutrition, and income security for vulnerable groups in Sub-Saharan Africa. World Med Health Policy 7(3):187–210. https://doi.org/10.1002/wmh3.148

Schreinemachers P, Sequeros T, Lukumay PJ (2017) International research on vegetable improvement in East and Southern Africa: adoption, impact and returns. Agricultural Economics. Available from: https://doi.org/10.1111/agec.12368

Shiferaw BA, Kebede TA, You L (2008) Technology adoption under seed access constraints and the economic impacts of improved pigeonpea varieties in Tanzania. Agric Econ 39(3):309–323. https://doi.org/10.1111/j.1574-0862.2008.00335.x

Springmann M, Godfray HC, Rayner M et al (2016) Analysis and valuation of the health and climate change cobenefits of dietary change. Proc Natl Acad Sci USA 113:4146–4151. https://doi.org/10.1073/pnas.1523119113

Stifel D, Minten B (2017) Market access, well-being, and nutrition: evidence from Ethiopia. World Dev 90:229–241

Taub DR, Miller B, Allen H (2008) Effects of elevated CO_2 on the protein concentration of food crops: a meta-analysis. Glob Chang Biol 14:565–575

Vermeulen SJ, Campbell BM, Ingram JSI (2012) Climate change and food systems. Annu Rev Environ Resour 37:195–222

WFP (2017) Home-grown school feeding: a framework to link school feeding with local agricultural production. World Food Programme, Rome Available from: http://documents.wfp.org/stellent/groups/public/documents/newsroom/wfp204291.pdf?_ga=2.196122148.945026317.1507775688-1019979156.1507775688

World Economic Forum (2017) Shaping the Future of Global Food Systems. World Economic forum, Geneva Available from:https://www.weforum.org/whitepapers/shaping-the-future-of-global-food-systems-a-scenarios-analysis

22

Climate-Smart Agriculture and Private-Sector Action on Climate Risks

Sonja Vermeulen

22.1 The Climate-Smart Agriculture Initiative of the World Business Council for Sustainable Development

Millions of small and large businesses interact to transform agricultural inputs into the food we consume each day. Agrifood value chains involve a wide range of business interests, including suppliers, farmers, logistics companies, manufacturers and processors, retailers, caterers, financial services, and researchers. Harnessing the collective interest and input of this diverse private sector will be critical to achieving a lasting impact at global scale in climate-smart agriculture and food systems.

Large companies with an international reach are leading the way on partnerships for large-scale action on climate risks. Central to this global effort is the Climate-Smart Agriculture initiative of the World Business Council for Sustainable Development (WBCSD). WBCSD is a membership organization of companies organized into 70 national councils across the world, working together to accelerate the transition to a sustainable world. The Climate-Smart Agriculture initiative involves self-selected WBCSD member companies that are active in agrifood, in all continents, with a focus on Brazil, Ghana, India, Southeast Asia, and the United States of America. Successes in these places, which are called the 'road-test countries', may be scaled up in future.

Announced at the 2015 United Nations Climate Change Conference in Paris, the WBCSD Statement of Ambition on Climate-Smart Agriculture (WBCSD 2015) draws on multiple sources—most importantly, the United Nation's Sustainable Development Goals coupled with regional consultations with farmers, businesses, governments, civil society organizations and research institutes. The Statement of

S. Vermeulen (✉)
Hoffmann Centre for Sustainable Resource Economy, London, UK
e-mail: sonja.vermeulen@wwf.org

Ambition sets out global targets for private-sector action by 2030, under each of the three pillars of climate-smart agriculture (CSA).[1]

22.2 The Three Pillars of Climate-Smart Agriculture

Pillar 1: Productivity ambition: "Increase global food security by making 50% more nutritional food available through increased production on existing land, protecting ecosystem services and biodiversity, bringing degraded land back into productive use and reducing food loss from field to shelf." The footnotes to this WBCSD Statement explain that nutritious food means a range of macro- and micro- nutrients; that 'ecosystem services' follow the definition of the Millennium Ecosystem Assessment and, thus, include cultural as well as ecological services; and that 'food loss' is up to the point of the consumer.

Pillar 2: Climate change resilience, incomes and livelihoods ambition: "Strengthen the climate resilience of agricultural landscapes and farming communities to successfully adapt to climate change through agro-ecological approaches appropriate for all scales of farming. Invest in rural communities to deliver improved and sustainable livelihoods necessary for the future of farmers, bringing prosperity through long-term relationships based on fairness, trust, women's empowerment and the transfer of skills and knowledge."

Pillar 3: Climate change mitigation ambition: "Reduce greenhouse gas (GHG) emissions by at least 30% of annual agricultural carbon dioxide equivalent (CO_2e) emissions against 2010 levels (aligned with a global 1.6 gigatonnes of carbon dioxide equivalent per year ($GtCO_2e$ yr) reduction by 2030). A substantial portion of these reductions will also be achieved through reducing food waste up to the point of sale to the end consumer, in line with WBCSD's Action 2020 to halve food waste. Further emissions reductions will come from elimination of GHG emissions from land-use change to commercial agriculture and land restoration under the WBCSD Land Degradation Neutrality initiative."

Achievement of the targets will constitute a major contribution to the Paris Agreement as well as wider societal goals enshrined in the Sustainable Development Goals (SDGs) and global environmental agreements. Getting there will entail profound change across agrifood chains if total emissions are to fall sharply while food productivity and farming livelihoods rise. WBCSD sought CGIAR partnership to improve businesses' ability to trace, measure and monitor progress on climate-smart agriculture (Vermeulen and Frid-Nielsen 2017). The aim was not to provide a comprehensive new protocol for member companies, but rather to support monitoring and evaluation by synthesising metrics that businesses and other entities collect already.

---

[1] Note: these differ slightly from other definitions of CSA, such as that used by the Food and Agriculture Organization (FAO) of the United Nations (FAO 2013).

Ahead of formal reporting by the WBCSD Climate-Smart Agriculture initiative, this chapter provides an initial analysis of progress towards the targets and the key potentials for—and obstacles to—measuring collective advances towards the global targets. These findings hopefully offer valuable early lessons to the increasing number of companies in agrifood value chains that are starting to plan and implement actions on climate risks, assisted by partners in governments, non-governmental organisations and research.

22.3 A Simple Framework and Method for Measuring Climate-Risk Actions in Agrifood Chains

Selection of appropriate global- and company-level indicators under each CSA pillar would try to: avoid new costs of measurement by piggy-backing on existing metrics and data collection; establish a baseline, and cover multiple subsequent years towards 2030; and provide a set of both activity and outcome indicators, linked by a testable theory of change.

The WBCSD Statement of Ambition for each CSA pillar includes both outcomes and activities towards achieving these objectives. These outcomes and activities are linked by an implicit theory of change—a hypothesis, or best bet, that the activities will deliver the outcomes (Vogel 2012). For example, the intended outcome of pillar two is to strengthen climate resilience of agricultural landscapes and farming communities. The stated activities, or best bet, to achieve this outcome include adopting agro-ecological approaches, investing in rural communities, and building long-term empowering relationships between farmers and industry.

Combining and triangulating information from WBCSD member companies with external global data sources is crucial. The WBCSD Statement of Ambition addresses the global agrifood sector, not just WBCSD members. Individual member companies will be seeking both to improve their own performance on various metrics and to track collective progress towards the global goal. They will also be looking to stimulate positive change across their own sub-sectors, as well as among governments, rural communities, consumers and other agents of change.

Among sources of global indicators, which include the World Bank, the International Fund for Agricultural Development and others, the Food and Agriculture Organization of the United Nations Statistics Division (FAOSTAT) provides global data sets on the quantity of food produced and yields (pillar 1), and direct agricultural emissions (pillar 3), but not on activities and outcomes for pillar two. Simple linear regression created business-as-usual projections and compared these with business-as-usual projections to target scenarios for 50% more food (pillar 1), and 30% fewer direct agricultural emissions (pillar 3). It was not possible to gauge progress relating to pillar two, since the pillar does not have a quantitative target and lacks global data to support a projection to 2030.

When it comes to company indicators, several companies report CSA-relevant indicators within annual reports, corporate social responsibility reports and sustainability reports, or under the Global Reporting Initiative of the Carbon Disclosure Project. Companies track progress differently, in terms of which indicators are used, how the indicators are measured (e.g. absolute versus relative progress), and how far back the reporting goes. There are major gaps in data availability, both across companies and for individual companies over time. At least five companies (approximately 40%) provided data for both 2010 and 2015 on five indicators: total waste to landfill (pillar 1), total water use (pillar 2); absolute Scope 1 emissions or Direct GHG i.e. from sources that are owned or controlled by the company, plus Scope 2 emissions or Energy Indirect GHG i.e. emissions from the consumption of purchased electricity, steam, or other sources of energy (e.g. chilled water) generated upstream from the company, and emissions intensity (pillar 3).

Current indicators at global and company levels are inadequate to provide a full picture of progress on climate-smart agriculture in line with the WBCSD definition and vision. At the company level, in particular, supplementing current data collection and reporting with additional indicators that resonate with both the SDGs and the Paris Agreement is recommended. Specific indicators are recommended in Vermeulen and Frid-Nielsen (2017).

22.4 Snapshot of Progress Towards the WBCSD Statement of Ambition

For pillar one on productivity, trends in global yield and production quantities from 2010 to 2014 indicate that we are on track to produce enough food to meet the WBCSD target of 50% more food by 2030, but without the guarantee that this food will meet food security needs or sustainability criteria such as zero land expansion. From 2010 to 2014, global average production quantity and yield of important food groups (cereals, vegetables, roots and tubers, fruit, meat, and milk) increased 10.8% and 2.7% respectively. To reach the 2030 target, food production must increase by approximately 1.9% per year. However, this trajectory is not certain; it is subject to multiple risks including climate, geo-politics and markets. We do not have evidence that this food will be nutritious or accessible to poorer consumers. Moreover, there is no data available on the wider WBCSD definition of pillar one, specifically whether this increased production is on existing land (thereby improving resource-use intensity), protects ecosystem services and biodiversity, brings degraded land back into productive use, or reduces food loss from field to shelf.

For pillar two on resilience, we know very little indeed. Neither companies nor global data sets are keeping track of the resilience and welfare of agricultural communities and landscapes under climate change. Global data sets on rural poverty can provide a metric towards the overall intended outcome of pillar two; but are not especially useful for the WBCSD CSA initiative, because they do not link to climate change or to private-sector activities. Company data, for example, on farmer training or contracts, is too patchy to aggregate into meaningful global statistics. Few companies report on resilience indicators, let alone in both 2010 and 2015. Similarly, FAOSTAT does not have relevant data that correspond to the WBCSD subcomponents of pillar two. For WBCSD members to demonstrate their collective progress towards building resilience in farming communities and landscapes, more companies will need to provide quantitative information on indicators that cover both activities (e.g. training; on-farm agro-ecological practices) and outcomes (e.g. women's share of assets and decisions; reductions in exposure to climate risks).

For pillar three on mitigation, the agrifood sector is already falling behind targets for agricultural and food system emissions. From 2010 to 2015, global agricultural emissions increased 3.3% (FAOSTAT 2017). If the trend continues, the 2030 goal of 30% emissions reductions compared to 2010 will not be met; a 2.4% decrease per year is required for that. While many WBCSD member companies have demonstrated some impressive improvements in emissions intensity per unit of revenue (Vermeulen and Frid-Nielsen 2017), increasing levels of production mean that absolute emissions are rising across the global agricultural sector (see also Bennetzen et al. 2015). Deforestation—a major source of global emissions associated with agriculture that is not included in FAOSTAT data—will also contribute to some companies' emissions. Impacts of waste reduction on emissions are not yet reported.

22.5 Challenges and Potentials for Tracking Global Private Sector Action on CSA

This early analysis reveals that gaps in data availability, transparency and standardisation create a major obstacle to demonstrating private-sector progress towards the WBCSD Statement of Ambition on Climate-Smart Agriculture. The companies involved are not yet measuring their own performance on the targets they have set themselves, nor is this information available from global data sets.

A deeper set of issues concerns the links among shared measurement, shared management and shared accountability. Food systems are complicated, with many interconnections and feedback loops. A full discussion of private incentives to provide public goods is beyond the scope of this paper, but negatives and positives are noted briefly. The logic of collaboration to meet collective environmental and social objectives may not be compelling to individual companies, particularly if these add to operational costs. On the other hand, many companies have embraced self-regulation and voluntary collaboration as a strategy to forestall increased regulation and public activism (Haufler 2013).

These issues raise questions on possible success factors for the WBCSD members that are active in the Climate-Smart Agriculture initiative. Reflecting on progress to date, three key areas of potential might be:

Amplifying complementary actions across a value chain: Coordinated initiatives across agrifood value chains have become a widespread approach to challenges such as the inclusion of small-scale producers in modern markets, exclusion of illegal practice, or achieving zero deforestation associated with specific commodities. Such initiatives, often linked to certification, have demonstrated positive outcomes but are not alone a sufficient solution to environmental and social challenges (DeVries et al. 2017). Climate change action faces multiple potentials for leakage, trade-offs or inequitable outcomes (Vermeulen et al. 2016), for which supply chain approaches may provide partial solutions. For a nutrient supply company, for example, helping to raise smallholder productivity might involve higher company-level emissions as more mineral fertiliser is manufactured to meet demand, but a value-chain and landscape approach might demonstrate how this is more than offset by gains in local livelihoods and resilience, coupled with reduced deforestation. For an insurance company, the returns to a crop weather insurance product might increase if issued with lower premiums for farmers who use agro-ecological approaches, climate-adapted breeds or other proven practices for climate adaptation or mitigation. Value chain initiatives can also broaden private-sector inclusion beyond multi-nationals to relevant national companies and small enterprises.

Balancing group versus individual accountability: Monitoring, reporting and accountability at the level of the value chain may be more meaningful and sensible than separate accounting by individual companies. On the other hand, targets and reporting that happen only at the group level may fail to provide incentives for action by the actual players involved, especially when the group is the whole of the global agrifood sector, as in the Statement of Ambition on Climate-Smart Agriculture. A mechanism to link individual and group accountability seems essential. The Paris Agreement itself provides one promising model, in which parties commit their individual contributions to an agreed global target. Likewise, individual companies and their alliances may find value in coordinating actions and reporting on the multiple standards and targets set by their regulators, shareholders, financiers and global agendas such as the SDGs, for reasons of efficiency and effectiveness. There may be a strong rationale to building climate-risk assessment into regular monitoring and evaluation protocols, not as standalone CSA initiatives, but through the integration of additional indicators into existing reporting and accountability on social and environmental performance.

Moving beyond dispersed local activities and outcomes to broader system-wide change: While much positive impact may come from the global sum of activities and outcomes at the level of individual farms, companies or value chains, more systemic action is likely necessary to achieve targets to reduce emissions by 30% while improving livelihoods and increasing food production by up to 50%. Impacts at this scale will arise from a mix of public- and private-sector action. Recognising the importance of the system-wide enabling environment, the WBCSD Climate-Smart Agriculture initiative includes an action on scaling-up investment. This aims to increase CSA-friendly financial products for farmers and small businesses, and to assess options for internal carbon pricing (WBCSD 2015). Measuring and reporting of progress on this action area, and on related efforts to tackle system-wide barriers, is another key opportunity for WBCSD and other alliances. Critical to this effort will be the assessment of equity in outcomes from CSA actions at different scales (Karlsson et al. 2017). This poses the question: are actions on climate risks in the agrifood sector really delivering the intended positive outcomes for those more disadvantaged within value chains, particularly small-scale farmers and low-income consumers?

22.6 Implications for Development

In short, much work needs to be done—on measurement but, more importantly, on action. WBCSD member companies have rightly set out an ambitious statement of intent to address the massive climate challenges that global society faces together. Lessons from this early analysis of progress can hopefully contribute to renewed impetus to scale-up action on climate risks and bring benefits to the more disadvantaged participants in agrifood value chains globally.

References

Bennetzen E, Smith P, Porter J (2015) Decoupling of greenhouse gas emissions from global agricultural production: 1970–2050. Global Chang Biol. https://doi.org/10.1111/gcb.13120

DeVries R, Fanzo J, Mondal P et al (2017) Is voluntary certification of tropical agricultural commodities achieving sustainability goals for small-scale producers? A review of the evidence. Environ Res Lett 12:033001

FAO (2013) Climate-Smart Agriculture sourcebook. Food and Agriculture Organization of the United Nations. http://www.fao.org/3/i3325e.pdf

FAOSTAT (2017) Food and Agriculture Organization of the United Nations Statistics. http://www.fao.org/faostat

Haufler V (2013) A public role for the private sector: industry self-regulation in a global economy. Carnegie Endowment for International Peace, Washington, DC, p 160

Karlsson L, Nightingale A, Naess LO et al (2017) 'Triple wins' or 'triple faults'? Analysing policy discourses on climate-smart agriculture, CCAFS Working Paper no.197. CGIAR Research Program on Climate Change, Agriculture and Food Security (CCAFS), Copenhagen

Vermeulen S, Frid-Nielsen S (2017) Measuring progress towards the World Business Council on Sustainable Development Statement of Ambition on Climate-Smart Agriculture, CCAFS Working Paper 199. CGIAR Research Program on Climate Change, Agriculture and Food Security (CCAFS), Copenhagen https://ccafs.cgiar.org/publications/measuring-progress-towards-wbcsd-statement-ambition-climate-smart-agriculture-improving#.WPhnEVLMyRs

Vermeulen S, Richards M, De Pinto A et al (2016) The economic advantage: assessing the value of climate-change actions in agriculture. International Fund for Agricultural Development, Rome

Vogel I (2012) Review of the use of 'theory of change' in international development. Department for International Development, London https://www.gov.uk/dfid-research-outputs/review-of-the-use-of-theory-of-change-in-international-development-review-report
World Business Council for Sustainable Development (2015) CSA action plan. WBCSD, Geneva http://lctpi.wbcsd.org/wp-content/uploads/2015/12/LCTPi-CSA-Action-Plan-Report.pdf

Part V
Scaling Climate Smart Agriculture

23

Multi-Stakeholder Platforms and Climate Change Policymaking

Mariola Acosta, Edidah Lubega Ampaire, Perez Muchunguzi, John Francis Okiror, Lucas Rutting, Caroline Mwongera, Jennifer Twyman, Kelvin M. Shikuku, Leigh Ann Winowiecki, Peter Läderach, Chris M. Mwungu, and Laurence Jassogne

23.1 Introduction

With climate change posing a rising threat to rural livelihoods in East Africa (Niang et al. 2014; Kahsay and Hansen 2016), the need for adaptation and mitigation strategies has gained increasing attention among policymakers (Liwenga et al. 2014). Although the region has made advances in building the relevant governance and policymaking systems, major challenges remain, including insufficient coordination between institutions and government levels; limited access of policymakers and

M. Acosta (✉) · E. L. Ampaire · P. Muchunguzi · J. F. Okiror · L. Jassogne
International Institute of Tropical Agriculture, Kampala, Uganda
e-mail: M.Acosta@cgiar.org; E.Ampaire@cgiar.org; P.Muchunguzi@cgiar.org; JF.Okiror@cgiar.org; L.Jassogne@cgiar.org

L. Rutting
Copernicus Institute of Sustainable Development, Utrecht University, Utrecht, The Netherlands
e-mail: l.rutting@uu.nl

C. Mwongera · K. M. Shikuku · C. M. Mwungu
International Centre for Tropical Agriculture, Africa Regional Office, Nairobi, Kenya
e-mail: C.Mwongera@cgiar.org; k.m.shikuku@cgiar.org; C.Mwungu@cgiar.org

J. Twyman
International Centre for Tropical Agriculture, Cali, Colombia
e-mail: J.Twyman@cgiar.org

L. A. Winowiecki
World Agroforestry Centre (ICRAF), Nairobi, Kenya
e-mail: l.a.winowiecki@cgiar.org

P. Läderach
International Center for Tropical Agriculture, Asia Regional Office c/o Agricultural Genetics Institute, Hanoi, Vietnam
e-mail: P.Laderach@cgiar.org

technical staff to empirical evidence; and insufficient funding (Minde et al. 2013; Asekenye et al. 2016; Ampaire et al. 2017).

Multi-stakeholder platforms (MSPs) bring together representatives from different interest groups to discuss shared challenges, opportunities, policy actions and advocacy strategies (Warner 2005). They have the potential to tackle complex development challenges and to assist in the scaling up of necessary innovations (Hermans et al. 2017). In the realm of agricultural development, MSPs have played a pivotal role in addressing many complex problems around the world (for a good overview and a selection of case studies see Dror et al. 2016). Recent studies also demonstrate MSPs' potential in addressing climate change (Pinkse and Kolk 2012).

With its three-part approach to climate change—mitigation, adaptation and food security—climate-smart agriculture (CSA) has been gaining increasing attention. While there has been considerable research on scaling up CSA practices, less attention has been given to assessing the policy environments most conducive to addressing climate change (Jordan and Huitema 2014). Such research is crucial, as the sustainable scaling up of CSA technologies can seldom be achieved without an enabling policy environment (Ampaire et al. 2015; Barnard et al. 2015).

The objective of this chapter is to examine the role of MSPs in facilitating climate change policymaking in East Africa through a case study of eight national and subnational MSPs in Uganda and Tanzania.

23.2 Methods

The Policy Action for Climate Change Adaptation (PACCA) project[1] (2014–2017) focused on building climate-resilient food systems in Uganda and Tanzania by coordinating policies and institutions at the local, regional and national levels. The empirical data for this chapter was collected through:

- *Participant observation and meeting minutes:* Between July 2014 and December 2017, principal members of the research team attended a total of 80 MSP meetings and events. Researchers took notes, made observations and reviewed meeting minutes.
- *Questionnaires:* Researchers administered a baseline questionnaire at the inception meetings of the national platforms to assess participants' knowledge, attitudes and skills regarding three main topics: (i) impacts of climate-change adaptation, (ii) available, locally appropriate adaptation options and (iii) policy formulation and implementation processes. Information was collected from 29 stakeholders in Tanzania (31% women, 69% men) and 39 in Uganda (38% women, 62% men).

[1] Initiative of the CGIAR Research Program on Climate Change, Agriculture and Food Security (CCAFS).

- *Social network analysis (SNA)* was conducted to collect information on the key organizations for knowledge exchange. Data were collected from participants using a multistep process during the launch of district platforms in Nwoya (n = 24) and Mbale (n = 21) in December 2015 and June 2016, respectively. Participants were first asked to list all the institutions they represented, then all the organizations with which they collaborated. Finally, from these lists of organizations, participants identified which they considered the most important for knowledge exchange. Analysis of the data was undertaken using Gephi 0.9.1 software.

23.3 Results and Discussion

23.3.1 Establishment and Operation of the MSPs

The climate change MSPs were established between 2014 and 2015. In Tanzania three were formed, one national and two subnational (in Lushoto and Kilolo districts). In Uganda five were established, one national and four subnational (in Nwoya, Rakai, Luwero and Mbale districts). The subnational platforms influenced district-level policymaking and informed the national platforms, which in turn influenced national policymaking through information-sharing with parliamentarians and national ministries. Having both subnational and national organizations facilitated a bidirectional flow of information. This integrated approach is important because, although the effects of climate change are felt locally and technologies must be context-specific, change happens most effectively within an enabling national policy environment.

While PACCA acted as the initiator of the platforms, funded some of their activities and remained a stakeholder member, the MSPs functioned largely as independent entities. The national platform in Uganda was hosted by the Climate Change Department of the Ministry of Water and Environment, and the one in Tanzania by the Environmental Management Unit of the Ministry of Agriculture Livestock and Fisheries (MALF-EMU). In the districts, the platforms were hosted by the national offices of environment and natural resources. Embedding the platforms within government structures provided those official bodies with convening power, a greater sense of ownership over the process and, ultimately, offered the platforms a pathway to sustainability. Facilitation of meetings was entrusted to members of the platform-hosting institutions who were recognized for their authority, their central role in local knowledge exchange and their credibility among other stakeholders.

Participant observation and an examination of minute meetings revealed that the platforms enabled their participants to share experiences and research findings on climate change. The PACCA project, as a member of the MSPs, contributed to the generation and dissemination of research findings on CSA and climate change adaptation (specifically on-farm trade-off and synergies for CSA, drivers for adoption of CSA, prioritization among CSA options for greater impact, scenario-guided policy development, policy-actor networks and gender-responsive policy-making), thereby contributing to an enhanced science-policy interface. This sharing of research evidence and experience became the basis for discussions and helped define the efforts by the MSPs to influence policy. Platform meetings, which generally took place quarterly, had two main sessions: the first featured sharing of research knowledge and experience, while in the second decisions were made in plenary through inclusive participatory processes, which normally involved working in groups followed by a plenary discussion. These processes of knowledge sharing contributed towards building trust between stakeholders and facilitated finding common goals and interests, which helped foster unified action.

23.3.2 The Role of MSPs in Promoting CSA

23.3.2.1 Knowledge Creation and Capacity Building

Initial knowledge levels about climate change and CSA varied widely among participants. The questionnaire revealed that stakeholders were generally familiar with the impacts of climate change, with 83% in Tanzania ($n = 29$) and 71% in Uganda ($n = 39$) reporting a high level of understanding. Knowledge of locally appropriate adaptation options was considerably lower, with 58% in Tanzania and 77% in Uganda reporting low or medium knowledge levels. Knowledge of policy processes was higher in Tanzania, where 41% and 45% rated their level of familiarity as high in policy formulation and implementation processes, respectively, as compared to 21% and 29% in Uganda. These differences can be explained by the actor composition of MSPs: Uganda had a higher proportion of representatives from non-state actors in their MSPs, whereas MSPs in Tanzania were disproportionately composed of government representatives who were familiar with policy formulation and implementation processes (Table 23.1).

Table 23.1 Composition of national MSPs at inception meetings (October–November 2014)

	Sample size	
Institution category	Uganda	Tanzania
Association	5	–
Academia	–	3
Consultant	–	1
Media	2	–
NGOs/CSOs	22	3
Government ministries	2	8
Government departments	1	6
Government agencies	2	5
Local government	4	2
Research	1	1
Total	**39**	**29**

Once the results of the questionnaires were presented to the MSPs, they made changes to their meeting structures in order to address the knowledge gap: all actors were invited to share their experiences with climate change adaptation projects, and experts regularly presented research-based evidence on the CSA technologies favorable for each region. This transfer of knowledge was expected to enhance the technical capacity of the platforms' members, which in turn was expected to translate into attitudinal and behavioral change both within each member's organization and in the actions of the platforms as a whole.

While an end-line study was not available at the time this chapter was written, limiting our ability to quantify the extent of participants' learning over time, there is evidence of the platforms' impact. In event evaluation forms, participants indicated that they shared their newly acquired knowledge with colleagues, politicians and community members. The role of the MSPs in the dissemination of knowledge was also publicly acknowledged by a representative of the Climate Change Department in Uganda, who stated that the MSPs "have improved the understanding of climate change and its impacts, thus enabling public institutions, individuals and non-state actors to tap into the opportunities and co-benefits arising from mitigation and adaptation actions" (Semambo 2017).

Results of the SNA showed that institutions in both districts were linked through information-sharing processes, but the relationships were not necessarily reciprocal. For example, in Nwoya the most important participants for knowledge exchange were the District Local Government and ZOA, a Dutch NGO, but there was no exchange between the two. We found a similar situation in Mbale (Fig. 23.1), where the organizations considered important for knowledge exchange were Mbale District Local Government (MDLG), National Forestry Authority (NFA) and

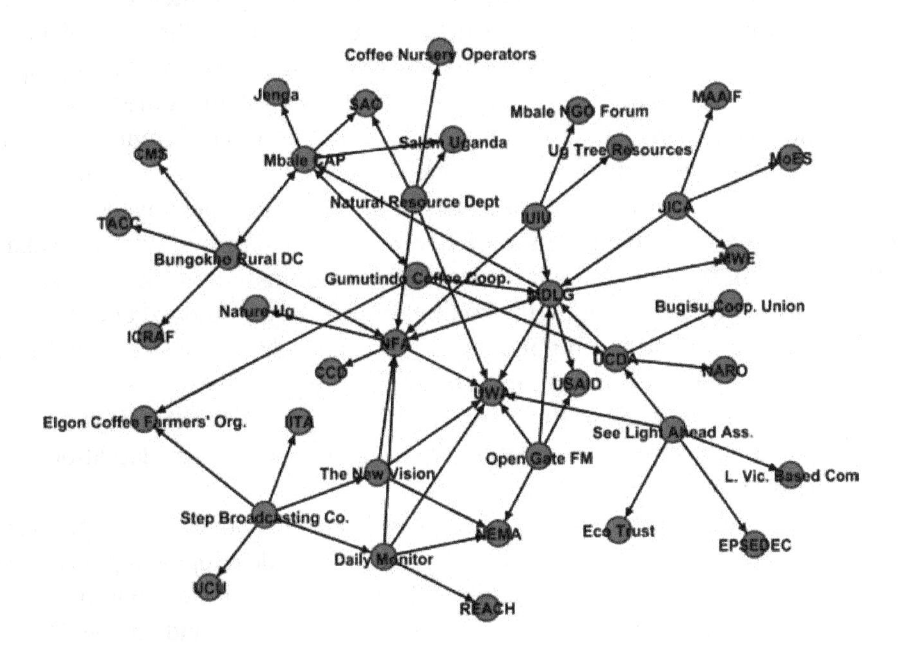

Fig. 23.1 Knowledge exchange sociogram for Mbale District

Uganda Wildlife Authority (UWA), but the knowledge exchange relationship existed only between NFA and MDLG, not with UWA. The SNA also identified institutions that acted as bridges for other institutions that would otherwise not be connected to the knowledge network (e.g., Bungokho Rural DC). In both MSPs, the district local governments were among the institutions better connected in terms of knowledge sharing. This further justifies the strategy of hosting the district MSPs within the local governments as a way to promote sustainability, knowledge exchange and coordination of local climate-change actors.

23.3.2.2 Influencing Subnational and National Policies

The platforms' meetings played a role in promoting collective action to influence national and subnational policies. In both Uganda and Tanzania, the platforms' actions were decided in plenary on the basis of the research-based evidence on CSA, responsive and equitable policymaking and climate change adaptation. MSPs were able to contribute and influence key national policies. In Uganda the Climate Change MSP was recognized as having influenced and complemented a number of policy reviews and strategic development plans (Semambo 2017). In both Uganda and Tanzania the national platforms were also active in influencing national climate change policy (Table 23.2).

Like the national platforms, the district platforms had regular meetings that involved the sharing of experiences with context-specific adaptation strategies and locally appropriate CSA options. Key representatives of the district platforms were also members of the national platforms, ensuring coordinated action. District platforms engaged in participatory zonal planning of their territories for the prioritization of adaptation investments—an example of the type of initiatives aimed at fostering a conducive policy environment for the scaling up of CSA practices which MSPs are especially well suited to address precisely because they require the collaboration of stakeholders from different sectors and across different scales. Stakeholders began by defining the zoning criteria and dividing the area into different zones based on what they perceived as locally important factors: the main source of livelihood and farming system (Rakai), altitude (Kilolo), rainfall gradient (Luwero) or a combination of these (Lushoto). The fact that districts differed in the zoning criteria they employed and the number of zones they identified highlights the fact that adaptation needs and local priorities are unique to each territory. The zoning was usually followed by stakeholder discussions on each of the zone's main enterprises, the effects of climate change on these enterprises, and the pertinent policy issues and adaptation measures needed to overcome these constraints.

In subsequent meetings, district officers and representatives of the platforms would prioritize the issues to be integrated in the district development plans. Since the formation of the platforms, there has been progress in incorporating CSA components in the District Development Plans in Uganda and in the District Agriculture Development Plans in Tanzania. For example, in Lushoto, Tanzania,

Table 23.2 Policy engagement activities of the national climate-change MSPs

Uganda national MSP activities	Tanzania national MSP activities
Scenario-guided policy review of the Uganda National Agricultural Sector Strategic Plan (ASSP)	Water-use technology study used in a policy engagement meeting with the National Irrigation Commission, Basin Water Boards and the Ministry of Agriculture, Food Security and Cooperatives. Recommendations given on the need to promote efficient water-use technologies and other CSA practices as a package, rather than individual technologies, for enhanced adoption of these technologies at large scale
Preparatory meetings to organize and ensure a coordinated approach of the Uganda position in the COP21	Scenario-guided policy review of the National Environmental Policy
Participation in the Joint Sector Reviews of the Ministry of Water and Environment (MWE) and the Ministry of Agriculture, Animal Industry and Fisheries (MAAIF)	Informing the development of the Intended Nationally Determined Contributions (INDCs)
Informing the draft irrigation policy	Participation in the development of the CSA Country Plan for Tanzania
Participation in a live national dialogue on climate change and women	
Participation in several climate-change workshops organized by other actors	

the district council allocated the equivalent of US$3800 to execute various CSA interventions for the financial year 2016–2017. In Luwero, Uganda, district officials prioritized working on the institutional framework for addressing climate change in the district, and in Rakai, Uganda, a District Climate Change Action Plan was created.

Conscious of the importance of using scientific evidence on gender, CSA and climate change to influence legislative decisions, the MSPs in Uganda undertook a National Reflection Workshop with members from civil society organizations, research institutions, local governments, ministries and the media. The evidence and messages that emerged from the event, together with information from the continued policy engagement that followed, were presented at a high-level event attended by members of the Uganda Parliamentary Forum on Climate Change and the Parliamentary Forum on Food Security, Population and Development. The event helped raise awareness and advocate for gender responsiveness in climate change adaptation among members of Parliament amidst discussions on the Uganda Climate Change Bill, the Biotechnology Bill and the pre-negotiations for the COP22 in Morocco. Inspired by the event in Uganda, the LA in Tanzania organized a sister event in 2017 with members of the Tanzanian Parliament, where evidence was presented to encourage legislators to ensure gender-responsive climate change policymaking. In addition to these parliamentarian engagements, representatives of the MSPs have also participated in other high-level policy engagements organized by partner organizations.

With growing evidence of their efficacy and acceptance by stakeholders, the MSPs have become increasingly institutionalized. In Uganda the national Climate Change Department is establishing a climate change MSP at the ministry, department and agency levels—independently of the PACCA project—to operationalize article 6 of the United Nation Framework Convention on Climate Change (UNFCCC) on capacity building. In Tanzania, the district government of Lushoto has formalized the incorporation of the MSPs into their district frame and has replicated the MSP model in villages, appointing "ambassadors" who monitor and report on their respective activities. Furthermore, officials from MALF-EMU have expressed interest in using the MSP model and acknowledged MSPs as central to national climate-change policy planning and to the scaling up of CSA in the country (Okiror and Cramer 2017). Further research will be needed to assess the levels of funding provided to these MSPs and what affects the availability of funds has on their operation, efficacy and perceived legitimacy.

23.4 Implications for Development

Through a case study in Uganda and Tanzania, this chapter has examined the role of MSPs in influencing climate change policy processes. MSPS foster the sharing of information among diverse stakeholders and allow participatory approaches for influencing policy recommendations across multiple governance levels. We argue that these MSP interventions can help build an enabling policy environment for climate - change adaptation and mitigation policy, as evidenced by the scenario-guided policy planning processes, CSA participatory zonal planning exercises and multiple policy reviews and consultations. With specific reference to the role of MSPs in fostering CSA science-policy dialogue, the results of the questionnaire highlighted the need for greater knowledge-sharing among stakeholders. Findings from the social network analysis suggest the importance of platform composition in the knowledge-exchange process. Furthermore, concrete policy action such as budgeting for tangible CSA projects at the local level (e.g., Lushoto, district MSPs), recommending specific packages of CSA water-efficient technologies for enhanced adoption (Tanzania, national MSPs) and mainstreaming CSA and climate change in district development plans (Uganda, Tanzania district MSPs) exemplifies the role that continuous science-policy interaction through MSPs can have in influencing policymaking.

While these MSP processes have succeeded in enhancing CSA science-policy dialogues and promoting evidence-based policy outcomes in East Africa, additional research is needed if the MSP model is to be successfully replicated elsewhere. Specifically, further context-specific studies are needed on the optimal balance between non-state actors (including the private sector) and government representatives in the platforms, as these case studies appear to suggest that an overrepresentation of either could hinder the ability to achieve policy change. End-line evaluation and follow-up studies will also be required to determine whether the degree and manner of the East African MSPs' embeddedness in local government structures was sufficient to maintain their financial sustainability over time while preserving their independence and participatory approach.

References

Ampaire EL, Providence H, Van Asten P, Radeny M (2015) The role of policy in facilitating adoption of climate-smart agriculture in Uganda. CGIAR Program on Climate Change, Agriculture and Food Security (CCAFS), Copenhagen

Ampaire EL, Jassogne L, Providence H et al (2017) Institutional challenges to climate change adaptation: a case study on policy action gaps in Uganda. Environ Sci Pol 75:81–90. https://doi.org/10.1016/j.envsci.2017.05.013

Asekenye C, Ampaire E, Epp MV, Van Asten P (2016) Climate Change Social Learning (CCSL) report for Uganda and Tanzania. Influencing and linking policies and institutions from national to local level for the development and adoption of climate-resilient food systems. International Institute of Tropical Agriculture (IITA), Kampala

Barnard J, Manyire H, Tambi E, Bangali S (2015) Barriers to scaling up/out climate smart agriculture and strategies to enhance adoption in Africa. Forum for Agricultural Research in Africa, Accra

Dror I, Cadilhon J-J, Schut M et al (eds) (2016) Innovation platforms for agricultural development: evaluating the mature innovation platforms landscape. Routledge, London

Hermans F, Sartas M, van Schagen B et al (2017) Social network analysis of multi-stakeholder platforms in agricultural research for development: opportunities and constraints for innovation and scaling. PLoS One 12:e0169634. https://doi.org/10.1371/journal.pone.0169634

Jordan A, Huitema D (2014) Policy innovation in a changing climate: sources, patterns and effects. Glob Environ Chang 29:387–394. https://doi.org/10.1016/j.gloenvcha.2014.09.005

Kahsay GA, Hansen LG (2016) The effect of climate change and adaptation policy on agricultural production in eastern Africa. Ecol Econ 121:54–64. https://doi.org/10.1016/j.ecolecon.2015.11.016

Liwenga ET, Jalloh A, Mogaka H (2014) Review of research and policies for climate change adaptation in the agriculture sector in East Africa. Future Agricultures Consortium, Brighton

Minde H, Kateka A, Tilley H, et al (2013) Tanzania national climate change finance analysis. Overseas Development Institute, London/The Centre for Climate Change Studies, University of Dar es Salaam, London

Niang I, Ruppel OC, Abdrabo MA et al (2014) Africa. In: Barros VR, Field CB, Dokken DJ et al (eds) Climate change 2014: impacts, adaptation, and vulnerability. Part B: regional aspects. Contribution of working group II to the fifth assessment report of the intergovernmental panel on climate change. Cambridge University Press, Cambridge, UK, pp 1199–1265

Okiror JF, Cramer L (2017) Lessons for successful scaling of climate-smart agriculture innovations. https://ccafs.cgiar.org/blog/lessons-successful-scaling-climate-smart-agriculture-innovations. Accessed 22 Dec 2017

Pinkse J, Kolk A (2012) Addressing the climate change—sustainable development nexus: the role of multistakeholder partnerships. Bus Soc 51:176–210

Semambo M (2017) PACCA project experience sharing. Ministry of Water and Environment (MWE). Climate Change Department. Kampala (Uganda)

Warner J (2005) Multi-stakeholder platforms: integrating society in water resource management? Ambient Soc 8:4–28

24

Boosting Climate-Smart Agriculture through Farmer-To-Farmer Extension

Steven Franzel, Evelyne Kiptot, and Ann Degrande

24.1 Introduction

The rise in importance of climate-smart agriculture (CSA) has been accompanied by increased concern over how CSA practices can be scaled up to reach millions of smallholder farmers. While climate-smart agricultural practices—like conventional ones—are sometimes complex and knowledge-intensive, public investment in extension services has been declining in many countries (Harvey et al. 2014). This situation makes it increasingly difficult for farmers to access the CSA information they need. McCarthy et al. (2011) cite the high cost of accessing information and the key role that extension can play in reducing the cost and the risks of adopting CSA practices. Lipper et al. (2014) claim that public financial support is needed to promote CSA through extension services and other types of information dissemination.

In the face of increased demand for agricultural information and the reduced capacity of extension systems, many extension providers have been using farmer-to-farmer extension (F2FE), which is defined as the provision of training by farmers to farmers, often through the creation of a structure of farmer-trainers (Scarbourough 1997). Surveys reveal that most farmers rely on other farmers as their primary source of information about new technologies. The F2FE approach therefore can be viewed as an extension of farmers' existing practices. We use "farmer-trainer" as a generic term, even though we recognize that different names (e.g., lead farmer, farmer-promoter, community knowledge worker) may imply somewhat different roles.

S. Franzel (✉) · E. Kiptot
World Agroforestry Centre (ICRAF), Nairobi, Kenya
e-mail: s.franzel@cgiar.org

A. Degrande
World Agroforestry Centre (ICRAF), Yaounde, Cameroon
e-mail: a.degrande@cgiar.org

F2FE programs have a long history, having been used in the Philippines since the 1950s and in Central America since the 1970s (Selener et al. 1997). They have grown tremendously in Africa in recent years (Simpson et al. 2015) and are now quite common, with 78% of a sample of 37 development organizations using the approach in Malawi (Masangano and Mthinda 2012) and 33% using it across seven regions of Cameroon (Tsafack et al. 2014). At least two public extension systems—those of the Rwanda Agricultural Board and the Ministry of Agriculture in Malawi—each work with over 12,000 farmer-trainers.

The objective of this paper is to assess the potential of F2FE to promote CSA, based on experiences in Cameroon, Kenya and Malawi. Specifically, we assess the approach based on the following four criteria:

- Extension program managers' perceptions of the approach
- The effectiveness of the approach as a means for training farmers
- The efficiency of the approach, as judged by comparing its costs per trained farmer with the costs of an approach in which extension agents train groups of farmers
- The potential of the approach for improving the proportion of female extension providers

First, we describe the methods used in the study and the main features of the F2FE approach as it is applied in the three countries. Next, we examine each of the four criteria in turn. Finally, we focus on the implications of the results for using the F2FE approach to promote CSA, especially those CSA practices that are complex and knowledge-intensive.

24.2 Methods

This study relied on semi-structured surveys of extension program managers to assess 80 development organizations' experiences with F2FE in the following countries:

- Cameroon (25 organizations operating in the 5 southernmost of the country's 8 regions) (Tsafack et al. 2014)
- Kenya (30 organizations in Rift Valley, Western and Nyanza provinces) (Franzel et al. 2014)
- Malawi (25 organizations across all three of the country's administrative regions) (Kundhlande et al. 2014)

The organizations included international non-governmental organizations (NGOs) (39%), national or local NGOs (35%), government agencies (14%), producer organizations (10%) and private companies (2%). Most of the sampled extension program managers were using not only F2FE but also three or four other approaches (such as farmer field schools, training groups, exchange visits and field days). Given this diversity of approaches, the managers were not likely to be biased in favor of F2FE, as they might have been had they been using F2FE alone. Sampling was done using the snowball sampling method, in which respondents directed interviewers to other potential respondents (Goodman 1961). This method was used because no lists of development organizations using the F2FE approach were available.

In Cameroon and Malawi, interviews were also conducted with 160 and 203 randomly selected farmer-trainers, respectively, who were working with the organizations visited (Tsafack et al. 2015; Khaila 2015). No such survey was conducted in Kenya because of the availability of a data set from interviews with 99 randomly selected farmer-trainers working with a dairy project (Kiptot and Franzel 2012). In Kenya, a survey also was conducted of 113 randomly selected trainees (farmers trained by farmer-trainers) to confirm that they had indeed been trained and to determine their level of knowledge about the improved practices and whether they were using them (Kiptot et al. 2013).

24.2.1 Main Features of the Approach

The organizations surveyed used farmer-trainers for a wide variety of enterprises and initiatives, including livestock, crops, agroforestry, nutrition and sustainable land management. Proportions of organizations promoting CSA (or some component of it, such as conservation agriculture) ranged from 4% in Cameroon to 23% in Kenya and 40% in Malawi. The most common terms for those doing the training were lead farmers, farmer-trainers, contact farmers and community facilitators. The uptake of the approach appears to be fairly recent: In all three of the countries, over half of the organizations reported having adopted it over the 7 years prior to 2014.

The trainers' main roles were to train farmers, to follow up with those farmers, and to mobilize them for meetings and training events. Most trainers (over 84% in all three countries) received an initial training (often 1 week in length), and nearly half received additional training after they had served for some time. Many hosted a demonstration plot. Over 72% of farmer-trainers in each country received training materials such as leaflets and brochures.

In most cases their key responsibility was to train other members of their own farmer groups, although sometimes they worked with neighboring groups as well. In only a few cases did they serve villages. In 25 of the organizations, the trainers received no compensation; in 43 they were compensated for some expenses such as communication and travel to meetings; and in 12 they received salaries or periodic allowances. There was little variation among countries in the proportions of farmer-trainers compensated in these various ways.

24.2.2 *Extension Program Managers' Perceptions of the Approach*

In the surveys, the managers of extension program expressed widespread satisfaction with F2FE as a means to promote the adoption of innovations (Simpson et al. 2015). Median scores across the three countries on the overall effectiveness of the approach ranged from 7.5 to 8 out of 10 in each of the 3 countries, with a score of 1 being not effective and 10 being extremely effective. Over 70% of respondents in each country gave the approach a score of 7 or 8. The main benefits of the F2FE approach, as perceived by over 60% of organizations using it, was that it boosted their ability to cover large areas and reach large numbers of farmers (Fig. 24.1). Many also cited the enhanced sustainability of extension efforts, because they believed that farmer-trainers would continue their training work even after the projects ended. In fact, there is considerable evidence that volunteer trainers continue working effectively after the projects supporting them come to an end—either because the government takes over support (Kiptot et al. 2016); because the trainers are accountable to local community structures (Lukuyu et al. 2012); or because producer organizations support them (Karanja et al. 2017). Many managers also felt that the approach helped increase adoption rates because farmers preferred to learn novel practices from their colleagues rather than from extension staff.

Organizations reported three main problems in implementing F2FE programs (Fig. 24.2). First, as reported by over 40% of organizations in Cameroon and over 20% in Kenya and Malawi, farmers sometimes had unreasonably high expectations in terms of financial and non-financial benefits, despite organizations' attempts to reduce such expectations. Unmet expectations could be a cause of high dropout

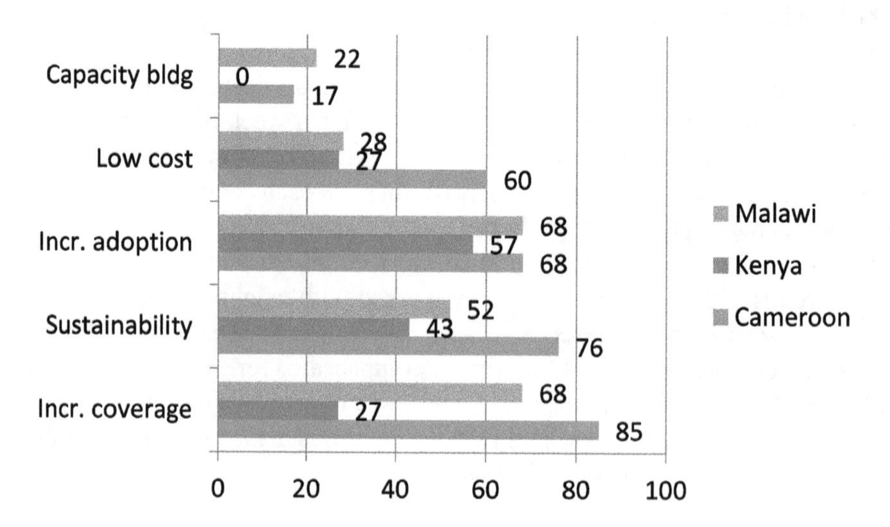

Fig. 24.1 Organizations' views of the main benefits of farmer-to-farmer extension programs (percent of organizations reporting)

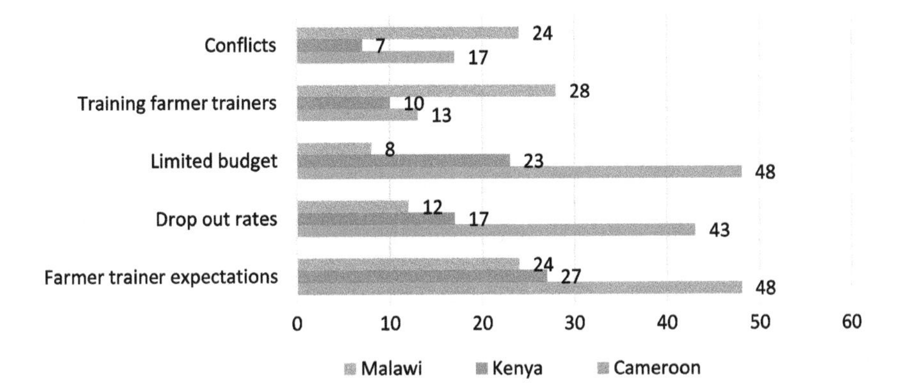

Fig. 24.2 Organizations' views of the main problems of farmer-to-farmer extension programs (percent of organizations reporting). Note: Conflicts refers to conflicts between organizations using the same volunteer farmers or problems arising when organizations compensate their farmer-trainers differently or offer contradictory recommendations to farmers

rates, which were also reported as a problem. Limited budgets for supporting farmer-trainers also created challenges.

24.2.3 Effectiveness of the Approach

Overall findings on effectiveness were positive. In Kenya, farmer-trainers each reported training an average of 201 farmers (median: 37) over the month prior to the interview. In Cameroon the average was 58 (median: 17) over the previous year (Tsafack et al. 2015), and in Malawi the number was 61 (median: 25) over the previous year. In all three countries, averages were skewed upward because a few of the farmer-trainers hosted large numbers of trainees who had been brought in by NGOs or other extension providers. The median number of farmers trained by typical farmer-trainers (17–37) approximates the number of members in the groups to which the farmer-trainers belonged and to whom their training was targeted.

In Kenya, the effectiveness of farmer-trainers was corroborated by a survey of 113 randomly selected trainees, who were found to be knowledgeable about the innovations on which they had received training, and most of whom were testing some of the new practices they had learned (Kiptot et al. 2013).

24.2.4 Efficiency of the Approach

An extension approach is deemed to be cost-efficient if the cost per farmer trained is lower than that of alternative approaches. Table 24.1 compares the cost efficiency of an F2FE program in Malawi with a conventional program in which extension

Table 24.1 Comparison of cost efficiency of an F2FE program in Malawi with a conventional program in which extension staff members directly train farmer groups

A. Cost of maintaining a front-line extension staff member (dollars/year)	
Salary	3600
Fringe benefits	1200
Fuel	200
Motorbike	1200 cost of 3000, depreciated over 3 years, plus $200 maint/year
Other (comm., training materials)	240 $20/month
Total	6440
B. Cost of maintaining a farmer trainer (dollars/year)	
Demo plot inputs	30
Bicycle	50 (cost of 150 depreciated over 3 years. No maintenance provided
Badge, t-shirt, cap, gum boots	30
Training costs 2 days per year	150 (residential training incl room ($30*2), board ($20*2), transp.($20)), trainers ($10), training materials ($20)
Total	260
C. Model 1 extension worker trains 10 farmer groups	
Each extension staff spent 10 days per month training	10 days per month
Each trains 10 farmers (1 club) per day of training	10 farmers trained per day of training
Farmers trained per month	100 farmers trained per month
Farmers trained per year	100 Same farmer groups visited once per month
Cost per farmer trained	65 costs are cost of extension staff +2 demos
D. Model 2. Farmer to farmer extension model	
Each extension staff spent 10 days per month training	10 days per month
Each extension staff trains 20 LFs per month	20 lead farmers
Each lead farmer trains 20 farmers	20 farmers trained
Total farmers trained per month	400 farmers trained
Cost per farmer trained	29.1 Costs are cost of extension staff + cost maintaining 20 farmer trainers
E. Ratio of costs per farmer in model 1 over model 2	2.2

The cost of training extension staff in the two approaches, F2FE and the group training, are considered to be the same and are thus not included in the model

staff members directly train farmer groups. The data in the model are from interviews with extension staff. The cost of a front-line extension staff member—which is a cost in both models—is $6440 per year, and the cost of a farmer-trainer is $260 per year. In model 1, the conventional approach, an extension worker trains 100 farmers per year at a cost of $65 per farmer. In model 2, an extension worker trains 20 farmer-trainers per year, each of whom trains 20 farmers, amounting to 400 farmers at a cost of $29 per farmer. The cost per farmer trained in the F2FE model is thus 55% lower than the cost in the conventional approach. If farmer-trainers trained only 9 farmers instead of 20, the 2 models would have the same costs per farmer trained. The analysis omits some costs (e.g., training of extension staff and administrative costs) that are not likely to vary between the different approaches.

24.2.5 The Approach's Potential for Increasing the Proportion of Female Extension Providers

Gender imbalance in agricultural extension has been widely recognized as an important challenge (World Bank, FAO, IFAD 2009; GFRAS 2014). The problem is generally attributed to two key facts: women make up only a small proportion of extension staff, and female farmers have less access to extension services than do male farmers. The low proportion of female staff has been linked to the fact that relatively few women enroll in the agricultural extension departments of universities and training institutes or choose agricultural extension as a career path (Simpson et al. 2012).

Figure 24.3 compares proportions of women in farmer-trainer programs to their proportions in professional frontline extension positions in the same organizations. These organizations included NGOs, government agencies, producer organizations and private companies. If the proportion of women among farmer-trainers is higher than the proportion among professional frontline extension staff, then farmer-trainer programs can be said to help increase the proportion of women providing extension services. Results on this issue are mixed.

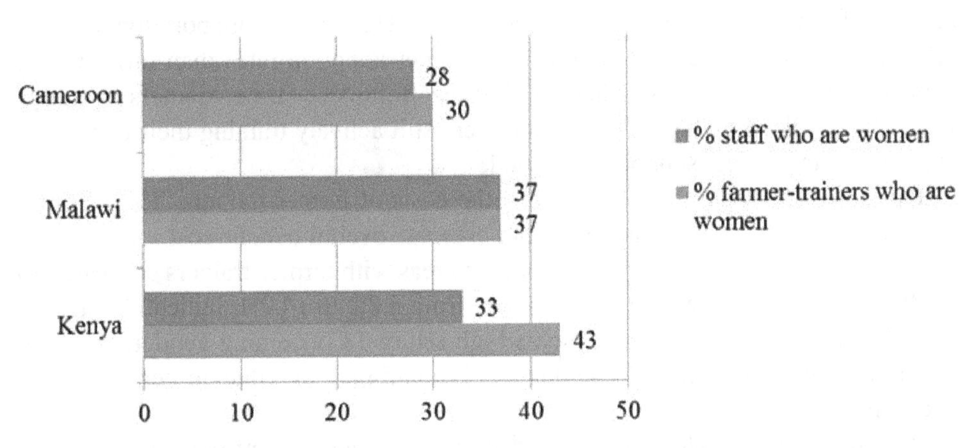

Fig. 24.3 Proportion of field staff and farmer-trainers who are women in organizations providing extension services. Note: Includes government, NGOs, private sector and farmer organizations

In Kenya, the mean proportion of farmer-trainers who were women across the F2FE programs of 30 organizations was 43%, while the mean proportion of field staff who were women in the same organizations was only 33%. Thus the proportion of female extension providers among farmer-trainers was about 30% higher than the proportion of women among field staff. In contrast, in Cameroon and Malawi, the mean proportion of farmer-trainers who were women was about the same as the mean proportion of field staff who were women (Fig. 24.3).

However, in certain organizations, the proportion of farmer-trainers who were women was dramatically higher than the proportion of female field staff. For example, in the East Africa Dairy Development Project in Kenya in 2013, only 10% of the professional trainers were women, whereas 28% of the 1141 farmer-trainers were women. In the Ministry of Agriculture of Malawi in 2013, only 21% of the field staff were women, whereas 40% of the 12,000 volunteer farmer-trainers were women. Organizations making special efforts to recruit female farmer-trainers, using such measures as targeting women's groups and providing childcare during training, were able to recruit high numbers of them. Nevertheless, low literacy levels among women limit the efforts being made to increase the number of women trainers because an ability to read and write is often a prerequisite for the position.

24.3 Discussion

These findings contribute to a growing body of knowledge on the effectiveness of farmer-trainers. Whereas no formal impact evaluations using randomized controlled trials were found, there have been several studies using quasi-experimental techniques or surveys of participants' perceptions and trainees' knowledge. Nakano et al. (2015) found that farmer-trainers in Tanzania were successful in disseminating knowledge about rice cultivation, and that the improved practices diffused to other farmers as well over a 5-year period. Wellard et al. (2013) found that farmers working with farmer-trainers in Ghana, Malawi and Uganda in collaboration with an NGO (Self-Help Africa) had significantly higher adoption rates than farmers in a control group. Lukuyu et al. (2012) found that volunteer farmer-trainers in Kenya were highly prized by their trainees and were still actively training their peers several years after project support had ended.

Only Wellard et al. (2013) examined the costs of farmer-trainers. They did not examine costs per farmer trained but did estimate overall benefit-cost ratios, based on increased adoption, yields and income in areas with farmer-trainers compared to controlled areas. Benefit-cost ratios ranged from 6.8:1 to 14.2:1, indicating that the investment in an F2FE program yields high returns. Concerning gender, no study was found that examined the role that F2FE could play in reducing gender imbalances in extension.

It is important to note that neither the above studies nor our own surveys dealt exclusively with CSA practices. One question, therefore, is whether findings concerning F2FE's performance in promoting a wide range of agricultural practices are also relevant for promoting CSA practices. An examination of a list of CSA practices, such as that provided by Harvey et al. (2014), reveals that it is not possible to generalize: CSA practices vary considerably in complexity; in whether new skills are required to apply them; and in the cost, resource intensity and length of time necessary to generate benefits (CIAT, BFS/USAID 2016). Given this complexity, perhaps a more relevant question is this: Are there types of practices that are not suitable to be promoted by farmer-to-farmer extension, and might these include some CSA practices? Franzel et al. (2015) reported that F2FE, though effective for a wide range of innovations, is less appropriate for complex practices (e.g., conservation agriculture), high-risk innovations where the cost of an error may be high (e.g., agrichemicals or artificial insemination) and for what are essentially permanent decisions (e.g., water retention ponds and dams). Degrande and Benoudji (2017) noted the need for close supervision of farmer-trainers by extension staff in a conservation agriculture initiative in Chad.

Franzel et al. (2015) also reported that F2FE may not perform well in certain situations, regardless of the type of practice promoted. For example, it does not do well in areas of low population density where transportation is a constraint, unless means of transport are provided. Performance may also be compromised in areas where farmers are not well organized. It also appears to be less suited to high-income commercial systems, where the opportunity cost of labor is high and social networks may be weak. It works best where farmer-trainers are serving members of a farmer group or a producer organization, because in these circumstances trainers already have a ready clientele.

24.4 Implications for Development

Among those using it, F2FE is perceived to be an effective approach for promoting a wide variety of practices—including CSA—under a wide range of circumstances. Scaling up of extension is essential for helping farmers adapt to climate change, and F2FE has great potential for helping in these efforts. The findings presented in this paper show that adding F2FE programs can help extension services increase the numbers of farmers they reach and promote uptake of CSA practices. F2FE, however, is not appropriate for all agricultural practices or situations. Moreover, whereas extension programs often find it difficult to recruit female professionals, some are able to significantly increase their proportion of female farmer-trainers by making special efforts to do so.

There are two other caveats to using the approach. First, F2FE can never be used to compensate for a poorly performing extension service. Farmer-trainers rely on extension staff for backstopping and education. As a result, if the extension staff are inadequately trained or lack transportation, farmer-trainers will not be effective. Second, neither F2FE nor any single extension approach on its own can scale up CSA to millions of farmers. Rather, F2FE needs to be combined with other complementary approaches such as extension campaigns, farmer field schools or information and communication technology (ICT) approaches.

Finally, more research is needed on whether F2FE is effective for promoting CSA. For example, it would be useful to compare F2FE with more conventional extension approaches (e.g., extension staff working directly with farmer groups) in their ability to cope with changes in weather (e.g., rising temperatures over time) and increased weather risks (e.g., a rising probability of drought). Such research questions are extremely difficult and expensive to answer as they would require randomized controlled trials in which researchers compare the adoption of CSA technologies in villages that have farmer-trainers with those that do not. Villages with farmer-trainers would have to be located at considerable distance from those without in order to prevent "leakage" of the training information from one village to the other. Such research studies would likely require large samples (several thousand farmers) to ensure statistically valid findings and would take several years to conduct. Both quantitative and qualitative assessments would be needed.

But certain proxy questions for assessing the relevance of F2FE to CSA could be answered more easily, such as whether farmers prefer—and are more willing—to discuss their farming risks and risk responses with farmer-trainers rather than with extension agents. One could also test whether farmer-trainers are more skilled at recognizing the weather risks characterizing their agro-ecologies compared to extension agents, who are often less likely to have extensive experience in the particular region. Given that so many extension services already use F2FE, it would also be useful to assess more generic questions about how to improve its effectiveness, such as how to choose farmer-trainers, how to train and motivate them, and how to ensure the sustainability of the program.

References

CIAT, BFS/USAID (2016) Climate-smart agriculture in Senegal. CSA country profiles for Africa series. International Center for Tropical Agriculture (CIAT), Bureau for Food Security, United States Agency for International Development (BFS/USAID), Washington, DC

Degrande A, Benoudji C (2017) L'Agriculture de conservation repensée: contextualiser l'innovation pour la résilience au Tchad et au Soudan. BRACED Innovation for resilience case studies

Franzel S, Degrande A, Kiptot E, et al (2015) Farmer-to-farmer extension. Global forum for rural advisory services good practice notes for extension and advisory services, no. 7. Global Forum for Rural Advisory Services, Lindau, Switzerland

Franzel S, Sinja J, Simpson B (2014) Farmer-to-farmer extension in Kenya: the perspectives of organizations using the approach. ICRAF Working Paper no. 181. World Agroforestry Centre, Nairobi. http://www.worldagroforestry.org/downloads/publications/PDFs/WP14380.PDF

GFRAS (2014) Gender equality in rural advisory services. Global Forum for Rural Advisory Services (GFRAS) brief no. 2. Global Forum for Rural Advisory Services, Lindau, Switzerland

Goodman LA (1961) Snowball sampling. Ann Math Stat 32:148–170

Harvey CA, Chacón M, Donatti CI et al (2014) Climate-smart landscapes: opportunities and challenges for integrating adaptation and mitigation in tropical agriculture. Conserv Lett 7:77–90. https://doi.org/10.1111/conl.12066

Karanja E, Kiptot E, Franzel S (2017) The volunteer farmer-trainer approach three years after the exit of the East Africa Dairy Development Project: a case study of four dairy producer organizations in Kenya. ICRAF, Nairobi http://www.worldagroforestry.org/output/volunteer-farmer-trainer-approach-three-years-after-exit-east-africa-dairy-development

Khaila S, Tchuwa F, Franzel S et al (2015) The farmer-to-farmer extension approach in Malawi: a survey of lead farmers, ICRAF Working Paper no. 189. World Agroforestry Centre, Nairobi http://www.worldagroforestry.org/downloads/publications/PDFs/WP14200.PDF

Kiptot E, Franzel S (2012) Effectiveness of the farmer-trainers approach in dissemination of livestock feed technologies: a survey of volunteer farmer-trainers in Kenya. East African Dairy Development Project, Nairobi https://cgspace.cgiar.org/bitstream/handle/10568/34449/Effectiveness%20of%20farmer%20trainer%20approach%20in%20disseminating%20feed%20technologies.pdf?sequence=1

Kiptot E, Franzel S, Karanja E et al (2013) Effectiveness of the volunteer farmer-trainer approach in dissemination of livestock feed technologies: a survey of farmer trainees in Kenya. East African Dairy Development Project, Nairobi

Kiptot E, Franzel S, Nzigamasabo P et al (2016) Farmer-to-farmer extension of livestock feed technologies in Rwanda: a survey of volunteer farmer-trainers and organizations, ICRAF working paper no. 221. World Agroforestry Centre, Nairobi. https://doi.org/10.5716/WP16005.PDF

Kundhlande G, Franzel S, Simpson B et al (2014) Farmer-to-farmer extension approach in Malawi: a survey of organizations, ICRAF working paper no. 183. World Agroforestry Centre, Nairobi http://www.worldagroforestry.org/downloads/publications/pdfs/WP14391.PDF

Lipper L, Thornton B, Campbell M et al (2014) Climate-smart agriculture for food security. Nat Clim Chang 4:1068–1107

Lukuyu B, Place F, Franzel S et al (2012) Disseminating improved practices: are volunteer farmer-trainers effective? J Agric Educ Ext 18(5):525–554

Masangano C, Mthinda C (2012) Pluralistic extension system in Malawi, IFPRI discussion paper no. 01171. IFPRI, Washington, DC

McCarthy N, Lipper L, Branca G (2011) Climate-smart agriculture: smallholder adoption and implications for climate change adaptation and mitigation. FAO, Rome

Nakano Y, Tsusaka TW, Aida T et al (2015) The impact of training on technology adoption and productivity of rice farming in Tanzania: is farmer-to-farmer extension effective? JICA-RI working paper no. 90. JICA Research Institute, Tokyo

Scarborough V, Killough S, Johnson DA et al (eds) (1997) Farmer-led extension: concepts and practices. Intermediate Technology Publications, London

Selener D, Chenier J, Zelaya R (1997) Farmer-to-farmer extension: lessons from the field. International Institute for Rural Reconstruction, New York

Simpson B, Franzel S, Degrande A et al (2015) Farmer-to-farmer extension: issues in planning and implementation, MEAS technical note. USAID Modernizing Extension and Advisory Services, Washington, DC http://agrilinks.org/sites/default/files/resource/files/MEAS%20TN%20Farmer%20to%20Farmer%20-%20Simpson%20et%20al%20-%20May%202015.pdf

Simpson BM, Heinrich G, Malindi G (2012) Strengthening pluralistic agricultural extension in Malawi. MEAS rapid scoping mission. USAID Modernizing Extension and Advisory Services, University of Illinois, Urbana

Tsafack S, Degrande A, Franzel S et al (2014) Farmer-to-farmer extension in Cameroon: a survey of extension organizations, ICRAF working paper no. 182. World Agroforestry Centre, Nairobi http://www.worldagroforestry.org/downloads/publications/PDFs/WP14383.PDF

Tsafack S, Degrande A, Franzel S et al (2015) Farmer-to-farmer extension: a survey of lead farmers in Cameroon, ICRAF working paper no. 195. World Agroforestry Centre, Nairobi http://www.worldagroforestry.org/downloads/Publications/PDFS/WP15009.pdf

Wellard K, Rafanomezana J, Nyirenda M et al (2013) A review of community extension approaches to innovation for improved livelihoods in Ghana, Uganda and Malawi. J Agric Educ Ext 19(1):21–35

World Bank, Food and Agricultural Organization of the United Nations, International Fund for Agricultural Development (WB, FAO, IFAD) (2009) Gender in agriculture sourcebook. World Bank, Washington, DC

25

CSA and Smallholder Farms: Role of Innovative Partnerships

Mariam A. T. J. Kadzamira and Oluyede C. Ajayi

25.1 Introduction

Southern Africa is particularly vulnerable to climate change because smallholder subsistence farmers, a majority in the region, rely almost entirely on rain-fed farming (Nhemachena et al. 2010). Weather patterns such as droughts, floods and erratic rainfall impact rural households' food security, nutrition and income, which is significant in a region where the majority of the population are poor and have the lowest adaptive capacity (Beegle et al. 2016; Nhemachena and Hassan 2007). Amongst the most vulnerable are smallholders, especially female farmers who have the least capacity to adapt, and are thus disproportionally impacted by climate change compared to their male counterparts (UN 2009). This high vulnerability to climate change causes food insecurity (recurrent swings between food scarcity and surplus) for up to six million people annually. For example, in the 2015–2016 agricultural season, countries including Lesotho, Malawi, Swaziland, Botswana, Namibia and Zimbabwe declared national emergencies because of drought (WFP 2016). There is an increasing call from farmers, development practitioners and policymakers to recognise drought as the "new normal" in the region, and to respond appropriately (Ajayi et al. 2007a, b). Part of the response is to shift from "relief efforts" (giving food aid to farmers after crop failure) to "production relief"—helping farmers to adopt practices that make them resilient, so they can continue to produce food despite climate change uncertainties.

Innovative partnerships are increasingly recognised as essential for addressing the negative consequences of climate change, however there is limited literature about the practicalities of putting into use innovative partnerships to scale up climate-resilient agricultural solutions in southern Africa (Surminski and Leck 2016;

M. A. T. J. Kadzamira (✉) · O. C. Ajayi
The Technical Centre for Agricultural and Rural Co-operation (CTA (ACP-EU)), Wageningen, The Netherlands
e-mail: kadzamira@cta.int

Andersson et al. 2016; Fünfgeld 2015; UNEP 2015; Forsyth 2010). The objective of this chapter is therefore to describe the processes and experiences of forming country project teams, partnership models and approaches to reach farmers in Zimbabwe, Zambia and Malawi. This will improve understanding of methods of setting up sustainable partnerships that exist beyond donor-funded projects.

25.2 Solutions for Scaling CSA

The specific climate-smart solutions for scaling up were selected over multiple phases, in consultation with farmers and a range of stakeholders including development workers and researchers. The process began with a call for proposals for climate-smart solutions, followed by an evaluation process that involved many experts from Europe, African, Caribbean and Pacific (ACP) countries (CTA 2015). The experts were six individuals from Europe, Africa and the Caribbean; an agro-ecologist, a climate change scientist, an international agricultural development specialist, and representatives from the farmers' organisation and the Forum for Agricultural Research in Africa (FARA). The top 15 solutions were documented and published to assess their respective development, adoption, impact and potential for being scaled up in other regions, after which four climate-smart solutions were selected (CTA 2015). The project adopts a "bundled solution" approach rather than a single technology. The four climate-smart solutions being scaled up are:

(i) Drought tolerant maize seeds
(ii) Information and communication technology (ICT) enabled weather information services for smallholder farmers
(iii) Weather based index insurance for smallholder farmers
(iv) Diversified options for livestock farmers

25.3 Partnerships for Scaling Up Climate Resilient Solutions: Country Case Studies

The scaling up project is currently implemented in three countries: Zimbabwe, Zambia and Malawi (Table 25.1).

25.3.1 Bilateral Partnership Model: Zimbabwe

The partnership in Zimbabwe is bilateral, with the two implementing partners, the Zimbabwe Farmers Union (ZFU) and Econet Wireless, sharing common economic interests and equal responsibility. Whilst ZFU acts as the aggregator to reach

Table 25.1 Summary of country projects

Country	Zimbabwe	Zambia	Malawi
Project goal	To contribute to climate resilient agrifood systems that improve food security, nutrition and income for smallholder farm households under climatic uncertainties		
Beneficiaries	140,000 farmers (approximately 40% female) over 2 years in Zimbabwe (30,000) Zambia (60,000) and Malawi (50,000)		
Regions	Mashonaland West, Masvingo, Midlands	Eastern, Central and Southern Provinces	Southern, Central and Northern Regions
Districts	Chegulu, Makonde, Zvimba, Hurungwe, Chivi, Masvingo, Zaka, Gokwe South, Shurungwe, Kwekwe	Lundazi, Chipata, Nyimba, Petauke, Chibombo, Mumbwa, Kapiri Mposhi, Serenje, Kalomo, Choma, Monze, Mazabuka	Zomba, Mchinji, Nkhotakota, Ntchisi, Mzimba
Key innovations promoted	Zimbabwe Farmers Union (ZFU) EcoFarmer Combo: A service bundle offering weather information (including advice for livestock farmers), weather index insurance, payment for ZFU membership and funeral cover	Market facilitation: Better price negotiations for farmers and links to seed producers, meat traders and processors to smallholders, training agro-dealers in CSA to enable them to provide advice at point of sale	Weather based index insurance
	Agronomic advisory services via SMS	Awareness campaigns for farmers to create demand for drought tolerant maize seed and weather based index insurance	ICT-enabled weather information services for smallholder farmers
	Dial-a-Mudhumeni: A phone-in facility for crop and livestock farmers to get extension advice	Agronomic and animal husbandry training for Lead Farmers[a]	Drought tolerant maize seeds
		Advisory services for integrated crop-livestock farming	
Implementing partners	Zimbabwe Farmers Union (ZFU), Econet Wireless	Zambia Open University, Musika Development Initiatives, Professional Insurance Company of Zambia (PICZ)	National Smallholder Farmers Association of Malawi (NASFAM)

(continued)

Table 25.1 (continued)

Country	Zimbabwe	Zambia	Malawi
Other collaborating partners	Public sector: Meteorological Services Department (MSD), Government Extension Service Department, Zimpost	Public sector: Ministry of Agriculture (MOA), National Agricultural Information Services (NAIS), MOA Block and Camp Agricultural Extension Officers	Public sector: Department of Climate Change and Meteorological Services (DCCMS)
	Private sector: Seed Co, Pannar Seeds, Klein Karoo (K2) Seeds, Agriseeds, aWhere	Private sector: Seed Companies	Private Sector: NICO General Insurance Limited, Risk Shield Consultants

[a]The lead farmer is the main contact for the project and partner organisations. He/she is selected from his/her peers in the community based on educational background as well as standing in the community and is trained (as an entry point to the community). Lead farmers are given training materials, a push bike and a schedule to train or disseminate the information to other farmers in their respective locality

farmers, Econet Wireless provides a platform to digitally register farmers and disseminate information to them. Their shared interest (i.e., reaching farmers with information) and mutually agreed delineation of responsibility (including management of project resources) allows autonomy and ease of operations when carrying out specialised project activities. Additionally, ZFU and Econet Wireless have an equal share of decisive power in project planning, programming and implementation, facilitated by regular project meetings, a joint project implementation plan, and a signed Memorandum of Understanding (MoU). This also supports the management and dissemination of profit from farmer subscriptions to the information services provided. The partners also share an online portal containing project statistics, including farmers' subscriptions to the insurance product, billings and payments.

This second-generation partnership—the two partners have previously worked together in partnership—in Zimbabwe was established in 2015 with the aim of providing farmers with highly valuable services at a minimum cost through ICT. The services, collectively referred to as the original ZFU EcoFarmer Combo, included: crop advice, weather index insurance, payment for ZFU membership and funeral cover. This original combo reached approximately 39,000 farmers by mid 2017.

The current ZFU EcoFarmer Combo costs 1 USD per month and includes all the services in the original version, as well as weather information in real time, toll-free phone information on drought tolerant seeds, and advice for crop and livestock farmers (via the *Dial-a-Mudhumeni* phone in extension service). With CTA support (especially regarding real-time weather data and increasing reach to farmers) this current combo has reached approximately 10,000 farmers in the first 2 months of mobilization.

The partnership is mutually beneficial. Organisational theorists have demonstrated that member organisations are most likely to survive if they are able to mobilise resources and demonstrate legitimacy, both of which ZFU achieved through this partnership (Walker and McCarthy 2010). As a member-based organisation which farmers subscribe to, ZFU relies on membership fees for continuity. Before the EcoFarmer Combo was introduced, these were collected manually (i.e., at the district level by ZFU employees and then remitted to ZFU central via a bank deposit), a process that was laborious, inefficient and ineffective. The EcoFarmer Combo package includes the ZFU membership fee and is bought by mobile payment, thereby resolving the issue for ZFU. The partnership also enables ZFU to demonstrate relevance to members through the delivery of tangible, valuable services at a relatively low cost, and reduces cost thanks to their discounted telecom rates from Econet Wireless.

The key benefit for Econet Wireless is that working with ZFU opens up a new clientele of smallholder farmers to which they can market their services, including their insurance products and ICT subscription services. Both organisations profit financially from farmer subscriptions to the EcoFarmer Combo, which ensures that ZFU can exist and that Econet Wireless achieve their commercial goals, both of which are key outcomes for the continuity for the partnership.

The public sector (specifically the Ministry of Agriculture) has also helped facilitate the scale-up of CSA innovations in Zimbabwe. Specifically, farmer recruitment to the ZFU EcoFarmer Combo within Municipal Wards (the smallest political demarcation of a district in Zimbabwe) is supported by local government extension staff. By providing skeptical members of the general public and local (political) leaders with project information, they have played a vital role in validating and legitimising the work of the partnership.

25.3.2 Multilateral Partnership Model: Zambia

In Zambia a multilateral partnership comprised of Zambia Open University, Musika Development Initiatives (Musika) and the Professional Insurance Company of Zambia (PICZ) works in collaboration with officers and field workers from the Ministry of Agriculture and Livestock to reduce the vulnerability of smallholder farmers. Specifically, the partnership is working to increase farmers' resilience through diversified, adaptive, climate resilient production systems. Each partner has equal influence in the implementation of the project, achieved through joint regular meetings, field implementation and monitoring. Zambia Open University, by leading the consortium, has ensured government engagement in the project.

Musika previously implemented the DFID-funded project *Vuna* to promote different aspects of CSA through the creation of a supportive policy environment. They trained farmers and agro-dealers to understand changes in the agricultural landscape, the benefits of using climate smart practices, and the use of pesticides, herbicides and post-harvest technology in an altered environment.

Musika has capitalised on this previous work in the current partnership by using agro-dealer networks already established and previously trained in CSA.

To ensure sustainability beyond the life of the project, each partner in the Zambian consortium is responsible for the areas in which they specialise. For example, PICZ developed the insurance product for farmers, and provide the digital platform for farmer registration. Musika is responsible for farmer mobilization (sensitising farmers so that they can register for the program) using existing staff and local structures, while Zambia Open University, apart from coordinating and leading project implementation, manages the research components of the project. As in the Zimbabwe partnership described above, there are multiple benefits to this approach for each partner; PICZ acquires new clientele in the form of smallholder farmers, Musika Initiatives (which uses a market facilitation approach to link farmers with agribusinesses) mobilises new farmers for their market linkages work. Zambia Open University benefits from capacity building for their students, field project management experience and research outcomes.

The consortium aims to reach 60,000 farmers in 2 years. Between 2013 and 2015, up to 75,000 smallholders in Zambia were introduced to weather index insurance. In early 2017, the Zambian Government announced a policy that made the purchase of weather index insurance compulsory for all farmers benefiting from the Farmer Input Support Programme (FISP).[1] Approximately 1.2 million farmers will subscribe to weather index insurance as a result of this policy. The Zambian Government has approached the project consortium to support the implementation of the new policy, and has adopted their training materials on CSA and weather index insurance for use by all front line government extension staff and farmers. Additionally, nationwide efforts to scale up weather index insurance will draw lessons from the consortium target areas. Government buy in, although not a panacea to low adoption and limited access to CSA innovations by smallholders, is of paramount importance in the scale up of climate resilient solutions. Furthermore, the public sector in Zambia plays a key role in the project via the National Agricultural Information Services (NAIS) - the source of all technical agronomic information that the consortium in Zambia intends to disseminate to farmers via mobile phone services during the cropping season. This creates synergies between on-going public services and upcoming private/academic initiatives, such as CTA's work to scale up CRS under discussion. Most importantly, the use of government-approved technical information ensures that farmers do not get conflicting extension service messages from different service providers.

[1] The FISP is the Zambian government's programme that distributes subsidised agricultural inputs to small-scale producers of the staple food crop, maize. Started in 2002, the programme benefitted approximately 1.6 Million farmers by 2017.

25.3.3 Unipolar Partnership Model: Malawi

Project implementation in Malawi began in the last quarter of 2017. The processes that have thus far taken place have drawn lessons from the Zambian and Zimbabwean consortiums, as summarised in Table 25.2. As the partnership unfolds, there are some insights for consideration in Malawi listed below, particularly for the National Smallholder Farmers Association of Malawi (NASFAM):

(i) Beyond moral persuasion, the opportunity for continuous scaling up of CSA on a sustained basis increases when there are stakeholders who have well defined economic interests and a sound business case to engage in scaling up efforts (i.e., the private sector).
(ii) Strong, vibrant partnerships require transparency, trust, shared influence and decision-making, mutual benefits (economic interests), commitment, recognition of partners' specialised roles and profit sharing.
(iii) Champions are needed to rally the private sector to develop market-driven climate resilient solutions; the government to create an enabling environment; and the farming community to raise awareness of CSA benefits. Strategically placed individuals or institutions in the climate change/agriculture/food security nexus are best suited to act as champions.
(iv) Using field based agricultural extension staff helps the case for scaling up climate resilient solutions. For this to be effective there is need to build capacity of the field staff in key technical areas related to CSA. In the case of NASFAM, this entails building the capacity of Association Field Officers in weather index insurance, drought tolerant seeds and ICT-enabled weather information services.

25.4 Implications for Development

The identification of the "best partnership type" for scaling up CSA in southern Africa is not yet conclusive, however some preliminary lessons for successful partnerships can be drawn: they must be inclusive and participatory, have clear mutual benefits, and ensure transparency in project operations. Other enabling factors needed for success include:

1. *Private sector involvement*: Bringing non-state actors on board effectively improves the chances of successfully scaling up proven innovations (such as CSA) in a sustainable manner. In Zimbabwe, farmers outside the ZFU project have access to similar innovations as the ZFU-EcoFarmer combo because Econet Wireless—as a private commercial entity—is working nationally. In Zambia, PICZ facilitates access to weather index insurance for farmers in the CTA project areas and nationally. Both the Zimbabwe and Zambia case studies show that

Table 25.2 Summary of partnerships in Zimbabwe, Zambia and Malawi

Country case study	Consortium	Partnerships	Relationships	Business	Government	Farmers	Value chain innovations	Outcomes (November 2017)
Zimbabwe	Farmer Organisation	Bilateral	Formalised (MoU)	Profit sharing	Informed at HQ level	Member of farmer organisation	Second generation partnership	10,000 farmers registered on digital platform (2nd generation farmers)
	Private Sector		Joint planning	Clear business case	Informed at field level (Provincial and District officers)	Non-member farmers inclusive of small, medium and large-scale farmers		134 insurance agents trained
			Joint implementation		Engaged at field level (sensitising general public and local leaders)			6 input suppliers engaged
			Project funds shared		Vets weather data sourced externally, before dissemination to farmers			66 farmers and stakeholders sensitised on the project

Zambia	Academia	Multilateral	Formalised (MoU)	Profit sharing	Engaged and consulted at HQ level	Any type of smallholder farmer in target districts	Musika Development Initiatives is a member of the consortium and has previously worked on scaling up CSA	13,000 farmers registered on digital platform
	Private Sector		Joint planning	Clear business case	Trained at field level (extension officers)		Consortium has been co-opted by government in scaling up some CSA innovations at the national level	3591 stakeholders participated in seed fairs to show case drought tolerant seeds
	NGO		Joint implementation		Involved in implementation (field extension staff)		Using Musika Development Initiatives's agro-dealer network to incorporate agro-dealers as point of information to farmers	331 government extension staff trained
					Source of technical agronomic data that is disseminated to farmers			132 agro-dealers trained
Malawi	Farmer Organisation	Unipolar	Informal	Non-profit	Informed at HQ level	Member of farmer organisation	Will engage farmers within affiliate district level farmer associations	
					Will be trained at local level			

bundled climate resilient solutions will be scaled up by the private sector provided they have commercial viability.

2. *Strong and charismatic institutional leadership*: this is key to galvanise others into action, and leverage financial incentives.
3. *Financial incentives*: key for private sector buy in and investment.
4. *Make use of existing value chain innovations*: partnerships must build on existing and successful mechanisms and processes (e.g., second-generation partnerships in Zimbabwe, engagement of Musika Development Initiatives after their successful implementation of the *Vuna* project in Zambia).

Beyond the lessons described here, further research is needed to critically assess the challenges associated with scaling up single solutions in relation to bundled solutions, and their impacts on the livelihoods of the poor in a changing climate. Additional research is also needed to better understand how partnerships can be flexible and adaptable in light of the dynamic nature of climate change, and to determine how to better provide clear evidence of a business case for the private sector to invest in scaling up climate resilient solutions. Additionally, action research is needed to monitor and evaluate the extent to which partnerships, such as those explored in this chapter, deliver results and achieve impact.

References

Ajayi OC, Akinnifesi FK, Gudeta S et al (2007a) Adoption of renewable soil fertility replenishment technologies in southern African region: lessons learnt and the way forward. Nat Resour Forum 31(4):306–317

Ajayi OC, Akinnifesi FK, Sileshi et al (2007b) Economic framework for integrating environmental stewardship into food security strategies in low-income countries: case of agroforestry in southern African region. Afr J Environ Sci Technol 1(4):59–67

Andersson E, Hermelyova S, Pedroni V (2016) Opportunities and challenges for multi-stakeholder partnerships: linking climate change with the post-2015 development agenda. Graduate Institute Geneva. Available from: http://gsogeneva.ch/wp-content/uploads/FINAL-REPORT-GSO.pdf

Beegle K, Christiaensen L, Dabalen A et al (2016) Poverty in a rising Africa. World Bank, Washington, DC

CTA (2015) Climate solutions that work for farmers. The Technical Centre for Agricultural and Rural Cooperation (CTA), Wageningen Available from: https://publications.cta.int/media/publications/downloads/1867_PDF_PBBJWiT.pdf

Forsyth T (2010) Panacea or paradox? Cross-sector partnerships, climate change and development. Wiley Interdiscip Rev Clim Chang 1(5):683–696

Füngled H (2015) Facilitating local climate change adaptation through transnational municipal networks. Curr Opin Environ Sustain 12:67–73

Nhemachena C, Hassan R (2007) Micro-level analysis of farmer's adaptation to climate change in Southern Africa, IFPRI Working Paper, vol 714. International Food Policy Research Institute, Washington, DC Available from: http://ebrary.ifpri.org/utils/getfile/collection/p15738coll2/id/39726/filename/39687.pdf

Nhemachena C, Hassan R, Kurukulasuriya P (2010) Measuring the economic impact of climate on African agricultural production systems. Clim Chang Econ 1(1):33–55

Surminski S, Leck H (2016) You never adapt alone – the role of multi-sectoral partnerships in addressing urban climate risks. Centre for climate change economics and policy working paper no. 262/Grantham Research Institute on Climate Change and the Environment Working Paper No. 232. Available from: http://www.lse.ac.uk/GranthamInstitute/wp-content/uploads/2016/03/Working-Paper-232-Surminski-and-Leck.pdf

UN (2009) Women, gender equality and climate change. Fact sheet. United Nations, New York Available from: http://www.un.org/womenwatch/feature/climate_change/downloads/Women_and_Climate_Change_Factsheet.pdf

UNEP (2015) Collaborating for resilience: partnerships that build disaster-resilient communities and economies. The PSI global resilience project. United Nations Environment Programme (UNEP) Finance Initiative, Geneva Available from: http://www.unepfi.org/psi/wp-content/uploads/2015/12/collaborating-for-resilience.pdf

Walker ET, McCarthy JD (2010) Legitimacy, strategy, and resources in the survival of communnity-based organizations. Soc Probl 57(3):315–340

WPF (2016) Southern Africa: food security crisis. World Food Programme, Rome Available from: https://www.wfp.org/stories/southern-africa-food-security-crisis-0

26

Adoption and Upscaling of CSA: Usage of Rural Finance Instruments

Ruerd Ruben, Cor Wattel, and Marcel van Asseldonk

26.1 Introduction

Agricultural development is strongly influenced by the availability of rural finance. Given the time lag between sowing and harvesting, upfront funding is generally required to enable input purchase before returns are realized. This time lag is even larger for perennial (tree) crops and for production practices that have a longer gestation period, such as irrigation, land consolidation and cover crops.

Access to credit is even more important in the adoption and subsequent upscaling of climate-smart agriculture (CSA) practices. CSA investments tend to be resource-intensive and can be recovered only over a long period of time. Farmers who make these investments often are motivated not only by direct costs and returns but also by the prospect of reduced volatility, increased resilience and a higher degree of certainty regarding future revenue streams.

Different types of financial services fulfil different functions in the production cycle. Whereas credit provision is most helpful for short-term input intensification and medium-term investments, market contracts and insurance (e.g., crop, health and life) provide coping strategies for risk-averse decision-makers. Furthermore, savings provide a way for farmers to both pay for inputs and ride out adversity. The effectiveness of these financial services depends on the availability of other non-financial services (such as training, extension and certification) and the incentives provided by the market (e.g., price premiums, input costs and payments for environmental services). The latter types of incentives may enhance the profitability of CSA investments and encourage farmers to adopt CSA practices (Long et al. 2016; Nyasimi et al. 2014).

R. Ruben (✉) · C. Wattel · M. van Asseldonk
Wageningen Economic Research (WEcR), Wageningen University & Research,
Den Haag, The Netherlands
e-mail: ruerd.ruben@wur.nl

There are also crucial differences among the types of agencies that promote climate-smart investments. While on-farm and value-chain investments are driven by private financial returns, public and civic agencies seek to support societal benefits such as sustainability, poverty reduction and inclusiveness. Public-private partnerships can therefore be helpful.

The purpose of this paper is to understand how rural finance instruments (credit, savings and insurance) can support the adoption and upscaling of CSA. We do not address targeted international financial mechanisms (such as Global Environment Facility and Green Climate Fund) that intend to create specific supportive investment conditions for climate-smart practices. Instead, we focus on methods of linking local financial markets with adaptive CSA practices, with the goal of identifying viable market-based pathways for bringing CSA systems to scale. Our study primarily addresses ways to enhance local adaptive capacity, since mitigation usually requires more global and long-term mechanisms.

In this paper, we first outline the theories of change underlying investments in CSA practices. Then we review the available empirical evidence from studies that analyse these pathways. We give special attention to integrated finance models that address critical complementarities among these pathways, and to different analytical approaches for assessing the impact of CSA-supportive financial policies.

26.2 Theories of Change

The term climate-smart agriculture describes systems designed to improve food security and rural livelihoods and to support climate-change adaptation and mitigation efforts. Mitigation refers to reducing greenhouse gas concentrations in the atmosphere, while adaptation—our focus in this paper— aims to reduce vulnerability to anticipated negative impacts of climate change such as rising temperatures, increases or decreases in precipitation, and changes in the timing of the rain season (UNFCCC 1992).

Meeting the financing requirements for implementing CSA is a significant challenge, since both technological innovations and socio-economic and institutional changes are required. There are three markedly different pathways for assessing CSA investments (Fig. 26.1):

- Direct pathway: Financial instruments for enhancing direct investments for climate-smart practices, ranging from short-term input loans to medium- and long-term loans (Pender and Gebremedhin 2008; Arimi 2014; Marenya et al. 2014; Nyong et al. 2007);
- Indirect pathway: Economic incentives for supporting farm-household incomes that generate expenditure effects in favour of climate-smart practices (Lopez-Ridaura et al. 2018; Ksoll et al. 2016; Jette-Nantel 2013; Wood 2011);

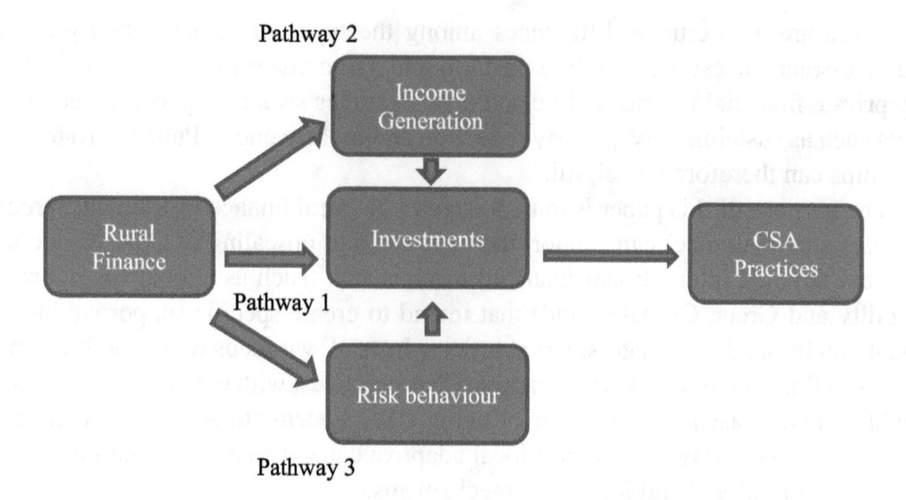

Fig. 26.1 Impact pathways for financing CSA practices. (Source: elaborated by the authors)

- Behavioural pathway: Incentive mechanisms that influence behaviour towards weather risks and enhance resilience of revenue streams generated by climate-smart practices (Dercon and Christiaensen 2011; Brick and Visser 2015).

In Fig. 26.1 we outline these three pathways but also indicate that there might be critical interactions among them. Improved input use and CSA investments (pathway 1) are likely to result in higher net incomes (pathway 2), thus reinforcing the opportunity for a (self-financed) investment pathway (Pender and Gebremedhin 2008). In a similar vein, if farmers become more tolerant of risk (pathway 3) they will be more inclined to intensify input use (pathway 1) (Arslan et al. 2016). And farmers with higher income (pathway 2) tend to become less risk-averse (pathway 3).

Disentangling these pathways is difficult but important. Most research on CSA investments has focused on the identification of supply-side financial services (Branca et al. 2012) that can best cover the costs of adaptation (see www.cgap.org/blog/series/climate-smart-financial-services). And it is true that in less-developed markets, lack of available financial services can be the major limiting factor. Far more often, however, the constraint is on the demand side: low-income smallholders often simply do not wish to borrow money to make CSA investments. Physical access to rural banking facilities is still very limited, resulting in high transaction costs for loans (Branca et al. 2012). Farmers also resist borrowing based on aversion to risk and transaction costs. Opportunity costs (of time and assets) can be barriers as well (McCarthy et al. 2011).

Effective financing of CSA requires business models with multiple market linkages—on both the input and output sides—and integrated contracts that simultaneously enable input intensification and enforce rewarding output market engagement (Hayami and Otsuka 1993; Ton et al. 2017). Creating complementarities, coherence and synergies between instruments and practices represents a major challenge for reaching CSA policy effectiveness. Therefore, interactions between the three CSA finance pathways are of critical importance.

26.3 Evidence Base

Few studies address the wide variety of barriers (financial, economic and behavioural) that stand in the way of CSA investments. The effectiveness of finance for CSA adoption can be judged by considering the net welfare effects at farm-household level (income, health and food) and the environmental effects at village/landscape level. This includes the simultaneous improvement of income/wealth and sustainability by reducing trade-offs and managing volatility. Many available impact studies (Norman et al. 2015) focus, however, on higher scale levels (village, region, district) and on single indicators (either socio-economic or sustainability outcomes).

Suitable finance depends on the type of CSA practices undertaken (see Table 26.1). Some require upfront investments in inputs (e.g., adapted seed varieties or integrated nutrient management), whereas others require longer-term investments (e.g., laser levelling, solar pumps, land-water conservation). Credit amounts involved and their impact on household risk and cash flow can differ widely, with consequences for the required financial products. Sometimes a CSA practice can be easily accommodated in the household production system, without the need for external finance (Asfaw et al. 2014; Di Falco et al. 2012; Yirga and Hassan 2006).

The impacts of different CSA practices can vary widely as well. Whereas CSA practices targeting improved water and nutrient management and diversified seed systems focus on input efficiency (pathway 1), weather-smart services may be particularly helpful in reducing risks (pathway 3), and market reforms deliver more potential for managing the vulnerability and composition of revenue streams (pathway 2). This also translates into different credit scores used by financial institutions, which tend to vary depending on the likelihood of reaching improved efficiency and/or higher resilience (Basak 2017).

We will briefly discuss some key finding from these field studies that address the three impact pathways. This also permits us to highlight major differences in the approaches to assessing impact.

26.3.1 Input Intensification and Investment Pathways

Many adoption studies point to rural finance as a key enabler of technology change (Feder et al. 1985; Feder and Umali 1993). The positive impact of credit use on CSA adoption has been confirmed in studies of highland crops in Ethiopia (Pender and Gebremedhin 2008), fisheries systems in Nigeria (Arimi 2014) and soil conservation in Malawi (Marenya et al. 2014).

To assess impact, these studies generally rely on cross-section regression for likelihood of adoption with a single binary dummy for access to credit services. Few studies rely on balanced samples or use sound counterfactual procedures for robust impact analysis. In fact, individual characteristics are highly correlated with access to credit, and therefore sample selection correction methods (Heckman procedure)

Table 26.1 Examples of evidence of the effect of financial services on the adoption of CSA practices for adaptation

CSA practice	CSA type[a]	Case	Evidence
Pathway 1: Input intensification and investment			
Land management practices: manure or compost, burning to clear the plot, contour ploughing, reduced tillage, intercropping or mixed cropping	Carbon-nutrient smart	Pender and Gebremedhin (2008) on smallholders in Ethiopia	Credit is not strongly associated with the use of land management practices
Adapting aquaculture practices (e.g., water management in ponds, shifting production calendar)	Water-smart	Arimi (2014) on fish farmers in Nigeria	Fish farmers with access to credit showed higher adoption rates of adaptation measures
Conservation practices that reduce soil erosion and increase yields	Carbon-nutrient smart	Marenya et al. (2014) on small farmers in Malawi	Most farmers preferred cash payments to index insurance contracts, even when the insurance contracts offered substantially higher expected returns. Further, more risk-averse farmers were more likely to prefer cash payments
Pathway 2: Income and expenditures			
Changing crop varieties, soil and water conservation measures, water harvesting, tree planting, change planting and harvesting dates	Seed-breed smart, carbon-nutrient smart, water-smart	Di Falco et al. (2012), cereal farmers in Ethiopia	Access to formal credit had a positive but not significant effect on the adoption of the practices
Planting of agro-forestry trees, change of date of planting, land terracing, construction of drainages, cover cropping, making ridges across slope, selling assets, borrowing loans, diversifying livelihoods, short-term migration, support from social network, compensation of losses from government	Weather-smart, water-smart, carbon-nutrient smart, institutional-market-smart	Enete et al. (2016) on flood-coping strategies of small farmers in Nigeria	Access to credit had a negative relationship with selling of assets and short-term migration, suggesting that farmers do not need to recur to more radical and expensive coping strategies when they have access to credit

(continued)

Table 26.1 (continued)

CSA practice	CSA type[a]	Case	Evidence
Maize-legume intercropping, soil and water conservation measures, organic fertilizer, inorganic fertilizer, high-yielding maize varieties	Carbon-nutrient-smart, seed-breed-smart	Arslan et al. (2016) on maize farmers in Tanzania	Positive effect of credit for practices that require liquidity (inorganic fertilizer, improved seeds). Negative effect of credit for intercropping, probably because intercropping is perceived as a way to compensate for lack of fertilizers. Credit appears to increase the use of modern inputs but decrease maize-legume intercropping, a practice that has which has longer-term benefits for soil health and adaptation
Pathway 3: Risk mitigation			
Diversity of climate change adaptation practices	Weather-smart, water-smart, seed/breed-smart, carbon-nutrient-smart, institutional/market-smart	Shackleton et al. (2015) reviewing evidence from 64 case studies worldwide	The cluster "financial, technical and infrastructural barriers" is the most cited barrier. This includes lack of cash and lack of credit, but also lack of inputs and poverty in general
Forest sequestration (mitigation), CSA (adaptation) and information-communication (disaster management)	All	Wong (2016) reviewing evidence from a variety of case studies worldwide	Women face more obstacles in accessing credit and cash, preventing them from adopting certain practices. Existing policies have not paid sufficient attention to the gender gap in access to land, capital and other productive resources. Engaging women in CSA without fully understanding the constraints they face risks reinforcing their subordinated positions
Crop diversification, adjustment of crop management practices or agricultural calendar, land use and land management, and other strategies	Weather-smart, carbon-nutrient-smart, institutional/market-smart	Yegbemey et al. (2014) on maize farmers in Benin	Access to credit enables the farmer to choose adaptation strategies that require investments (larger doses of fertiliser, purchase of better seeds, etc.)

[a]We use the CSA typology from CCAFS (2016)

should be used for unbiased impact assessment (Lipper et al. 2018). It must also be noted that little distinction is made between different types of loans (formal vs. informal) and their terms and conditions (such as loan size, interest rate, collateral requirements and duration). Loans can serve rather different purposes (e.g., a micro-credit loan for a woman's trading activity plays an entirely different role in the household economy than a crop input loan) and will thus have different effects on resource-management practices and CSA outcomes.

The overall evidence base supporting the idea that lack of available credit limits CSA adoption is therefore rather weak. Sometimes access to credit can even lead to land-use specialization and intensification at the expense of climate-friendly technologies. For resource-poor farmers, credit constraints can support the adoption of more labour-intensive climate mitigation practices as an alternative to more expensive external input technologies (Rioux et al. 2016). As Arslan et al. (2016) demonstrated for Tanzania, improving access to credit is likely not only to increase the adoption of modern inputs (such as high-yielding maize varieties and inorganic fertilizer) but also to decrease maize-legume intercropping practices that have longer-run benefits for soil health and adaptation. There are thus important trade-offs to be considered among different intensification strategies that are supported through access to finance.

26.3.2 Income and Expenditures Pathway

For investing in CSA, access to savings and financial services such as insurance, transfers and remittances may be as important as access to credit. Poor farmers who wish to avoid debt may prefer to invest using funding from their own non-farm or off-farm income. An indirect pathway may work best: helping farmers to build a larger household income derived from a variety of resources may allow them to make investments in CSA practices.

Based on integrated farm-household models that combine production and expenditure decisions (Singh et al. 1986), smallholder farmers who have solid expectations for stable revenue streams (even at low levels) are more likely to invest in CSA practices (Lopez-Ridaura et al. 2018; Ruben et al. 2007). Such models also enable assessment of the likely impact of different policy incentives on the allocation of resources within the farm household. Analytical simulation modelling suggests that risk-sharing instruments (e.g., risk-bearing credit, input dealers' risk sharing, voluntary cost sharing and hired-labour risk sharing)[1] can lead to higher CSA adoption rates compared to subsidized loans. In fact, offering low-interest credit appears to be a relatively ineffective strategy for encouraging the adoption of agricultural technologies (Feder and Umali 1993). Instead, activity diversification has repeatedly been shown to encourage both savings and investing in strategies that help cope

---

[1] Much of the risk modelling takes place within the framework of the AgMip programme (https://www.agmip.org/).

with risk (Ksoll et al. 2016). Moreover, rural households are more likely to make in-depth CSA investments if they either receive remittances or are engaged in off-farm employment, since these give access to more stable revenue streams (Jette-Nantel 2013).[2]

26.3.3 Risk Mitigation Pathway

Investment in CSA is closely related to perceptions of risk. Dercon and Christiaensen (2011) show with panel data from Ethiopia that households have different tolerances for taking on risky production technologies based on their fears of poor harvests. In this situation, CSA adoption is discouraged not just by lack of credit but also by a lack of insurance or other risk-mitigating measures. Either indemnity-based or index-based insurance might help (Ndagijimana et al. 2017; Brick and Visser 2015).

CSA practices may also require that farmers have access to specific inputs, such as tree seedlings, seeds or fertilizers. Many farmers lack access to fertilizer, which is a key determinant of productivity and efficient resource use. Duflo et al. (2011) have shown with experiments in Kenya that innovative means of input delivery—including those that rely on mobile phones—can improve the certainty of input available and thus enhance CSA use.

Intra-household decision-making can also play an important role in risk mitigation and resource allocation. Women tend to be somewhat more risk-averse but are also more likely to invest in activities with a longer gestation period (Wong 2016). Consequently, gender-transformative rural finance strategies are likely to be better able to overcome trade-offs between short-term (consumption) and longer-term (resource conservation investment) goals (World Bank et al. 2015).

26.4 Complementarities

Our discussion of these different pathways may give the impression that we are dealing with fully separate activities. That, of course, is not the reality of rural finance, where financial products and services tend to be linked to several farm household activities. The connected nature of financial services and activities can offer advantages. Since rural households simultaneously target a number of different objectives (like nutrition, resilience and resource-use sustainability), it is important to build synergies between instruments that contribute to climate-change mitigation, adaptation and food security.

---

[2] There is also some contrary evidence in the sense that remittances are not used to invest but rather to increase consumption or reduce labour supply (Lartey 2013).

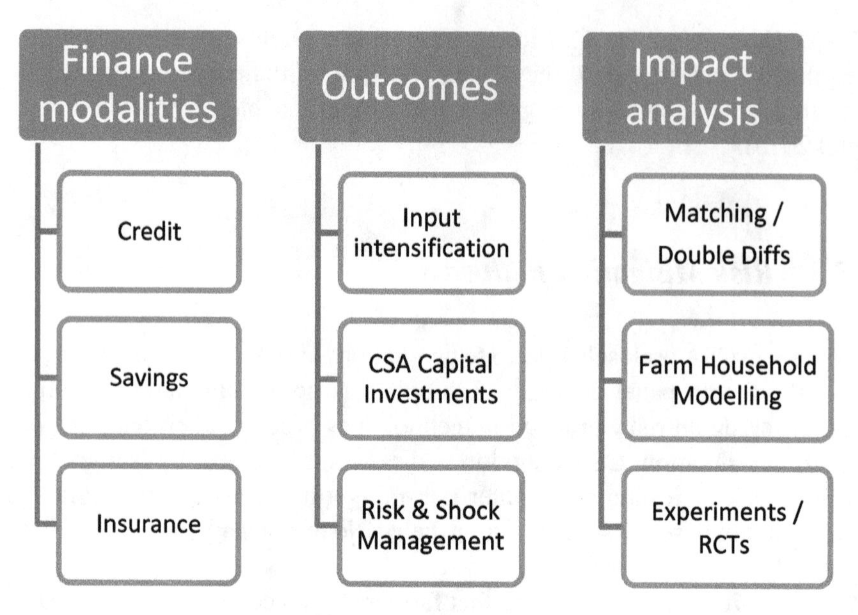

Fig. 26.2 Rural finance CSA outcomes and impact analysis framework. (Source: elaborated by the authors)

Understanding the different aspects of rural finance—credit, savings and insurance—allows a better understanding of its contribution to CSA upscaling. While each instrument may be useful in its own right, a combination offers the greatest benefits (see Fig. 26.2). Adoption of water- and nutrient-smart practices supported by CSA credit schemes will be enhanced if greater certainty on expected revenue streams can be guaranteed (Yegbemey et al. 2014). Investments in CSA infrastructure also can be supported through interlinked insurance systems (Brick and Visser 2015). For some CSA interventions, such as drought-tolerant maize varieties, widespread adoption may well occur without much development of financial markets.

Rural finance for CSA upscaling is heavily dependent upon the combined offer of finance, technical assistance and business support services. It also requires improved financial literacy among farmers. Smallholders usually lack bookkeeping and business-planning skills that would enable them to make more informed investment decisions. They also tend to lack the knowledge of insurance products that would allow them to manage production at the required scale (Branca et al. 2012).

Investments also must be made at different scales (farm, village, landscape and region). Whereas many CSA investments take place at the farm level, it is equally important to support infrastructure (in roads, irrigation, energy, etc.) that benefit larger communities and thus create the enabling conditions for CSA upscaling (Lipper et al. 2018). Also crucial is a supportive business environment that protects property rights and market incentives (Branca et al. 2012). Moreover, participation by different social actors and engagement of multi-stakeholder networks in the development of inclusive financial services is a key condition for broadly anchoring these CSA initiatives.

The interconnectedness of finance modalities (credit, savings, insurance) supports the different pathways towards CSA adoption. These linkages also might help ascertain the real impact of local rural finance on CSA outcomes (see Fig. 26.2). Common approaches for measuring the net effects of climate-smart input intensification (pathway 1) are usually based on the comparison of sites with and without CSA credit by means of the matching of households and assessing income and yield differences. Assessment of CSA capital investments (pathway 2) could be complemented by more dynamic analyses of household resource allocation (using bio-economic farm household simulation models) that provide insights into the implications of changes in the expenditure patterns for CSA investments. This could also inform decisions on resource allocation in non/off-farm activities that result in improved expenditures and enhanced farm-level investments in CSA practices. Finally, changes in risk behaviour (pathway 3) can be analysed by using experimental designs that offer insights into farmers' willingness to invest in CSA practices. A more forward-looking analysis of finance impact would also need to consider changes in attitudes toward risk that influence both intra-household resource allocation and extra-household supply-chain linkages. These behavioural changes are considered to be crucial for supporting long-term CSA resilience.

26.5 Implications for Development

A broad and scalable process of climate adaptation will require comprehensive interventions that both improve income and change attitudes toward risk. At the regional level, it is important to make use of the identified complementarities in rural finance systems that generate multipliers through simultaneous changes in expenditure patterns and risk attitudes. The latter changes tend to have more long-term implications and favour continuous engagement by farmers in CSA investments beyond increasing short-run profitability. Upscaling of CSA practices can thus be encouraged by systematically linking credit, savings and insurance products that influence different dimensions of the rural farm household decision-making structure. In addition, the effectiveness of finance products in fostering CSA adaptation depends heavily on the wider institutional, policy and market environments (Ruben et al. 2007). Therefore, local rural finance should be embedded in a framework of financial intermediation involving multiple stakeholders.

Our literature review suggests a number of strategies for better tailoring local rural financial products and services towards CSA anchoring:

- A broad use of CSA practices cannot be based on single financial products but requires the development of integrated financial markets.
- Effectiveness of rural finance for CSA upscaling is heavily supported by the combined offer of finance with technical assistance and business support.
- Prospects for scaling of CSA practices increase alongside coherent public investments in market development and institutional arrangements.

- Combined public and private sector engagement in CSA investments may have an additional payoff since such joint efforts also contribute to a more resilient business climate.
- Blended finance can take shape in the form of softer financing conditions for climate-smart investments (e.g., risk-sharing, risk layering, interest rates rebates and longer repayment periods), performance clauses and prohibitions, and combinations of finance with subsidized interventions (e.g., training, technical assistance, business development services and certification).

Rural finance plays a double role in CSA anchoring, both supporting individual farm-households as they adopt CSA practices and encouraging the local and regional business climate to favour CSA production systems that deliver credible outcomes.

References

Arimi K (2014) Determinants of climate change adaptation strategies used by fish farmers in Epe Local Government Area of Lagos State. Nigeria J Sci Food Agric 94(7):1470–1476

Arslan A, Belotti F, Lipper L (2016) Smallholder productivity under climatic variability: adoption and impact of widely promoted agricultural practices in Tanzania. ESA Working Paper No. 16–03. FAO, Rome

Asfaw, S, Davis, B, Dewbre, J. et al. (2014) Cash transfer programmes, productive activities and labour supply: evidence from randomized experiment in Kenya. J Dev Stud 50(8):1172–1196

Basak R (2017) Credit scoring and climate-smart agriculture. World Bank AgriFIN Working Paper, Washington, DC

Branca B, Tennigkeit T, Mann W et al (2012) Identifying opportunities for climate-smart agriculture investments in Africa. FAO: Economics and Policy Innovations for Climate-smart Agriculture, Rome

Brick K, Visser M (2015) Risk preferences, technology adoption and insurance uptake: a framed experiment. J Econ Behav Organ 118:383–396. https://doi.org/10.1016/j.jebo.2015.02.010

CCAFS (2016) Climate-smart villages: an AR4D approach to scale up climate-smart agriculture. CGIAR Research Program on Climate Change, Agriculture and Food Security (CCAFS), Copenhagen

Dercon S, Christiaensen L (2011) Consumption risk, technology adoption and poverty traps: evidence from Ethiopia. J Dev Econ 96:159–173

Di Falco S, Yesuf M, Kohlin G et al (2012) Estimating the impact of climate change on agriculture in low-income countries: household level evidence from the Nile Basin. Ethiop Environ Resour Econ 52(4):457–478

Duflo E, Kremer M, Robinson J (2011) Nudging farmers to use fertilizer: theory and experimental evidence from Kenya. Am Econ Rev 101:2350–2390

Enete A, Obi J, Ozor N et al (2016) Socioeconomic assessment of flooding among farm households in Anambra state, Nigeria. Int J Clim Change Str 8:96–111

Feder G, Umali D (1993) The adoption of agricultural innovations. Technol Forecast Soc Change 43:215–239

Feder G, Just R, Zilberman D (1985) Adoption of agricultural innovations in developing countries: a survey. Econ Dev Cult Chang 33:255–298

Hayami Y, Otsuka K (1993) The economics of contract choice: an agrarian perspective. Clarendon Press, Oxford

Jette-Nantel S (2013) Implications of off-farm income for farm income stabilization policies. Dissertation, University of Kentucky

Ksoll C, Lilleør H, Lønborg J et al (2016) Impact of village savings and loan associations: evidence from a cluster randomized trial. J Dev Econ 120:70–85

Lartey E (2013) Remittances, investment and growth in sub-Saharan Africa. J Int Trade Econ Devel 22:1038–1058

Lipper L, McCarthy N, Zilberman D et al (2018) Climate smart agriculture: building resilience to climate change. Springer, Berlin

Long T, Blok V, Poldner K (2016) Business models for maximising the diffusion of technological innovations for climate-smart agriculture. Int Food Agribus Man 20:5–23. https://doi.org/10.22434/IFAMR2016.0081

Lopez-Ridaura S, Frelata R, van Wijk M et al (2018) Climate smart agriculture, farm household typologies and food security: an ex-ante assessment from Eastern India. Agric Syst 159:57–68

Marenya P, Smith V, Nkonya E (2014) Relative preferences for soil conservation incentives among smallholder farmers: evidence from Malawi. Am J Agric Econ 96(3):690–710. https://doi.org/10.1093/ajae/aat117

McCarthy N, Lipper L, Branca G (2011) Climate smart agriculture: smallholder adoption and implications for climate change adaptation and mitigation. Mitigation of Climate Change in Agriculture Working Paper 3

Ndagijimana M, van Asseldonk M, Kessler A et al (2017) Facing climate change in Burundi with an integrated agricultural and health insurance approach. Microfinance Insur Netw 3:32–37

Norman M, Nakhooda S, Canales N et al (2015) Climate finance: how are dedicated climate funds progressing towards impact? ODI Report, London

Nyasimi M, Amwata D, Hove L et al (2014) Evidence of impact: climate-smart agriculture in Africa. CGIAR Research Program on Climate Change, Agriculture and Food Security (CCAFS) and the Technical Centre for Agricultural and Rural Cooperation (CTA), Wageningen

Nyong A, Adesina F, Osman Elasha B (2007) The value of indigenous knowledge in climate change mitigation and adaptation strategies in the African Sahel. Mitig Adapt Strat Glob Change 12:787–797

Pender J, Gebremedhin B (2008) Determinants of agricultural and land management practices and impacts on crop production and household income in the highlands of Tigray, Ethiopia. J Afr Econ 17(3):395–450

Rioux J, San Juan MG, Neely C et al (2016) Planning, implementing and evaluating climate-smart agriculture in smallholder farming systems. FAO, Rome

Ruben R, Pender J, Kuyvenhoven A (2007) Sustainable poverty reduction in less-favoured areas. CABI, Wallingford, UK

Shackleton S, Ziervogel G, Sallu S et al (2015) Why is socially-just climate change adaptation in sub-Saharan Africa so challenging? A review of barriers identified from empirical cases. Wiley Interdiscip Rev Clim Chang 6(3):321–344

Singh I, Squire L, Strauss J (eds) (1986) Agricultural household models: extensions, applications, and policy. Johns Hopkins University Press, Baltimore

Ton G, Desiere S, Vellema W (2017) The effectiveness of contract farming in improving smallholder income and food security in low- and middle-income countries: a mixed-method systematic review. International Initiative for Impact Evaluation (3ie), London

UNFCCC (1992) The United Nations Framework Convention on Climate Change. New York

Wong S (2016) Can climate finance contribute to gender equity in developing countries? J Int Dev 28(3):428–444

Wood RG (2011) Is there a role for cash transfers in climate change adaptation? IDS Bull 42:79–85

World Bank, FAO and IFAD (2015) Gender in climate-smart agriculture. IBRD, FAO and IFAD, Rome

Yegbemey RN, Yabi JA, Aihounton GB et al (2014) Simultaneous modelling of the perception of and adaptation to climate change: the case of the maize producers in northern Benin. Cah Agric 23(3):177–187

Yirga C, Hassan RM (2006) Poverty soil conservation efforts among smallholder farmers in the central highlands of Ethiopia. South Afr J Econ Man Sci 9(2):244–261

Permissions

All chapters in this book were first published in TCSAPIBPRLEF, by Springer; hereby published with permission under the Creative Commons Attribution License or equivalent. Every chapter published in this book has been scrutinized by our experts. Their significance has been extensively debated. The topics covered herein carry significant information for a comprehensive understanding. They may even be implemented as practical applications or may be referred to as a beginning point for further studies.

The contributors of this book come from diverse backgrounds, making this book a truly international effort. We would like to thank all the contributing authors for lending their expertise to make the book truly unique. They have played a crucial role in the development of this book. Without their invaluable contributions this book wouldn't have been possible. They have made vital efforts to compile up to date information on the varied aspects of this subject to make this book a valuable addition to the collection of many professionals and students.

This book was conceptualized with the vision of imparting up-to-date and integrated information in this field. To ensure the same, a matchless editorial board was set up. Every individual on the board went through rigorous rounds of assessment to prove their worth. After which they invested a large part of their time researching and compiling the most relevant data for our readers.

The editorial board has been involved in producing this book since its inception. They have spent rigorous hours researching and exploring the diverse topics which have resulted in the successful publishing of this book. They have passed on their knowledge of decades through this book. To expedite this challenging task, the publisher supported the team at every step. A small team of assistant editors was also appointed to further simplify the editing procedure and attain best results for the readers.

Apart from the editorial board, the designing team has also invested a significant amount of their time in understanding the subject and creating the most relevant covers. They scrutinized every image to scout for the most suitable representation of the subject and create an appropriate cover for the book.

The publishing team has been an ardent support to the editorial, designing and production team. Their endless efforts to recruit the best for this project, has resulted in the accomplishment of this book. They are a veteran in the field of academics and their pool of knowledge is as vast as their experience in printing. Their expertise and guidance has proved useful at every step. Their uncompromising quality standards have made this book an exceptional effort. Their encouragement from time to time has been an inspiration for everyone.

The publisher and the editorial board hope that this book will prove to be a valuable piece of knowledge for students, practitioners and scholars across the globe.

Index